T0214141

Lecture Notes in Computer Science 12608

More information about this subseries at http://www.springer.com/series/7409

Meikang Qiu (Ed.)

Smart Computing and Communication

5th International Conference, SmartCom 2020
Paris, France, December 29–31, 2020
Proceedings

 Springer

Editor
Meikang Qiu
Texas A&M University – Commerce
Commerce, TX, USA

ISSN 0302-9743 ISSN 1611-3349 (electronic)
Lecture Notes in Computer Science
ISBN 978-3-030-74716-9 ISBN 978-3-030-74717-6 (eBook)
https://doi.org/10.1007/978-3-030-74717-6

LNCS Sublibrary: SL3 – Information Systems and Applications, incl. Internet/Web, and HCI

This Springer imprint is published by the registered company Springer Nature Switzerland AG
The registered company address is: Gewerbestrasse 11, 6330 Cham, Switzerland

Preface

This volume contains 29 papers, and all papers were presented at SmartCom 2020 (the Fifth International Conference on Smart Computing and Communication), December 29–31, 2020, in Paris, France.

There were 162 initial submissions. Each submission was reviewed by at least three reviewers, and on average, each paper is reviewed by about 3.5 Program Committee members. Finally, the Program Chairs and technical committees decided to accept 29 papers.

Recent booming developments in Web-based technologies and mobile applications have facilitated a dramatic growth in the implementation of new techniques, such as edge computing, cloud computing, big data, pervasive computing, Internet of Things, social cyber-physical systems, and so on. Enabling a smart life has become a popular research topic with an urgent demand. Therefore, SmartCom 2020 focused on both smart computing and communications fields, especially, the smart technology in communication fields. This conference aimed to collect recent academic work to improve the research and practical applications in these fields.

The scope of SmartCom 2020 was broad, from smart data to smart communications, from smart cloud computing to smart edge computing. The conference gathered all high-quality academic or industrial papers related to smart computing and communications, and also aimed at proposing a reference guideline for further research. SmartCom 2020 was held in Paris, France, and the publisher of this conference proceeding is Springer.

SmartCom 2020 continued to be one of the series of successful academic gatherings, following SmartCom 2019 (Birmingham, UK), SmartCom 2018 (Tokyo, Japan), SmartCom 2017 (Shenzhen, China) and SmartCom 2016 (Shenzhen, China).

We would like to thank the conference sponsors: Springer LNCS, The Alliance of Emerging Engineering Education for Information Technologies, China Computer Federation, North America Chinese Talents Association, and Longxiang High Tech Group Inc.

December 2020

Meikang Qiu
General Chair

Gerard Memmi
Neal Xiong
Zhongming Fei
Keke Gai
Program Chairs

Organization

General Chair

Meikang Qiu   Texas A&M University-Commerce, USA

Program Chairs

Gerard Memmi   Telecom Paris, France
Neal Xiong   Northeastern State University, USA
Zhongming Fei   University of Kentucky, USA
Keke Gai   Beijing Institute of Technology, China

Local Chairs

Han Qiu   Telecom Paris, France
Xiangyu Gao   New York University, USA

Publicity Chairs

Tao Ren   Beihang University, China
Yuanchao Shu   Microsoft Research Asia, China
Wei Liang   Hunan University, China
Fu Chen   Central University of Finance and Economics, China
Xiang He   Birmingham City University, UK
Jinguang Gu   Wuhan University of Science and Technology, China
Junwei Zhang   Xidian University, China
Md Ali   Rider University, USA
Li Bo   Beihang University, China
Meng Ma   Peking University, China

Technical Committee

Yue Hu   Louisiana State University, USA
Aniello Castiglione   University of Salerno, Italy
Maribel Fernandez   King's College London, UK
Hao Hu   Nanjing University, China
Oluwaseyi Oginni   Birmingham City University, UK
Thomas Austin   San Jose State University, USA
Zhiyuan Tan   Edinburgh Napier University, Scotland
Peter Bull   QA Ltd, UK
Hendri Murfi   Universitas Indonesia, Indonesia
Zengpeng Li   Lancaster University, UK

Contents

A Post-processing Trajectory Publication Method Under Differential Privacy

Sijie Jiang[1], Junguo Liao[1(✉)], Shaobo Zhang[1], Gengming Zhu[1], Su Wang[1], and Wei Liang[2]

[1] School of Computer Science and Engineering, Hunan University of Science and Technology, Hunan 411201, China
jgliao@hnust.edu.cn
[2] Hunan University, Hunan 410082, China

Abstract. Differential privacy is an effective measure of privacy protection in data analysis. We propose a differential private trajectory data publication method based on consistency constraints in road network space to significantly improve the accuracy of a general class of trajectory statistics queries. First of all, Laplace noise is injected into statistical data of each road segment. And then, in the post-processing phase, consistency constraint is employed to hold over the noisy output. Based on both synthetic datasets, we do experiments to evaluate the performance of the proposed method. The experimental results show that the proposed method achieves high availability and efficiency.

Keywords: Trajectory · Post-processing · Consistency constraint · Differential privacy · Privacy protection

1 Introduction

With the development of location-aware devices and the Internet of Things, studying the trajectories of people has received growing interest [1–3]. While the benefits provided by location-based services (LBSs) are indisputable, they, unfortunately, pose a considerable threat to privacy [4, 5]. Differentially private has received increased attention because it provides a rigorous theoretical basis and can realize data privacy-preserving by adding noise. Since standard differential privacy is a strong privacy protection model. Many existing techniques [6–8] add more noise than is strictly necessary to ensure differential privacy.

In this paper, we propose a publishing algorithm of the trajectory which satisfies the differential privacy. Furthermore, to address the problem of data inconsistency caused by adding noise to the statistics of the original trajectory data, we propose a post-processing adjustment approach, which can achieve the requirement of differential privacy. Finally, experiments on two virtual data sets proved that our mechanism can better protected users' trajectory privacy and improves higher trajectory quality.

In the early stage of differential privacy, C. Dwork [9]. proposed two definitions of differential privacy, event-level privacy and user-level privacy, aiming at the privacy

M. Qiu (Ed.): SmartCom 2020, LNCS 12608, pp. 1–8, 2021.
https://doi.org/10.1007/978-3-030-74717-6_1

protection of data streams. The technology first appeared in the field of the statistical database and then extended to other fields.

There is prior work focusing on the continual real-time release of aggregated statistics from streams. R. Chen et al. [10] consider applying differential privacy to trajectory data, proposed a variable-length n-gram model to realize differential privacy protection for trajectory statistical results. Y. Chen et al. [11] put forward a privacy-preserving mechanism with differential privacy, called PeGaSus, which can simultaneously support a variety of stream processing tasks. However, their algorithm cannot provide good scalability, there exist some parameters in the model required to be set. Z. Ma et al. [12] raised a mechanism for a real-time trajectory data release with differential privacy called RPTR. F. Y. Li et al. [13] proposed a new real-time trajectory data publishing privacy preserving method E_Group. Little work connects trajectory statistics in road network space and differential privacy.

2   Preliminaries

We denote scalar variables by normal letters, vectors by bold lowercase letters, and matrices by bold capital letters. Such as x_i to denote the i-th row element of X, $x_{i,j}$ to denote the i-th row and the j-th column element of X, x^T to denote the transpose of vector x. We use $\|\cdot\|_p$ to denote the l_p norm.

2.1   Differential Privacy

Differential Privacy is a notion of privacy from the area of statistical databases [14]. It places this constraint by requiring the mechanism to behave almost identically on any two datasets D and D' that are sufficiently close. D and D' are neighboring databases, they differing on at most one record.

Definition 1 (ε-differential privacy). Given a random algorithm S, if the output result $R, R \in Range(S)$, of the algorithm S satisfies the follow in g inequality, the algorithm S is said to satisfy ε-DP.

$$Pr[S(D) \in R] \leq exp(\varepsilon) \times Pr[S(D') \in R] \tag{1}$$

where ε denotes the privacy budget, represents the degree of privacy offered.

Definition 2 (Global sensitivity). Let $f : D \to \mathbb{R}^d$, and let $\|\cdot\|_1$ be the usual L1 norm. Define Δf as the global sensitivity of f, for two neighboring datasets D and D', the global sensitivity of f is

$$\Delta f = \max_{D,D'} f(D) - f(D') \tag{2}$$

Definition 3 (Laplace Mechanism): Let $f : D \to \mathbb{R}^d$ and $\varepsilon > 0$, and let $(Lap(\sigma))^d$ denote a d—length vector of i.i.d. samples from a Laplace with scale $\sigma (\sigma = \Delta f / \varepsilon)$. Define randomized algorithm Q as

$$Q(D) = f(D) + (Lap(\sigma))^d \tag{3}$$

then Q satisfies ε-differential privacy. ε is inversely proportional to privacy preserving intensity.

2.2 Data Model

Our model is represented by a directed graph G = (V, E) where V and E are the set of nodes and edges, respectively.

Users' location points from initial moment to later point composes a trajectory, which can be abstractly defined as Lr. Let us assume the length of release time is $T = \{t_1, t_2, \ldots, t_n\}$. Let $Lr = \{v_1, v_2, \ldots, v_n | 1 \leq i \leq n, \forall v_i \in V, v_i v_{i+1} \in E\}$ be all possible values of user's data in a timed sequence.

2.3 Inconsistency Problem

Computing the number of trajectories of a particular road. Random independent noise is added to each answer in the set through the Laplace mechanism, where the data owner scales the noise based on the sensitivity of the query set. Figure 1 shows the common application scenarios of trajectory.

Suppose that each user appears at only one location at each time point. The in-degree and out-degree of node D are both of 4. According to the Laplace mechanism, adding Laplace noise to perturb each count in Fig. 1(a). The node D in Fig. 1(b) has in-degree 3 and out-degree 4. It means that the number of users arriving at intersection D is equal to the leaving. However, the number of users coming in and going out after noise is added is different, resulting in the inconsistency of track flow statistics, which may lead to privacy breaches.

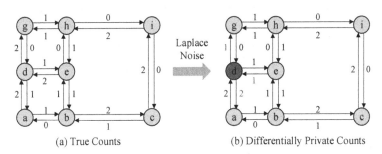

(a) True Counts   (b) Differentially Private Counts

Fig. 1. Trajectory application scenarios

3 Proposed Methods

We are now ready to introduce the trajectory flow data publishing method and consistency processing algorithm that satisfy differential privacy.

The query set is sent to the data owner, to obtain statistical data on the original trajectory data of a section. The set of noisy answers $\tilde{m}$ is sent to the analyst. The set of noisy answers $\tilde{m}$ returned may be inconsistent. Therefore, we propose a post-processing adjustment approach for resolving inconsistencies. This algorithm ensures that a new set of answers $\bar{m}$ are approximately equal to the noise value $\tilde{m}$ previously published by constrained inference.

3.1 Differential Privacy Publishing of Trajectory

We apply the Laplace technique to the query. For the sake of simplicity, we introduce a virtual node k_v which connected to each other nodes in the graph G, let k_v be the starting and ending points for all trajectories. Algorithm 1 shows a trajectory data publication method under differential privacy.

In Algorithm 1, we use the matrix A that represents the adjacency matrix of graph G, M represents the original trajectory flow graph, $\tilde{M}$ represents the trajectory flow matrix after adding Laplace noise. Line 3 is to add a virtual node k_v, forming a loop of head-to-tail.

Algorithm 1 Differential Privacy Publishing of Trajectory

Input: The original trajectory databases D, the matrix A, ε

Output: $M, \tilde{M}$

1:   Initialize $M = 0, \tilde{M} = 0$ # 0 is a null matrix
2:   FOR $Lr \in D$ DO
3:       Add a virtual node k_v, make k_v as head and tail of Lr
4:       FOR each $(v_i, v_j) \in Lr$ DO
5:           Calculated statistical values in $m_{i,j}$
6:       END FOR
7:   END FOR
8:   FOR each $(v_i \rightarrow v_j)$ DO
9:       IF $Lr \in D, a_{i,j} = 1$
10:          Add random Laplace noise to the trajectory statistical values $\tilde{m}_{i,j}$
11: END FOR
12: RETURN $M, \tilde{M}$

3.2 Constrained Inference

We first transform this problem into the optimal solution of convex quadratic programming under a uniform constraint problem. By exploiting the constraints in this problem, we then design a polynomial algorithm to calculate it in Eq. (5).

Formally, given $\tilde{M}$, the objective is to find an s that minimizes $\bar{M} - \tilde{M}_2$ and also satisfies the consistency constraints $\bar{M}I_n = \bar{M}^T I_n$. $\tilde{M}$ represents the trajectory flow matrix after constraint adjustment. In the following, we demonstrate the calculation of $L(\bar{M}, \mu)$.

$$L(\bar{M}, \mu) = \frac{1}{2}\left(\bar{M} - \tilde{M}_2\right)^2 + \mu^T\left(\bar{M}^T I_n - \bar{M}I_n\right) \tag{4}$$

Suppose the function f has a parameter μ, which associates the constraint function with the objective function $\min f(\bar{M})$, and a punishment coefficient α. Let Q is a penalty coefficient matrix, $q_{i,j} = \begin{cases} \alpha, & if \, m_{i,j} = 0 \\ 1, & if \, m_{i,j} \neq 0 \end{cases}$, the Hadamard product operator is $\circ$, then Eq. (5) is deduced.

$$L(\bar{M}, \mu) = \frac{1}{2} tr\left(\bar{M}^\circ Q - \tilde{M}^\circ Q\right)^T \left(\bar{M}^\circ Q - \tilde{M}^\circ Q\right) + \mu^T \left(\bar{M}^T I_n - \bar{M} I_n\right) \quad (5)$$

Let matrix A is taking the inverse of all the elements of Q, i.e. $a_{i,j} = q_{i,j}^{-1}$.

If $\alpha \to +\infty$, $a_{i,j} = \begin{cases} 0, & if\ m_{i,j} = 0 \\ 1, & if\ m_{i,j} \neq 0 \end{cases}$. Let $X = \bar{M}^\circ Q$, $\tilde{X} = \tilde{M}^\circ Q$ that $\bar{M} = X \circ A$,

$\tilde{M} = \tilde{X}^\circ A$, by Eq. (5), then the Eq. (6) is deduced.

$$L(X, \mu) = \frac{1}{2} tr\left(X - \tilde{X}\right)^T \left(X - \tilde{X}\right) + tr\left(A diag(\mu) \tilde{X}^T - A^T diag(\mu) X\right) \quad (6)$$

The idea of conjugate gradient path in unconstrained optimization irradiates us to use this method for solving the linear equality optimization subject to bounds on variables. For Eq. (7), the Lagrange multiplier method is used to obtain the optimal solution. In the end, we can solve the post-processing trajectory flow matrix $\bar{M}$ and objective function (i.e., the minimum error) $minf\left(\bar{M}\right)$ as follows:

$$\begin{cases} \bar{M} = \tilde{M} - A diag(\mu) + diag(\mu) A \\ minf\left(\bar{M}\right) = \frac{1}{2}\left(\sum_{i=1}^{n}\sum_{j=1}^{n}\left(\bar{m}_{i,j} - \tilde{m}_{i,j}\right)^2\right) \end{cases} \quad (7)$$

Based on the above analysis, we design Algorithm 2 for computing $\bar{M}$. The idea of Algorithm 2 is the following: Lines 2~5 are calculated μ, Lines 6~7 are calculated the trajectory flow graph matrix $\bar{M}$.

Algorithm 2 the post-processing to find an optimal solution $\bar{M}$

Input: $A, \tilde{M}$

Output: $\bar{M}$

1:    Initialize n $= size(A)$

2:    $H \leftarrow diag\left((A + A^T) I_n - (A + A^T)\right)$, H is a positive semi-definite matrix

3:    Make $K = H(1:n-1, 1:n-1)$, K is a positive definite matrix

4:    $\mu' \leftarrow K\mu' = \left(\tilde{M}^T - \tilde{M}\right) I_{n-1}$

5:    $\mu \leftarrow \mu = (E_{n-1}, o_{n-1})^T \mu'$

6:    Calculated $\bar{M}$, $\bar{M} = \tilde{M} - A diag(\mu) + diag(\mu) A$

7:    Remove the virtual nodes k_v in matrix $\bar{M}$

8:    RETURN $\bar{M}$

---

4 Evaluation

In the experiments, each configuration is repeated 200 times and the average result is reported. We used two synthetic datasets, it has been generated using Brinkhoff's

[15] network-based generators of moving objects. The former dataset is referred to as Oldenburg and the latter as San Joaquin for convenience. In Table 1 we report the characteristics of the two datasets.

The experiment evaluates the difference between real statistical data and noise statistics data. For the change of different privacy budget ε, Algorithm 1 and Algorithm 2 are compared and analyzed based on the Oldenburg and San Joaquin datasets. To evaluate the utility of an estimator, we measure its root mean square error (RMSE) [16]. The RMSE calculation formula is as follows:

$$\text{RMSE} = \sqrt{\sum_{i=1}^{n}\sum_{j=1}^{n}\left(\hat{m}_{i,j} - m_{i,j}\right)^2}$$

Table 1. Datasets used in the experiments.

D	Nodes	Edges	|D|
Oldenburg	6015	14058	1500
San Joaquin	18496	48061	10000

Figure 2 shows the results of the experiment. The results indicate that compared with Algorithm 1, the RMSE of Algorithm 2 is lower, which shows that under the same privacy budget, the post-adjusted algorithm (Algorithm 2) can reduce RMSE and improve the availability of published results. By comparing Fig. 2(a) and Fig. 2(b), under the same privacy budget, the RMSE of San Joaquin is larger than that of Oldenburg, because the road network size of San Joaquin is larger than Oldenburg. Because the error is only related to ε and the edge numbers, i.e. |E|, in the road map.

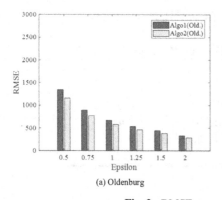
(a) Oldenburg

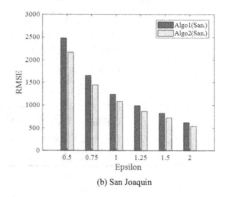
(b) San Joaquin

Fig. 2. RMSE across varying datasets and ε

5 Conclusion

This paper proposes a differentially private trajectory data publication method base on consistency constraints in road network space. The experiments show that bound by the incoming and outgoing flow of the nodes so that it is consistent, we can boost accuracy.

In our future work, we would also like to consider extensions of our strategy that would incorporate the exact timestamp of location points along with their relative time and extend our algorithm to publish trajectories with exact temporal features. Some results show that conventional differential privacy approaches can add more noise than is strictly required by the privacy condition. Therefore, how to reduce the impact of privacy protection on data availability is also an important research direction.

Acknowledgments. This work was supported in part by the Hunan Provincial Natural Science Foundation of China (Grant number 2020JJ4317). And the Hunan Provincial Science Popularization Project of China (Grant number 2020ZK4032).

References

1. Gai, K., Qiu, M., Zhao, H.: Security-aware efficient mass distributed storage approach for Cloud systems in big data. In: Proceedings of 2016 IEEE 2nd International Conference on Big Data Security on Cloud, pp. 140–145, New York (2016)
2. Dai, W.Y., Qiu, M., Qiu, L.F., Chen, L.B., Wu, A.: Who moved my data? privacy protection in smartphones. IEEE Commun. Mag. **55**(1), 20–25 (2017)
3. Zhang, S.B., Mao, X.J., Choo, K.R., Peng, T.: A trajectory privacy-preserving scheme based on a dual-K mechanism for continuous location-based services. Inf. Sci. **527**, 406–419 (2020)
4. Liang, W., Long, J., Li, K., Xu, J., Ma, N., Lei, X.: A fast defogging image recognition algorithm based on bilateral hybrid filtering. ACM Trans. Multimedia Comput. Commun. Appl. https://doi.org/10.1145/3391297
5. Liang, W., Zhang, D., Lei, X., Tang, M., Li, K., Zomaya, A.: Circuit copyright blockchain: blockchain-based homomorphic encryption for IP circuit protection. IEEE Trans. Emerg. Top. Comput. https://doi.org/10.1109/TETC.2020.2993032.2020.3
6. Erlingsson, Ú., Feldman, V., Mironov, I., Raghunathan, A., Talwar, K., Thakurta, A.: Amplification by shuffling: from local to central differential privacy via anonymity. In: Proceedings of the Thirtieth Annual ACM-SIAM Symposium on Discrete Algorithms, pp. 2468–2479, San Diego, CA, USA (2019)
7. Liu, F.: Generalized gaussian mechanism for differential privacy. IEEE Trans. Knowl. Data Eng. **31**(4), 747–756 (2018)
8. Chen, R., Acs, G., Castelluccia, C.: Differentially private sequential data publication via variable-length n-grams. In: Proceedings of the 2012 ACM Conference on Computer and Communications Security, pp. 638–649, Raleigh, NC, USA (2012)
9. Dwork, C.: A firm foundation for private data analysis. Commun. ACM **54**(1), 86–95 (2011)
10. Chen, R., Acs, G., Castelluccia, C.: Differentially private sequential data publication via variable-length n-grams. In: Proceedings of the 2012 ACM Conference on Computer and Communications Security, pp. 638–649, Raleigh, NC, USA (2012)
11. Chen, Y., Machanavajjhala, A., Hay, M., Miklau, G.: PeGaSus: data-adaptive differentially private stream processing. In: Proceedings of the 2017 ACM SIGSAC Conference on Computer and Communications Security, pp. 1375–1388, Dallas, TX, USA (2017)

12. Ma, Z., Zhang, T., Liu, X.M.: Real-time privacy-preserving data release over vehicle trajectory. IEEE Trans. Veh. Technol. **68**(8), 8091–8102 (2019)
13. Li, F.Y., Yang, J.H., Xue, L.F., Sun, D.: Real-time trajectory data publishing method with differential privacy. In: Proceedings of 2018 14th International Conference on Mobile Ad-Hoc and Sensor Networks, pp. 177–182, Shenyang, China (2018)
14. Hay, M., Rastogi, V., Miklau, G., Suciu, D.: Boosting the accuracy of differentially private histograms through consistency. Proc. VLDB Endowment **3**(1), 1021–1032 (2010)
15. Brinkhoff, T.: A framework for generating network-based moving objects. Geoinformatica **6**(2), 153–180 (2002)
16. Zheng, S.F.: A fast algorithm for training support vector regression via smoothed primal function minimization. Int. J. Mach. Learn. Cybern. **6**(1), 155–166 (2015)

Chinese Clinical Named Entity Recognition Based on Stroke-Level and Radical-Level Features

Feng Zhou[1], Xuming Han[2], Qiaoming Liu[3(✉)], Mingyang Li[4(✉)], and Yong Li[1]

[1] School of Computer Science and Engineering, Changchun University of Technology, Changchun 130012, China
[2] College of Information Science and Technology, Jinan University, Guangzhou 510632, China
[3] School of Computer Science and Technology, Harbin Institute of Technology, Harbin 150006, China
`cslqm@hit.edu.cn`
[4] School of Management, Jilin University, Changchun 130022, China

Abstract. Clinical Named Entity Recognition (CNER) is an important step for mining clini-cal text. Aiming at the problem of insufficient representation of potential Chinese features, we propose the Chinese clinical named entity recognition model based on stroke level and radical level features. The model leverages Bidirectional Long Short-term Memory (BiLSTM) neural network to extract the internal semantic in-formation of Chinese characters (i.e., strokes and radicals). Our method can not only capture the dependence of the internal strokes of Chinese characters, but also enhance the semantic representation of Chinese characters, thereby improving the entity recognition ability of the model. Experimental results show that the accuracy of the model on the CCKS-2017 task 2 benchmark data set reaches 93.66%, and the F1 score reaches 94.70%. Comp ared with the basic BiLSTM-CRF mod-el, the precision of model is increased by 3.38%, the recall is increased by 1.05% and F1 value is increased by 1.91%.

Keywords: CNER · Strokes · Radicals · BiLSTM · Internal semantic information · Chinese features

1 Introduction

In order to computerize health records, avoid serious medical accidents, reduce the growth of medical expenses, and improve medical standards, electronic medical rec-ords have been widely used worldwide. With the integration of medicine and comput-er science and the implementation of "medical reform", my country's electronic med-ical records have been comprehensively promoted, and the

---

F. Zhou, X. Han—These authors contributed equally

© Springer Nature Switzerland AG 2021
M. Qiu (Ed.): SmartCom 2020, LNCS 12608, pp. 9–18, 2021.
https://doi.org/10.1007/978-3-030-74717-6_2

number of electronic medical records is increasing. CNER aims to automatically identify and classify clinical entities in unstructured Electronic Medical Record (EMR), such as body parts, diseases, treatments, symptoms and others. Medical institutions use electronic medical records for outpatients, inpatients (or health care objects). Digital medical services record clinical diagnosis, treatment and guiding interventions. It contains rich health data and important clinical evidence, which help to support clinical decision-making and disease surveillance [1].

In recent years, Chinese clinical named entity recognition has been widely used in large-scale medical data analysis, but that is an important and difficult task. For example, large-scale medical literature databases are used to construct symptom-based human disease networks, and to study the links between the clinical manifesta-tions of diseases and potential molecular interactions [2]. The recognition of named entities in Chinese electronic medical records has achieved some results [3–5].

However, due to the structural complexity and diversity of Chinese characters themselves (for example, there are no fixed rules for medical entities; there is no fixed boundary in Chinese; and there are multiple meanings in Chinese), if introducing only one kind of internal structural features of Chinese characters into the traditional model, the model can not accurately capture the semantic information of Chinese characters. Chinese character strokes contain rich semantic information of Chinese characters, and have been used many times in research, but they still cannot better represent the semantic information of Chinese characters. For example, the two words "开", "井" are semantically unrelated, and the stroke sequenc-es are all "横"(一), "横"(一), "撇"(丿) and "竖"(丨). Chinese characters with the same stroke sequence have completely different semantics, so the stroke sequence of a Chinese character cannot uniquely identify a Chinese character. However, when we add the radicals of the Chinese characters on the basis of the strokes of the Chinese characters, The radical of "井" is '廾", and the radical of "开" is "二". we can distin-guish the Chinese characters based on the two characteristics of the strokes and the radicals.

Aiming at the problem of insufficient representation of Chinese potential features, we propose a BiLSTM-CRF model based on stroke level and radical level in this paper.

The main contributions of this article are as follows:

- Aiming at the problem of insufficient representation of potential Chinese features, this paper proposes to we propose the BiLSTM-CRF model based on stroke level and radical level to fully express the semantic information of Chinese characters.
- The evaluation results show that the model in this paper has achieved good per-formance on the CCKS-2017 Task 2 benchmark data set.

The remainder of this paper is structured as follows. Section 2 introduces the related work of Chinese CNER. Section 3 gives a detailed description of the proposed model. Section 4 presents extensive experiments to verify the effectiveness of our proposal, and Sect. 5 summarizes this work.

2   Related Work

In early research, the rule-based and dictionary-based methods were widely used in clinical named entity recognition tasks [5–8]. Ye et al. first used a Conditional Ran-dom Field (CRF) model incorporating dictionary information to identify named enti-ties in Chinese EMR [9]. Zeng et al. and Savova et al. used heuristics and manual rules to identify clinical entities and achieved effective results [10, 11]. Because the method based on machine learning has stronger adaptability and better performance than the method based on rules and dictionary, Classical ma-chine learning methods (such as based on hidden Markov model (HMM), maximum entropy Markov model (MEMM) and conditional random field (CRF)) are widely used in CNER tasks [12–14]. However, the method based on machine learning relies on a large number of feature engineering, which is too cumbersome and time-consuming.

With the rapid development of deep learning in recent years, researchers have widely applied deep learning to CNER tasks. In particular, the BiLSTM-CRF model [15–18] has achieved good results in CNER tasks. Wu et al. proposed a BiLSTM-CRF model with self-attention mechanism suitable (Att-BiLSTMCRF) for Chinese CNER tasks and proposed a fine-grained Chinese character-level rep-resentation method to understand the semantic information of Chinesecharacters better and introduced part of speech (POS) tag information into the model [19]. Yin et al. proposed a BiLSTM-CRF model based on radical-level features and self-attention mechanism to mine the deep semantic information of Chinese char-acters and capture the long-term dependence between characters [20]. Hu et al. proposed a hybrid system that combines four methods: rule, CRF, BiLSTM, and BiLSTM with features, and added a voting mechanism at the end [21]. Qiu et al. proposed a residual diffusion convolutional neural network (RD-CNN-CRF) with a conditional random field and projected Chinese characters and dictionary fea-tures into a dense vector representation, and then input them into the residual dilated CNN to capture contextual features [22]. Zhao et al. proposed a lattice LSTM-CRF based on adversarial training for Chinese CNER and the lattice LSTM is used to capture richer information in electronic medical records (EMR) so that the robustness of the model can be improved through adversarial training [23]. Although the above methods have achieved good results, they all use only the se-mantic information of the strokes or radicals of the Chinese characters and cannot fully express the semantic information of the Chinese characters. Therefore, we pro-pose a Chinese clinical named entity recognition model based on stroke-level and radical-level features. In our scheme, leverages BiLSTM neu-ral network to extract the internal semantic in-formation of Chinese characters (i.e., strokes and radicals) This method captures the dependence between inter-nal features while enhancing the se-mantic representation of Chinese characters Experimental results show that compared with the traditional Chinese named entity recognition model, this method can effectively improve the performance of Chinese clinical named entity recognition.

3   Hierarchical Structure of Proposed Model

In this section, we introduce the hierarchical structure of the BiLSTM-CRF model based on stroke level and radical level features. As shown in Fig. 1, the model mainly includes three components: the embedding layer, the BiL-STM layer and the CRF layer. Next, We will introduce the components of the BiLSTM-CRF Chinese clinical named entity recognition based on stroke-level and radical-level features from bottom to top, as shown below.

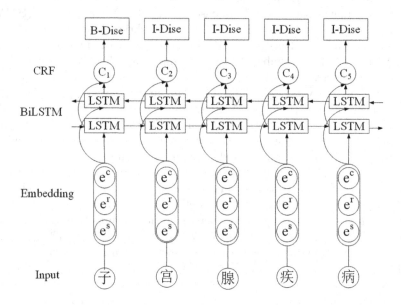

Fig. 1. BiLSTM-CRF model of based on stroke level and radical level features.

3.1   Embedding Layer

For a given sentence sequence $x = (c_1, c_2, c_3, \cdots c_n)$,the embedding vector is jointly created by Chinese characteristics $e_c \in R^{d_c}$ radical characteristic $e_r \in R^{d_r}$ and stroke characteristic $e_s \in R^{d_s}$. Formally, the embedding vector of each character C_i can be expressed as

$$e_i = e_i^c \oplus e_i^s \oplus e_i^r \tag{1}$$

where $i \in \{1, 2, 3, \ldots \ldots, n\}$, e_i^c, e_i^r, e_i^s respectively represent Chinese character embedding, radical embedding, stroke embedding, $\oplus$ means concatenation.

Radical Features. The radical of a Chinese character is the most important and basic unit, and has a very important impact on the semantics of Chinese characters. This paper uses the BiLSTM network to extract the semantic information of the corresponding radicals of Chinese characters. Figure 2 shows the overall structure of the model in details. The expression is as follows:

$$e_r = \left\lfloor \overrightarrow{h_t}; \overleftarrow{h_t} \right\rfloor \tag{2}$$

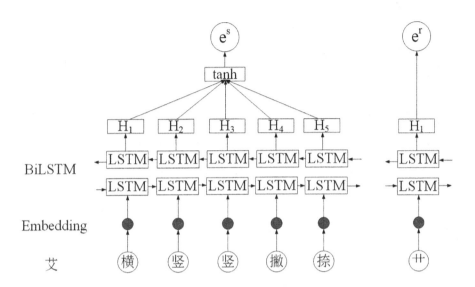

Fig. 2. Calculate the stroke and radical feature.

In the Formula (2), $\left\lfloor \overrightarrow{h_t}; \overleftarrow{h_t} \right\rfloor$ is the hidden layer vector obtained by training BiLSTM network.

Stroke Features. The stroke is the smallest continuous stroke unit that constitutes a Chinese character. As shown in Table 1, we divide the strokes into five types, and the corresponding numbers are 1 to 5.

Table 1. Stroke type and number.

Stroke name	horizontal	vertical	left falling stroke	right falling stroke	Others
Type	一(横)	｜(竖)	ノ(撇)	丶(捺)	〈, ﾚ, ㇄
number	1	2	3	4	5

The Chinese character writing system provides a guide for the stroke order of each Chinese character. These sequence information can be used when learning the internal semantic information of Chinese characters. Therefore, this paper uses the BiLSTM network to extract the contextual semantic information of strokes. Figure 2 shows the structure of propsed BiLSTM network. More Chinese character graphic features can be learned by this method. The expression is as follows:

$$e_s = \tanh \left(\sum_{j=1}^{N} H_{i,j} \right) \tag{3}$$

where $H_{i,j}$ is the j-th stroke feature vector of the i-th Chinese character. tanh represents the hyperbolic tangent function

3.2   BiLSTM Layer

Long Short-term Memory network (LSTM) is a special recurrent neural network. To cope with the problem of gradient explosion or disappearance of RNN model, LSTM introduces cell state, and uses input-gate, forget-gate and output-gate to maintain and control information. The formulas for updating the LSTM model at time t are as follows:

$$i_t = \sigma \left(W_i \left[h_{t-1}, x_t \right] + b_i \right) \tag{4}$$

$$f_t = \sigma \left(W_f \left[h_{t-1}, x_t \right] + b_f \right) \tag{5}$$

$$o_t = \sigma \left(W_o \left[h_{t-1}, x_t \right] + b_o \right) \tag{6}$$

$$\widetilde{c}_t = \tanh \left(W_c \left[h_{t-1}, x_t \right] + b_c \right) \tag{7}$$

$$c_t = i_t * \widetilde{c}_t + f_t * c_{t-1} \tag{8}$$

$$h_t = o_t * \tanh \left(c_t \right) \tag{9}$$

Where σ represents the sigmod activation function. tanh represents the hyperbolic tangent function. X_t represents unit input. i_t, f_t, o_t represent the input gate, forget gate and output gate at time t.W and b respectively represent the weight and deviation of the input gate, forget gate and output gate. $\widetilde{c}_t$ represents the current state of the input. c_t represents the update status at t. h_t represents the output at t.

3.3   CRF Layer

In the sequence labelling problem, the traditional seq2seq model does not consider the dependency between tags. For example, the "I-LOC" label cannot follow "B-PER". Therefore, this paper chooses the CRF model. The CRF model combines the characteristics of the maximum entropy model and the hidden Markov model to solve the problem. In the BiLSTM-CRF model, the input of CRF is the context feature vector learned from the BiLSTM layer. For the input text sentence

$$x = (x_1, x_2, x_3, \cdots\cdots x_n) \tag{10}$$

Let $P_{i,j}$ denote the probability score of the j-th label of the i-th Chinese character in the sentence. For a prediction sequence $y = (y_1, y_2, y_3, \cdots\cdots y_n)$, the CRF score can be defined as:

$$f(x, y) = \sum_{n=0}^{n+1} M_{y_i, y_{i+1}} + \sum_{i=1}^{n} P_{i, y_i} \tag{11}$$

Where M is the transition matrix, and $M_{i,j}$ represents the transition score from label i to j. y_0 and y_{n+1} represent the start and end tags, respectively. Finally, we use softmax to calculate the probability of the sequence y as follows:

$$P(y \mid x) = \frac{e^{f(x,y)}}{\sum_{\tilde{y} \in Y_x} e^{f(x, \tilde{y})}} \tag{12}$$

During the training process, maximize the log probability of the correct label sequence:

$$\log(P(x, y)) = f(x, y) - \log\left(\sum_{\tilde{y} \in Y_x} e^{f(x, \tilde{y})}\right) \tag{13}$$

In the decoding stage, we predict that the maximum score obtained by the output sequence is:

$$y^* = \underset{\tilde{y} \in Y_x}{\mathrm{argmax}} f(x, \tilde{y}) \tag{14}$$

In the prediction stage, the dynamic programming algorithm Viterbi is used to solve the optimal sequence.

4   Experiments and Results

4.1   Experimental Data and Evaluation Indicators

To evaluate the model proposed in this paper on the task of Chinese named entity recognition, we conduct experiments on the CCKS-2017 task 2 benchmark data set. CCKS-2017 Task 2 Benchmark Data Set: released by the 2017 China Knowledge Graph and Semantic Computing Conference for clinical entity recognition tasks. The data set contains 1,198 tagged files and five entity types, including diseases, exami-nations, symptoms, treatments, and body parts.

In order to fully evaluate the performance of the model, we use accuracy (Preci-sion, P), recall (Recall, R), and harmonic average F1 (F1-score) as the evaluation criteria for model performance. They are defined as follows:

$$P = \frac{TP}{TP + FP} \times 100\% \tag{15}$$

$$R = \frac{TP}{TP + FN} \times 100\% \tag{16}$$

$$F1 = \frac{2 * P * R}{P + R} \times 100\% \tag{17}$$

TP (True Positives) indicates the correct number of samples in the positive examples, FP (False Positives) indicates the number of incorrect samples in the negative examples, and FN (False Negatives) indicates the number of incorrect samples in the positive examples.

4.2 Model Building and Parameter Setting

The experimental codes are built using Pytorch. The experimental parameters are set as follows: the embedding dimension (embedding_dim) is 300, the input dimension (max_length) is 80, batch_size of the training set is set to 100, and the training learning rate is set to 0.001. In order to prevent overfitting during training, we set the weight decay factor (weight_decay) to $5e^{-4}$, if simultaneously using dropout technology to prevent overfitting, the value is set to 0.5.

4.3 Experimental Results

The experimental results are shown in Table 2. The proposed model in this paper achieves the highest F1 score among all models. Hu et al. used self-training to add training data to improve performance, which inevitably introduces noise [21]. Our proposed based on stroke-level and radical-level features method can mine the semantic information of this article without using external information. The performance of the model proposed by Zhao et al. [23] is the worst, because Lattice-LSTM introduces various segmenta-tion results into the model, which inevitably led to error segmentation. In order to avoid this problem, this paper treats Chinese named entity recognition as a character-level sequence labelling problem. Qiu and Wang et al. introduced dictionaries and human knowledge constructed from certain medical literature into deep neural networks, which required high-quality medical named entity dictionaries covering most entities [18,22]. Although Wu et al. also achieved good performance, but they only used one type of internal semantic information of Chinese characters, and there may be multiple Chinese characters with the same radical information, still can't get a better word representation [19]. In order to avoid this problem, this paper proposes a BiLSTM-CRF model based on stroke level and radical level features, and experimental results show that the F1 value of our model reaches 94.70% and is superior to other models.

Table 2. Comparison results between the model in this paper and the basic model.

Model	P	R	F1
Hu [21]	94.49	87.79	91.02
Wang [18]	90.83	91.64	91.24
Qin [22]	90.63	92.02	91.32
Zhao [23]	88.98	90.28	89.64
Wu [19]	92.04	90.68	91.35
Ours	**95.07**	**94.33**	**94.70**

5 Conclusion

Clinical named entity recognition is the key task of extracting patient information from clinical electronic medical records. In this paper, we propose a BiLSTM-CRF model based on stroke level and radical level features, which is suitable for Chinese clinical named entity recognition. In order to better capture the in-depth semantic information of Chinese characters and better enhance the representation of Chinese characters, we use BiLSTM to learn the internal strokes and radical features of Chi-nese characters. Compared with the BiLSTM-CRF model, the BiLSTM-CR Chinese clinical named entity recognition model based on stroke level and radical level fea-tures in this paper improves the precision by 3.38%, the F1-score by 1.91% and the recall by 1.05% on the CCKS-2017 Task2 dataset. Our model got the better F1 score than other models. Experimental results prove the effectiveness of strokes and radi-cal in clinical named entity recognition tasks.

Acknowledgments. The research was supported by the National Science Foundation of China under grant No. 61572225 and 61472049, the Foundation of Jilin Provincial Education Department under grant under grant No. JJKH20190724KJ, the Jilin Province Science & Technology Department Foundation under grant No.20190302071GX and 20200201164JC, the Development and Reform Commission Foundation of Jilin province under grant No. 2019C053-11.

References

1. Lossio-Ventura, J.A., et al.: Towards an obesity-cancer knowledge base: biomedical entity identification and relation detection. In: IEEE International Conference on Bioinformatics & Biomedicine IEEE (2016)
2. Habibi, M., Weber, L., Neves, M., Wiegandt, D.L., Leser, U.: Deep learning with word embeddings improves biomedical named entity recognition. Bioinformatics **33**(14), i37–i48 (2017)
3. Wang, Y., Ananiadou, S., Tsujii, J.: Improving clinical named entity recognition in Chinese using the graphical and phonetic feature. BMC Med. Inform. Decis. Mak. **19**(7), 1–7 (2019)
4. Cai, X., Dong, S., Jinlong, H.: A deep learning model incorporating part of speech and self-matching attention for named entity recognition of Chinese electronic medical records. BMC Med. Infor. Decis. Mak. **19**(2), 101–109 (2019)
5. Ji, B., Liu, R., Li, S., Jie, Yu., Qingbo, W., Tan, Y., Jiaju, W.: A hybrid approach for named entity recognition in Chinese electronic medical record. BMC Med. Inform. Decis. Mak. **19**(2), 149–158 (2019)
6. Coden, A., et al.: Automatically extracting cancer disease characteristics from pathology reports into a disease knowledge representation model. J. Biomed. Inform. **42**(5), 937–949 (2009)
7. Song, M., Yu, H., Han, W.S.: Developing a hybrid dictionary-based bio-entity recognition technique. BMC Med. Inform. Decis. Mak. **15**(1), S9 (2015)
8. Friedman, C., Alderson, P.O., Austin, J.H.M., Cimino, J.J., Johnson, S.B.: A general natural-language text processor for clinical radiology. J. Am. Med. Inform. Assoc. **1**(2), 161–174 (1994)

9. Feng, Y., Ying-Ying, C., Gen-Gui, Z., Hao-Min, L., Ying, L.: Intelligent recognition of named entity in electronic medical records. Chinese J. Biomed. Eng. **30**, 256–262 (2011)
10. Zeng, Q.T., Goryachev, S., Weiss, S., Sordo, M., Murphy, S.N., Lazarus, R.: Extracting principal diagnosis, co-morbidity and smoking status for asthma research: evaluation of a natural language processing system. BMC Med. Inform. Decis. Mak. **6**(1), 30 (2006)
11. Savova, G.K., et al.: Mayo clinical text analysis and knowledge extraction system (ctakes): architecture, component evaluation and applications. J. Am. Med. Inform. Assoc. **17**(5), 507–513 (2010)
12. McCallum, A., Li, W.: Early results for named entity recognition with conditional random fields, feature induction and web-enhanced lexicons (2003)
13. McCallum, A., Freitag, D., Pereira, F.C.N.: Maximum entropy Markov models for information extraction and segmentation. Icml **17**(2000), 591–598 (2000)
14. Zhang, J., Shen, D., Zhou, G., Jian, S., Tan, C.-L.: Enhancing HMM-based biomedical named entity recognition by studying special phenomena. J. Biomed. Inform. **37**(6), 411–422 (2004)
15. Xia, Y., Wang, Q.: Clinical named entity recognition: ECUST in the CCKS-2017 shared task 2. CEUR Workshop Proc. **2017**, 43–48 (1976)
16. Ouyang, E., Li, Y., Jin, L., Li, Z., Zhang, X.: Exploring n-gram character presentation in bidirectional RNN-CRF for Chinese clinical named entity recognition. CEUR Workshop Proc. **2017**, 37–42 (1976)
17. Li, Z., Zhang, Q., Liu, Y., Feng, D., Huang, Z.: Recurrent neural networks with specialized word embedding for Chinese clinical named entity recognition. CEUR Workshop Proc. **2017**, 55–60 (1976)
18. Wang, Q., Zhou, Y., Ruan, T., Gao, D., Xia, Y., He, P.: Incorporating dictionaries into deep neural networks for the Chinese clinical named entity recognition. J. Biomed. Inform. **92**, 103133 (2019)
19. Guohua, W., Tang, G., Wang, Z., Zhang, Z., Wang, Z.: An attention-based BiLSTM-CRF model for chinese clinic named entity recognition. IEEE Access **7**, 113942–113949 (2019)
20. Yin, M., Mou, C., Xiong, K., Ren, J.: Chinese clinical named entity recognition with radical-level feature and self-attention mechanism. J. Biomed. Inform. **98**, 103289 (2019)
21. Hu, J., Shi, X., Liu, Z., Wang, X., Chen, Q., Tang, B.: HITSZ CNER: a hybrid system for entity recognition from Chinese clinical text. CEUR workshop proc. 1976 (2017)
22. Qiu, J., Wang, Q., Zhou, Y., Ruan, T., Gao, J.: Fast and accurate recognition of Chinese clinical named entities with residual dilated convolutions. In: IEEE International Conference on Bioinformatics and Biomedicine (BIBM) vol. 2018, pp. 935–942 (2018)
23. Zhao, S., Cai, Z., Chen, H., Wang, Y., Liu, F., Liu, A.: Adversarial training based lattice LSTM for Chinese clinical named entity recognition. J. Biomed. Inform. **99**, 103290 (2019)

Vision-Based Autonomous Driving for Smart City: A Case for End-to-End Learning Utilizing Temporal Information

Dapeng Guo⬥, Melody Moh⬥, and Teng-Sheng Moh(✉)⬥

Department of Computer Science, San Jose State University, San Jose, CA, USA
teng.moh@sjsu.edu

Abstract. End-to-End learning models trained with conditional imitation learning (CIL) have demonstrated their capabilities in autonomous driving. In this work, we explore the use of temporal information with a recurrent network to improve driving performance especially in dynamic environments. We propose the TCIL (Temporal CIL) model that combines an efficient, pre-trained, deep convolutional neural network to better capture image features with a long short-term memory network to better explore temporal information. Experimental results in the CARLA benchmark indicate that the proposed model achieves performance gain in most tasks. Comparing with other CIL-based models in the most challenging task, navigation in dynamic environments, it achieves a 96% success rate while other CIL-based models had 82–92% in training conditions; it is also competitively by achieving 88% while other CIL-based models were at 42–90% in the new town and new weather conditions. We believe that this work contributes significantly towards safe, efficient, clean autonomous driving for future smart cities.

Keywords: Autonomous driving · Conditional Imitation Learning (CIL) · Convolutional Neural Network (CNN) · End-to-End learning · Long Short-Term Memory (LSTM)

1 Introduction

Autonomous driving has received increasing attention in recent years. With the rapid development of deep learning and Internet of Things (IoT) technologies, autonomous driving has evolved from a science-fiction concept to part of the Internet of Autonomous Things (IoAT) systems that are crucial to smart cities and smart societies [1].

The benefits of autonomous driving may be described below. First, it improves road traffic safety by removing human factors which are accountable for 94% of serious crashes [2]. Then, it subsequently reduces economic loss as billions of dollars have been lost each year because of road traffic accidents, in the form of loss of life, workplace productivity, etc. Moreover, it improves efficiency; autonomous driving can drive at the optimal speed and route to reduce traffic congestion while any small error of a driver can cause unnecessary congestion in dense traffic. Furthermore, it enables new applications for vehicles, which brings convenience to daily life [3]. Most notably, it provides people

© Springer Nature Switzerland AG 2021
M. Qiu (Ed.): SmartCom 2020, LNCS 12608, pp. 19–29, 2021.
https://doi.org/10.1007/978-3-030-74717-6_3

who are not fit to drive, such as the seniors or the visually impaired people, a way for mobility.

With decades of advancement in sensor technologies, computing hardware, and deep learning methods, significant progress has been made towards autonomous driving, especially the ones utilizing end-to-end deep learning [4, 5]. However, current autonomous driving systems are still unreliable in dynamic or unfamiliar circumstances.

The end-to-end learning model proposed in this paper aims to improve autonomous driving in dynamic environments. It is vision-based, accepts navigational commands for controllable routing, utilizes temporal information through a recurrent network to better control speed and distance, and uses transfer learning on a pre-trained, deep, efficient image module that greatly improves its generalization capability.

The rest of the paper is organized as follows. Section 2 describes the background and related works. Section 3 and 4 cover the proposed method and the proposed model, respectively. Section 5 and 6 present our experiment design and the results, respectively. Finally, Sect. 7 concludes the paper including future works.

2   Background and Related Works

The end-to-end learning models use one deep learning network that handles all self-driving tasks. Such a model does not have any human-defined intermediate steps. The assumption is that, with the appropriate deep learning model and training method, it will learn the internal features that are most suitable and optimize all the processing steps simultaneously. With the advancement of deep learning, the simple end-to-end learning models have achieved comparable performance to the complex modular pipeline models.

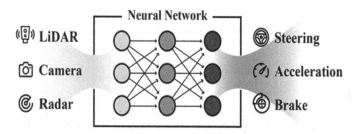

Fig. 1. An end-to-end model

A generalized end-to-end learning model is shown in Fig. 1. The model uses LiDARs, cameras, and ultrasonic radars as the perceptual sensors. The inputs are directly mapped to vehicle control signals using a deep neural network. There is great potential in end-to-end learning models, and models with significance are covered here [6–17].

The surveyed models share much in common. For instance, variations of Convolutional Neural Network (CNN) are used as the image module and imitation learning is utilized as a part of the training process [6–17]. Early works focus on autonomously lane keeping, and the vehicles make random turns at intersections [6–8, 14]. In [10], the author argues that autonomous vehicles need to accept navigational commands and make meaningful turns. Therefore, a high-level-command (HLC) is used to indicate navigational intentions.

CNN has been widely adopted in pattern recognition tasks. Nvidia published the research [6, 7] to demonstrate that CNN can be extended to autonomously steering a vehicle using camera images in an end-to-end fashion. The [6, 7] help shape the recent research trend in autonomous driving as we see an increasing number of researchers are utilizing the end-to-end approach. However, this model does not accept navigational command, thus cannot make meaningful turns. The research [10] proposed the conditional imitation learning (CIL) models which incorporates the HLC into the model so the vehicle can be navigated. Although the proposed models take navigational command into account, they use a simple CNN as the image module, which does not generalize well to the new environment.

In [11], the result suggests using Mean Absolute Error (MAE) of the trained model during offline evaluation has a higher correlation to the online performance comparing to using Mean Squared Error (MSE), therefore, MAE should be used as the loss function. In [12], depth information is used to supplement the RGB image, and the early fusion scheme which improves upon CIL produces a better result compared to using RGB image alone. Some works incorporate temporal information which is important for improving the performance [14, 15]. In [16], the author proposed to use pre-trained ResNet as the image module and utilize speed regularization to further improve the performance. In [17], the author proposed to perform reinforcement learning using the trained weight from imitation learning as the initial weight, which will greatly reduce the time required to converge.

Although multiple enhancements have been made to the basic model of end-to-end learning [6, 7]. We find none of the models satisfies all our design goals, including an image module that generalizes well to new environments, accepting HLC for navigation, and utilizing temporal information to make a better decision in a dynamic environment. Therefore, our proposed model is built upon the command input model in CIL [10]. We incorporate the idea of using pre-trained and deeper CNN [16] to better capture image features and generalize to the new environment. However, instead of using ResNet in [16], we use MobileNet which is more efficient. We are also inspired by [14, 15] to explore temporal information using the Long Short-Term Memory (LSTM). Furthermore, we adopt the idea proposed by [16] to perform speed regularization to force the image module to better learn speed-related features. Finally, as it is suggested in [11], we use the MAE as the loss function for a stronger correlation between online and offline evaluation results.

3 Proposed Method: Temporal Conditional Imitation Learning (TCIL)

In this work, we propose the Temporal Conditional Imitation Learning (TCIL) model. The features of the model are described below.

(1) **Vision-based**: The proposed model is vision-based where the primary perception source is a center-mounted RGB camera. By using RGB cameras, we can keep sensor cost and complexity low, which makes the model more feasible to deploy in the real world.

Furthermore, processing 2D RGB images require less computing power compared to 3D point cloud data, which compensates for the additional computation introduced by the LSTM network.

(2) **End-to-End Learning Paradigm**: The model follows the end-to-end learning paradigm. In this work, we improved upon the command input model in [10], and a CNN-LSTM network is used to map image, speed, and HLC inputs directly into vehicle control signals.

(3) **Command Input Model**: Accepting navigational command is a key part to achieve autonomous driving. From our preliminary study, we find using HLC in the form of one-hot encoding as input is sufficient to communicate the navigational intention to the model, therefore, we design our model based on the command input model proposed in [10].

(4) **Utilizing Temporal Information with LSTM**: Our proposed model utilizes the LSTM network to explore the temporal information which exists in a sequence of observations. Temporal information is critical to improving the reliability of the autonomous driving model as it helps the model to learn the sense of speed and relative movement which is especially important in avoiding the collision.

(5) **Deep, Efficient, Pre-Trained Image Module**: The proposed model uses MobileNet as the image module. It is designed to run on low-powered devices such as mobile devices with little impact on accuracy. We strive to keep the computation requirement of the image module low as the LSTM network requires the image module to be executed multiple times for each timestep. Furthermore, the MobileNet network is pre-trained on the ImageNet dataset, by using transfer learning, we can further improve the generalization of our model.

(6) **Speed Prediction Regularization and Loss Function**: The proposed model incorporates speed regularization, which uses a separate branch for vehicle speed prediction based on the features from the image module [16]. It forces the image module to learn speed-related features, therefore, the overall model better understands the vehicle dynamic and controls the speed without solely relying on the speed input. Our proposed model also adopts using MAE as the loss function during the training instead of MSE as MAE yields a stronger correlation between the offline evaluation result and online simulation evaluation result. It gives us more confidence in selecting the best-trained model for further evaluation according to offline evaluation results only.

(7) **System Structure during Offline Training**: The overall system structure during offline training is shown in Fig. 2. Expert driving demonstration contains sequences of observations of camera images and the synced control signals, vehicle speeds, and HLCs. During training, we sample from the original sequences to form short and fixed-length observation sequences to be consumed by the TCIL model. The model predicts the control signals of the vehicle with steering, throttle, and brake, as well as the vehicle speed. By using MAE as the loss function, we perform backpropagate on the model to update the weights.

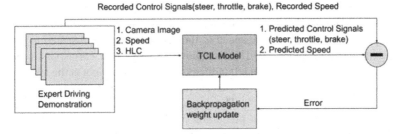

Fig. 2. System structure during offline training

4 Proposed Model: CNN-LSTM Network

The structure of the TCIL is shown in Fig. 3. The overall model is based on CNN-LSTM network structure, which is often used in sequence prediction tasks such as video sentiment classification. The model is inspired by the command input model in [10] where the HLC is used as a part of the input, and a single action prediction module is used for all HLC types. Furthermore, the model incorporates speed prediction regularization features proposed in [16], and the LSTM network to explore temporal information [15].

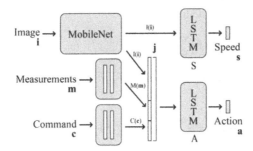

Fig. 3. Temporal conditional lmitation learning model

The proposed model includes the following, and described in detail below:

- Three input modules, namely, the image module **I**, the measurement module **M**, and the command module **C**.
- Two prediction branches, including speed prediction branch **S**, and action prediction branch **A**.

(1) The Image Module **I:** It is mainly responsible for learning image features. We find MobileNet works well for our model as it balances efficiency and performance. We use the full-sized MobileNet pre-trained on the ImageNet dataset. All trainable layers are unfrozen during training to enable transfer learning which lets the pre-trained MobileNet learn driving-related features. The image input **i** is represented as 200×88 pixels RGB image in our experiment.

(2) The Measurement Module **M** and Command Module **C:** They share the same structure. Two fully connected layers of size 128 are used in our experiment. Currently, the measurement **m** consists of the speed of the vehicle and is represented as a scalar value. The command **c** is represented by a one-hot encoding vector, corresponding to four HLC types: following the lane, turn left at the intersection, turn right at the intersection, and go straight. ReLU nonlinearity is used as the activation function.

(3) Action Prediction Module **A:** The outputs of the three input modules are concatenated, then fed into the action prediction module **A**. We use an LSTM network with 64 nodes, which allows us to explore the temporal information that exists in input sequences [15]. We organize the input **i, m, c** into sequences equally sampled from previous timesteps, and apply modules **I, M, C** in a timely distributed manner.

(4) Speed Prediction Module **S:** We jointly trained an LSTM speed prediction module **S** with 64 nodes connected to the image module. The prediction is also used to regulate the speed of the agent vehicle during simulation tests, such as avoiding the vehicle being stopped due to causal confusion.

(5) TCIL as Controller **F:** At each time step **t**, we use an input sequence consisting of **n** observations equally sampled from current and previous timesteps, denoted as images $\mathbf{i^n_t}$, speed measurement $\mathbf{m^n_t}$, and HLC $\mathbf{c^n_t}$, to make each action prediction $\mathbf{A(i^n_t, m^n_t, c^n_t)}$ and speed prediction $\mathbf{S(i^n_t)}$. The training dataset consists of such observations and ground truth action pairs recorded from the experts, and is denoted as $\mathbf{D = \{((i^n_j, m^n_j, c^n_j), (a_j, s_j))\}^N_{j=1}}$. The TCIL model acts as a controller **F**; the goal is to find trained weights θ that will minimize the loss:

$$\underset{\theta}{\text{argmin}} \sum_j L(F(i^n_j, m^n_j, c^n_j), (a_j, s_j)) \tag{1}$$

(6) Loss Function MAE: In our work, we use MAE as the loss function, which is defined as:

$$L = MAE = \frac{1}{n} \sum \left(\lambda |a - a_{gt}| + (1 - \lambda)|s - s_{gt}| \right) \tag{2}$$

5 Experiment

5.1 Dataset

The dataset consists of 40 GB of expert demonstration including RGB camera images, steer, brake, throttle signals, speed measurements, and HLCs collected from the CARLA simulator [10]. The data is recorded from map Town 1 with three different weather conditions. The map town 2 is exclusively used for testing. Furthermore, this dataset is consistent with the conditions in the CoRL2017 CARLA benchmark, thus, we can directly compare the performance of our model with other previous works. There are both human and machine driving footage included, and the vehicles are driven at a speed below 60 km/h while avoiding collision with obstacles. Traffic rules such as traffic lights, and stop signs are ignored in the dataset, therefore, our trained model will not be able to follow traffic lights or stop at stop signs.

5.2 Training

The original dataset was processed by the data generator which is responsible for splitting the data into a training set (80%) and a validation set (20%), generating an input sequence of length 5 with a sample interval of 3 for the CNN-LSTM network, scaling the input values including RGB images and speed measurements, and augment the RGB images for better generalization.

For the training parameters, we use Adam optimizer with an initial learning rate of 0.0002 and MAE as the loss function. We use a batch size of 64 and train the model in an episodic way where, in each epoch, all training data is used. Each model is trained for 60 epochs, and the network weights are saved after each epoch.

5.3 Evaluation

We evaluate the model with the smallest validation loss on the CoRL2017 CARLA benchmark to get a set of benchmark scores. The CoRL2017 CARLA benchmark uses the same sensor setup as our training dataset and consists of 4 tasks to be performed in a combination of seen and unseen weather and maps. The CoRL2017 CARLA benchmark has been used in multiple previous works; therefore, we can have a direct comparison.

6 Results

6.1 Comparison with the State-of-the-Art

Table 1. Success rate % comparison with state-of-the-art on CARLA benchmark.

Tasks	Training conditions					New town and New weather				
	CIL [10]	CIRL [17]	CILRS [16]	AEF [12]	TCIL (new)	CIL [10]	CIRL [17]	CILRS [16]	AEF [12]	TCIL (new)
Straight	98	98	96	**99**	98	80	**98**	96	96	96
One Turn	89	97	92	**99**	94	48	80	92	84	**98**
Nav	86	93	**95**	92	93	44	68	92	90	**94**
Nav. Dyn	83	82	92	89	**96**	42	62	90	**94**	88

We start analyzing our results in terms of the success rate on the CoRL2017 CARLA benchmark. In each task of the benchmark, there are multiple tracks for the agent to finish. The success rate measures the percentage of tracks the agent successfully finished on average among different weathers. Here, we report the best result in 5 runs. Our proposed model is referred to as TCIL, and it is compared with the state-of-the-art models including CIL [10], CIRL [17], CILRS [16], AEF [12]. To better utilize the speed prediction, we use the predicted speed to regulate our agent vehicle speed to prevent the vehicle stops due to causal confusion, and to prevent the vehicle over speed. We also limit the top speed of our agent to 45 km/h. The results are shown in Table 1.

There are four tasks in the CoRL2017, namely, straight, one turn, navigation, navigation in dynamic traffic. And there are two sets of evaluation conditions we are interested in. The training condition only contains the map and weather that are included in the training data, while the new town and new weather condition only contain the map and weather that never show up in the training data. We interpret the result under training conditions mostly as the model's capability in driving autonomously and the result under the new condition mostly as the model's capability in generalization to the new environment.

Training Conditions
In the straight task, which is the easiest one, the AEF model performs the best at 99% while our model is second-high at 98%. For the one-turn task, the agent is required to navigate a track including making one single turn at the intersection. The AEF model still takes the lead with a 99% success rate while our model is not far behind at 94%. In navigation tasks, the agent vehicle is required to navigate tracks with multiple turns, including driving straight in the multiple way intersections. The CILRS model takes the lead at a 95% success rate while the ours model has the second-high success rate at 93%. Finally, in navigation in a dynamic environment, our model performs the best with a 96% success rate while the second highest is CILRS at 92%.

Although our model does not top at every task, our model still has advantages over different models. TCIL outperforms the base CIL model in every task, and the improved CILRS model in three out of the four tasks. The reason we directly compare our model with CIL and CILRS is that all three models are trained similarly with imitation learning and take the same kind of input. The AEF model uses additional depth input, therefore, it has a slight advantage over our model in terms of input, while our model still outperforms it in the most difficult navigation in dynamic environment task. Although the CIRL is designed to improve both the driving performance and generalization, and we are only able to beat this model in navigation in the dynamic environment task, the CIRL does not generalize well in new environments. The result shows, our model has an edge in handling complex and dynamic driving environments.

New Town and New Weather Conditions
In the new town and new weather conditions, the results show the generalization capability of the models. In the navigation task, our model tops at 94% with CILRS and AEF keeping up at 90% and 92%. We observe a large performance drop for CIL and CIRL, which means both models do not generalize well to new environments, and this trend also applies to the navigation in dynamic environment tasks. In navigation in dynamic environment tasks, the AEF model impressively outperforms the rest at a 94% success rate, while our model is third highest at 88%. We see a trend that the models without pre-trained and deeper image modules, including CIL, CIRL lose performance dramatically in complex tasks in the new environment. And with a deeper pre-trained image network and LSTM for decision network, our TCIL tops at two out of four tasks. It shows our model generalizes relatively well to new environments, although we still need to work

on the performance and generalization issue in the navigation in dynamic environment tasks. In our case, the failure comes from the vehicle being confused by the observation and it stops in place until the episode times out.

6.2 Ablation Study

In addition to the comparison with state-of-the-art models, we also want to show the effect of different components of our model. To examine the effectiveness of each component, we remove the target component from the complete TCIL, then, train and evaluate the reduced model. The success rate of the CoRL2017 CARLA benchmark in the ablation study is shown in Table 2.

Table 2. Success rate % among ablated versions.

Tasks	Training conditions				New town and New weather			
	TCIL	TCIL w/o LSTM	TCIL w/o Speed	TCIL w/o MobileNet	TCIL	TCIL w/o LSTM	TCIL w/o Speed	TCIL w/o MobileNet
Nav.	**93**	84	79	88	**94**	76	78	52
Nav. Dyn..	**96**	88	76	87	**88**	84	70	44

Without LSTM: We replace the LSTM action prediction branch and speed prediction branch with fully connected networks in the same structure as in the CIL paper [10]. The results show, without the LSTM network, both the driving performance and generalization capability are lowered, especially in complex tasks. We also observe that the agent vehicle does not control its speed well where the vehicle is more likely to reach the speed limit, and crash in turns due to the vehicle speed being too high. Therefore, we consider the LSTM network has a positive impact on improving the model in both driving performance and generalization capability.

Without Speed Prediction Regulation: We remove the speed prediction branch and train the model. The results suggest the model experienced a large performance drop. We also observe that the vehicle stops more frequently due to casual confusion. We conclude that speed regularization can greatly reduce the effect of casual confusion. Incorporating speed regularization helps in the overall performance of our model.

Without MobileNet: Lastly, we replace the MobileNet with the CNN used in [10]. We observe a big performance drop in complex tasks, especially in the new town and new weather conditions. We observe many failed cases in bad weather, and the vehicle gets in the wrong lane and sidewalk frequently. The results suggest that the pre-trained MobileNet better captures driving-related features and contributes to both driving performance and generalization capability.

7 Conclusion

Autonomous driving is indispensable in future smart cities. We proposed the TCIL (Temporal Conditional Imitation Learning) model. The TCIL model aims to improve both driving performance and generalization capability, especially in dynamic navigations, by exploring temporal information via the CNN-LSTM structure, incorporating speed prediction and regularization, and using MobileNet pre-trained on ImageNet data. Performance evaluation based on the CARLA benchmark has shown that the proposed model is competitive compared to the state-of-the-art models; outperforms them in most tasks especially in navigating dynamic environments and in the new town and new weather conditions. Future works include improving the dataset, utilizing depth information, and exploring reinforcement learning to further enhance the generalization to a more variety of driving scenarios.

Acknowledgment. Melody Moh and Teng Moh are supported in part by a research grant from Dell and by SJSU RSCA Awards (2018–2023).

References

1. Marisi, A.C., Lovas, R., Kisari, A., Simonyi, E.: A novel IoT platform for the era of connected cars. In: Proceedings of 2018 IEEE International Conference on Future IoT Technologies, January 2018
2. National Highway Traffic Safety Administration (NHTSA), "Automated Vehicles for Safety", 15 June 2020. www.nhtsa.gov/technology-innovation/automated-vehicles-safety.
3. Zhu, M., et al.: Public vehicles for future urban transportation. IEEE Trans. Intell. Transp. Syst. **17**(12), 3344–3353 (2016). https://doi.org/10.1109/TITS.2016.2543263
4. Rao, Q., Frtunikj, J.: Deep learning for self-driving cars: chances and challenges. In: 2018 IEEE/ACM 1st International Workshop on Software Engineering for AI in Autonomous Systems (SEFAIAS), Gothenburg, pp. 35–38 (2018)
5. Janai, J., Güney, F., Behl, A., Geiger, A.: Computer vision for autonomous vehicles: problems, datasets and state-of-the-art. Comput. Res. Repository (2021). arXiv:1704.05519
6. Bojarski, M., et al.: End to End Learning for Self-Driving Cars, preprint submitted to arXiv, Nvidia, 25 April 2016
7. Bojarski, M., et al.: Explaining How a Deep Neural Network Trained with End-to-End Learning Steers a Car, Nvidia, 25 April 2017
8. Chen, Z., Huang, X.: End-to-end learning for lane keeping of self-driving cars. In: 2017 IEEE Intelligent Vehicles Symposium (IV), Los Angeles, CA, pp. 1856–1860 (2017)
9. Alexey, D.: CARLA: an open urban driving simulator. In: Proceedings of the 1st Annual Conference on Robot Learning, pp. 1–16 (2017)
10. Codevilla, F., Müller, M., López, A., Koltun, V., Dosovitskiy, A.; End-to-end driving via conditional imitation learning. In: 2018 IEEE International Conference on Robotics and Automation (ICRA), Brisbane, QLD, pp. 4693–4700 (2018)
11. Codevilla, F., López, A.M., Koltun, V., Dosovitskiy, A.: On offline evaluation of vision-based driving models. In: Ferrari, V., Hebert, M., Sminchisescu, C., Weiss, Y. (eds.) ECCV 2018. LNCS, vol. 11219, pp. 246–262. Springer, Cham (2018). https://doi.org/10.1007/978-3-030-01267-0_15

12. Xiao, Y., Codevilla, F., Gurram, A., Urfalioglu, O., López, A.M.: Multimodal end-to-end autonomous driving. IEEE Trans. Intell. Transp. Syst. (2020). https://doi.org/10.1109/TITS.2020.3013234

13. Wang, Q., Chen, L., Tian, B., Tian, W., Li, L., Cao, D.: End-to-end autonomous driving: an angle branched network approach. IEEE Trans. Veh. Technol. **68**(12), 11599–11610 (2019). https://doi.org/10.1109/TVT.2019.2921918

14. Chi, L., Mu. Y.: Learning end-to-end autonomous steering model from spatial and temporal visual cues. In: Proceedings of the Workshop on Visual Analysis in Smart and Connected Communities - VSCC 2017 (2017). https://doi.org/10.1145/3132734.3132737

15. Haavaldsen, H., Aasbø, M., Lindseth, F.: Autonomous vehicle control: end-to-end learning in simulated urban environments. In: Bach, K., Ruocco, M. (eds.) NAIS 2019. CCIS, vol. 1056, pp. 40–51. Springer, Cham (2019). https://doi.org/10.1007/978-3-030-35664-4_4

16. Codevilla, F., et al.: Exploring the limitations of behavior cloning for autonomous driving. In: 2019 IEEE/CVF International Conference on Computer Vision (ICCV) (2019)

17. Liang, X., Wang, T., Yang, L., Xing, E.: CIRL: controllable imitative reinforcement learning for vision-based self-driving. In: Ferrari, V., Hebert, M., Sminchisescu, C., Weiss, Y. (eds.) ECCV 2018. LNCS, vol. 11211, pp. 604–620. Springer, Cham (2018). https://doi.org/10.1007/978-3-030-01234-2_36

A Novel Estimation Method for the State of Charge of Lithium-Ion Battery Using Temporal Convolutional Network Under Multiple Working Conditions

Yuefeng Liu[1]($\boxtimes$) (ID), Jiaqi Li[1], and Neal N. Xiong[2]

[1] School of Information Engineering, Inner Mongolia University of Science and Technology, Baotou 014010, Inner Mongolia, China
liuyuefeng@imust.edu.cn
[2] Department of Mathematics and Computer Science, Northeastern State University, Tulsa, OK 74152, USA
xiong31@nsuok.edu

Abstract. Battery state of charge (SOC) is the available capacity of a battery expressed as a percentage of its nominal capacity. It is essential to estimate the SOC accurately for the normal use of Li-ion batteries equipment. However, SOC cannot be measured directly, but can only be estimated indirectly by measurable variables. Both the traditional methods and the methods based on adaptive filtering algorithms have some limitations. According to the nonlinear characteristics of Li-ion batteries under actual working conditions, this paper applies a new deep learning method, temporal convolutional network (TCN), which is firstly used to estimate the SOC of Li-ion batteries. This method can directly map the voltage, current, and temperature that can be observed and measured during the operation of Li-ion batteries into the SOC without using the battery model or setting additional parameters. Moreover, it can learn and update parameters by itself during the training process. Only one model is needed to estimate the SOC under different temperatures and working conditions. The experimental results show that the proposed method achieves a low average absolute error of 0.82% at a fixed temperature and 0.67% at multiple temperatures with relatively simple model complexity.

Keywords: Li-ion batteries · Electric vehicles · State of charge estimation · Neural networks · Temporal Convolutional Network (TCN)

1 Introduction

Under the dual drive of improving climate change and promoting sustainable growth, the global energy transformation is underway. With the continuous development of renewable energy technology, the power sector has witnessed impressive rapid development. Electrification is becoming a key solution to reduce emissions [1]. Due to its high energy density, low self-discharge performance, close to zero memory effect, high open-circuit

© Springer Nature Switzerland AG 2021
M. Qiu (Ed.): SmartCom 2020, LNCS 12608, pp. 30–39, 2021.
https://doi.org/10.1007/978-3-030-74717-6_4

voltage, and long service life, the lithium-ion battery has attracted extensive attention as a promising energy storage technology. In recent years, lithium-ion battery with high energy density is considered an ideal power supply for electric vehicles (EV) and hybrid electric vehicles (HEV) in the automotive industry [2].

SOC is one of the important characteristics of a battery. It is defined as the ratio of the residual capacity of the battery to its nominal capacity. Usually, it can be used to describe the current state of the battery in use, and it is an important part of the battery management system (BMS). However, the battery SOC cannot be directly measured. It needs to be indirectly estimated by the observable variables of the battery, such as voltage, current, and temperature. And it is affected by many uncertain factors including battery aging, environmental temperature change, and various vehicle driving conditions. Therefore, how to accurately estimate the battery SOC is a complex task [3].

With the development of artificial intelligence technology [4, 5], there are a number of studies that suggest that machine learning and deep learning techniques can be well applied to estimate the SOC of Li-ion batteries with higher accuracy and stronger robustness, such as support vector machine [6–8], and neural network [9–12]. Most researches on time series tasks have been carried out about the structures of Recurrent Neural Network (RNN), such as Long Short-Term Memory (LSTM) and Gate Recurrent Unit (GRU), which always been the first choice for most researchers [13]. In the paper [9], the LSTM-RNN model is used to accurately estimate the SOC of Li-ion batteries. In papers [10], the GRU-RNN model is used. However, some researches have established that Convolutional Neural Network (CNN) also showed good advantages in some time series tasks [14–16]. Although CNN was originally designed for computer vision tasks, its ability to extract high-level features from data through the convolutional structure is also applicable to time series data. At the same time, the advantage of CNN parallel speed is more suitable for real-time applications.

Recently, a new neural network TCN that combines the advantages of RNN and CNN has been applied to time series tasks. With its specially designed convolution structure, it has been used in forecasting energy-related time series [17], urban traffic prediction [18] which shows surprising results.

In this paper, we will introduce how the TCN model can be self-learning and estimate SOC according to measurable parameters of Li-ion battery. Specifically, the main contributions of this paper are as follows:

1. Using the TCN model, the voltage, current, and temperature measured during the use of a lithium-ion battery can be directly mapped to the SOC without using a battery model or performing a lot of calculations.
2. Through the back-propagation algorithm, the TCN model can learn and update parameters by itself during the training process, without the need for complex manual design and parameter settings.
3. Through training, only one TCN model is needed to estimate the SOC under different temperatures and working conditions. And there is no need to design new battery models or adjust parameters according to different conditions.

In the rest of this article, the second part introduces the theory of TCN and how dilated causal convolution can effectively capture long-term historical data information. The third part will introduce the data set and TCN model architecture we use, and then verify the performance of TCN by the tests including the data recorded at fixed as well as at varying ambient temperatures. Meanwhile, it is also compared with mainstream deep learning methods.

2   Our Proposed TCN for SOC Estimation

The large amount of measurement data collected during battery usage can provide valuable information about battery status. The battery management system can use this data to estimate battery SOC and adjust battery usage strategy. Currently, many deep learning models (such as RNN or CNN) have been proposed to deal with time series forecasting problems. And these methods can automatically capture complex patterns in time series by learning training data. Bai et al. [19] analyzed the above two kinds of networks and combined the advantages of the two networks, and proposed the TCN. Through the specially designed dilated causal convolution structure, TCN has the advantages of parallelism, flexible receiving area size, stable gradual change, low memory required for training, and variable-length input.

2.1   Causal Convolutions

The TCN network adopts a zero-filled one-dimensional full convolutional network (FCN) structure to ensure that the input and output lengths of each hidden layer are the same so that the model can be adjusted to any length as needed and can receive historical time data of different sizes. As well, in order to ensure that the past or future battery information will not be disclosed, in causal convolutions, the output at time t is convolved only with elements from time t and earlier in the previous layer. The formula is shown in Eq. (1), and Fig. 1 (a) shows a causal convolution block.

$$TCN = 1D\,FCN + causal\ convolutions \tag{1}$$

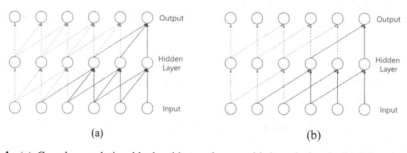

(a)   (b)

Fig. 1. (a) Causal convolution block with two layers with kernel size 3, (b) Dilated causal convolution block with two layers with kernel size 2, dilation rate 2

2.2 Dilated Convolutions

TCN model can capture information from time series for estimating SOC. A long time series often contains more information. However, longer time series means that the model needs a larger receptive field to capture information. When dealing with a large receptive field, usually it is necessary to increase the number of layers of the model or the size of the filter. But for TCN, holes can be injected into the standard convolution to increase the receptive field. In Fig. 1(b), a dilated causal convolution block is shown.

For a one-dimensional input sequence $x \in X$ and filter $f : \{0, \cdots, k-1\} \to \mathbb{R}$, the extended convolution operation F of sequence element s is defined as:

$$F(s) = (x *_d f)(s) = \sum_{i=0}^{k-1} f(i) \cdot x_{s-d \cdot i} \tag{2}$$

Where d is the dilate factor, k is the filter size, and $s - d \cdot i$ represents the past direction. Therefore, dilation is equivalent to introducing a fixed step between each two adjacent filter taps. When $d = 1$, the expansion convolution becomes regular convolution. By using larger expansion, the output of the top layer can represent a larger range of inputs without the loss of pooling information, thus effectively expanding the receiving range of the convolution network, and give it the ability to learn longer historical battery data.

The introduction of dilated causal convolution also makes the TCN model have a smaller network size when there is a large receptive field, which is suitable for the vehicle system.

2.3 Residual Block

Since the receptive field of the TCN model depends on the network depth, filter size, and dilate factor, it is important to stabilize the deeper and larger TCN model. The residual connection can eliminate the influence of partial gradient disappearance and explosion of the deep network to a certain extent, and it makes the network transmit information across layers, which is beneficial to a very deep network.

The residual block includes a branch, and its conversion result is added to input x by a transformation F:

$$o = Activation(x + F(x)) \tag{3}$$

We employ a generic residual module in place of a convolutional layer. The input of the residual block in TCN goes through a residual function F(x) which consist of two rounds of dilated causal convolution, weight normalization, activation function, and dropout. When the input and output of the residual block are of different dimensions, it will be transformed by a 1D convolution. Figure 2 shows the residual block of TCN.

2.4 Model Overall Structure

Figure 3 shows the model architecture used in this article. The model starts with a sequence input layer, where the input vector is the measured signals including voltage, current, and temperature: $x(k) = [V(k), I(k), T(k)]$. Then, the TCN layer which can contain multiple stacks is followed to learn the dependence on input. Meanwhile, in the TCN layer, the residual module including a 1D convolution is employed to replace the convolutional layer. Finally, a fully connected layer is used to estimate the SOC: $y(k) = [SOC(k)]$. The input sequence is processed by dilated causal convolution and extracted into more advanced features, and identity mappings aid training is introduced.

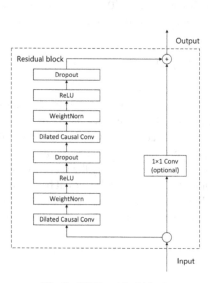

Fig. 2. TCN residual block   **Fig. 3.** Architecture of the TCN model

3 Performance Analysis

3.1 Dataset

The Panasonic 18650PF Li-ion battery dataset [20] is used in this paper. The nominal voltage and capacity of the Panasonic 18650PF battery are 3.6 V and 2.9 ah respectively. The data of nine driving cycles measured by the battery at 25 °C, 10 °C, and 0 °C were collected in the dataset, which can provide a wide range of battery information for the training of the model. Among them, US06 Supplemental FTP Driving Schedule (US06), Highway Fuel Economy Driving Schedule (HWFET), Urban Dynamometer Driving Schedule (UDDS), and LA92 Dynamometer Driving Schedule (LA92) are the four common driving cycles of electric vehicles. Cycle 1–4 consist of a random mix of US06, HWFET, UDDS, LA92, and the Neural Network driving cycles (Cycle NN) consists of a combination of portions of US06 and LA92 driving cycles and was designed to have some additional dynamics which may be useful for training neural networks. The recorded data includes the voltage, current, capacity, and battery temperature measured under the above conditions.

3.2 Experimental Settings

The proposed method is implemented by python and tensorflow keras, and the experiments were performed with an NVIDIA Tesla V100 32G.

TCN model uses 240 time-steps as historical data, the input vector of this model is defined as $[V(k), I(k), T(k)]$, where $V(k)$, $I(k)$, $T(k)$ represent voltage, current, and temperature at time k respectively. The architecture of the TCN model is shown in Fig. 3. It has 6 residual blocks with 4 convolution layers of kernel size 6, 64 filters, and the dilation of each layer is 1, 2, 4, 8. Finally, a single SOC value is an output through a fully connected layer. The model uses the Adam optimization algorithm and keras ReduceLROnPlateau learning rate reduction strategy during training. Using Adam to optimize the network parameters, the learning rate is 0.001, the β1 is 0.9, and the β2 is 0.999. The index of learning rate decline strategy monitoring is MAE, the learning rate reduction factor is 0.5, and patience is 5. The batch size and epoch are 32 and 100 respectively. At the same time, in order to facilitate the model training, we normalized data to the range $[-1,1]$ as shown in Eq. (4):

$$x_{norm} = \frac{2(x - x_{min})}{x_{max} - x_{min}} - 1 \tag{4}$$

where x_{max} and x_{min} are the maximum and the minimum values of data, x represents initial data, and x_{norm} represents data after normalization.

In order to evaluate the estimation performance of the model, three error functions Mean Absolute Error (MAE), Root Mean Squared Error (RMSE), and maximum error (MAX Error) are used to verify the estimation effect of the model.

3.3 SOC Estimation at Fixed Ambient Temperatures

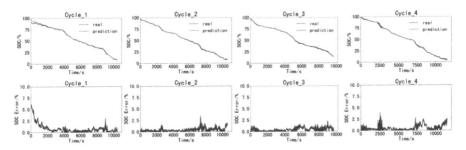

Fig. 4. Estimation results of fixed-temperature model at 25 °C

In this section, we use the dataset collected at 25 °C, 10 °C, and 0 °C to train models respectively, and evaluate the performance of these. For each temperature, the model was trained by US06, HWFET, UDDS, LA92, and Cycle NN, and tested on four hybrid driving cycle data include Cycle1 to Cycle4. The SOC estimated by the model trained at a single temperature is shown in Table 1 and Fig. 4 shows the estimation of SOC by

Table 1. Estimation results at fixed ambient temperatures

Temperature	Condition	MAE (%)	RMSE (%)	MAX Error (%)
25 °C	Cycle1	0.78	1.27	6.19
	Cycle2	0.40	0.54	3.44
	Cycle3	0.41	0.54	2.23
	Cycle4	0.55	0.77	3.97
10 °C	Cycle1	0.55	0.72	3.61
	Cycle2	0.82	0.95	5.37
	Cycle3	0.57	0.74	2.90
	Cycle4	0.61	0.83	3.31
0 °C	Cycle1	1.71	2.48	7.34
	Cycle2	1.57	2.24	7.68
	Cycle3	1.07	1.21	3.08
	Cycle4	0.81	0.99	3.48

TCN model and the error over time at 25 °C. Each subgraph contains the estimation curve and error curve of SOC.

The experiment shows that the average MAE and RMSE of SOC estimation results were 0.54%, 0.64%, 1.29% and 0.78%, 0.81%, 1.73% at three fixed temperatures of 25 °C, 10 °C, and 0 °C.

3.4 SOC Estimation at Varying Ambient Temperatures

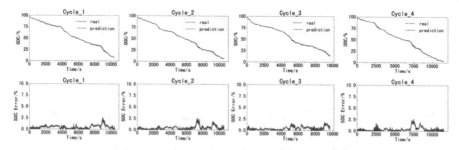

Fig. 5. Estimation results of multi-temperature model at 25 °C

In this section, the model is trained with a larger dataset, including 27 driving cycles collected at 25 °C, 10 °C, and 0 °C. Among them, all the US06, HWFET, UDDS, LA92, Cycle NN driving cycle data (15 driving cycles) were used as train dataset; all the cycle 1 to cycle 4 hybrid driving cycles (12 driving cycles) were used as test dataset. Unlike the

Table 2. Estimation results at varying ambient temperatures

Temperature	Condition	MAE (%)	RMSE (%)	MAX Error (%)
25 °C	Cycle1	0.47	0.60	2.61
	Cycle2	0.44	0.60	2.57
	Cycle3	0.43	0.56	1.84
	Cycle4	0.36	0.57	2.67
10 °C	Cycle1	0.59	0.82	4.22
	Cycle2	0.76	0.90	3.42
	Cycle3	0.67	0.80	2.23
	Cycle4	0.60	0.88	3.55
0 °C	Cycle1	1.15	1.57	4.09
	Cycle2	1.02	1.28	3.86
	Cycle3	0.88	1.09	3.46
	Cycle4	0.64	0.76	1.88

previous section, in this section, we hope to estimate the SOC of the battery at different temperatures using only one model.

The results of model estimation after training are shown in Table 2 and Fig. 5 shows the estimation at 25 °C. The MAE and RMSE estimated by the model at three temperatures are 0.43%, 0.66%, 0.92%, and 0.59%, 0.85%, 1.18%, respectively. In total, the average MAE and RMSE are 0.67% and 0.87%. The results show that the nonlinear relationship between the measured parameters and SOC at different temperatures can be learned by only one TCN model.

3.5   Comparative Analysis of Experimental Results

We compare the LSTM model and the GRU model which two also use the Panasonic 18650PF Li-ion battery dataset. Similar to the method described in this paper, both the LSTM model and the GRU model take voltage, current, and temperature as input parameters to directly estimate SOC value. And both of them adopt multiple driving cycles under three temperature conditions of 25, 10, and 0 for model training, and can estimate the SOC under multiple working conditions through a single model.

Table 3 Shows the estimation of various temperature and working conditions by the three models. MAE and RMSE are the average values of test results in different working conditions under the three temperature conditions, and MAX is the maximum Error of all the tests. The results show that the TCN model proposed in this paper performs better than the two RNN models.

Table 3. Comparison of estimation results with other neural network model methods

Method	Error
LSTM-RNN [9]	MAE = 1.21%, RMSE = 1.52%, MAX = 6.687%
GRU-RNN [10]	MAE = 0.86%, MAX = 7.59%
TCN (estimator in this paper)	MAE = 0.67%, RMSE = 0.87%, MAX = 4.22%

4   Conclusion

This paper describes how the voltage, current, and temperature of the TCN model are directly mapped to the SOC of the battery. Compared with other methods mentioned in literatures, the TCN model shows good advantages. During the training process, the model can learn and update parameters by itself, and the model has good accuracy and robustness. Only need one model can estimate the SOC of Li-ion batteries under different temperatures and operating conditions. Compared with the traditional methods, this method does not need to manually establish the battery model or determine and calculate a large number of parameters, but directly maps of battery measurement parameters to SOC which is more suitable for the vehicle system. At the same time, the performance of the model is verified by a public dataset. In conclusion, TCN has proved to be a powerful deep learning network, which can be applied to estimate SOC of Li-ion batteries and other similar time series tasks. For the future work, we need to study the dependence of the model on the amount of data, and how to further compress the network, reduce the network size without affecting its estimation performance, so as to make it more suitable for onboard real-time estimation.

Acknowledgement. This work was supported by Inner Mongolia Natural Science Foundation (2018MS06019).

References

1. Gielen, D., Boshell, F., Saygin, D., Bazilian, M.D., Wagner, N., Gorini, R.: The role of renewable energy in the global energy transformation. Energy Strat. Rev. **24**, 38–50 (2019)
2. Kim, T., Song, W., Son, D.-Y., Ono, L.K., Qi, Y.: Lithium-ion batteries: outlook on present, future, and hybridized technologies. J. Mater. Chem. A. **7**(7), 2942–2964 (2019)
3. Waag, W., Fleischer, C., Sauer, D.U.: Critical review of the methods for monitoring of lithium-ion batteries in electric and hybrid vehicles. J. Power Sources **258**, 321–339 (2014)
4. Song, Y., Li, Y., Jia, L., Qiu, M.: Retraining strategy-based domain adaption network for intelligent fault diagnosis. IEEE Trans. Industr. Inf. **16**(9), 6163–6171 (2019)
5. Chen, M., Zhang, Y., Qiu, M., Guizani, N., Hao, Y.: SPHA: smart personal health advisor based on deep analytics. IEEE Commun. Mag. **56**(3), 164–169 (2018)
6. Anton, J.C.A., Nieto, P.J.G., Viejo, C.B., Vilán, J.A.V.: Support vector machines used to estimate the battery state of charge. IEEE Trans. Power Electron. **28**(12), 5919–5926 (2013)
7. Sahinoglu, G.O., Pajovic, M., Sahinoglu, Z., Wang, Y., Orlik, P.V., Wada, T.: Battery state-of-charge estimation based on regular/recurrent Gaussian process regression. IEEE Trans. Industr. Electron. **65**(5), 4311–4321 (2017)

8. Sheng, H., Xiao, J.: Electric vehicle state of charge estimation: nonlinear correlation and fuzzy support vector machine. J. Power Sources **281**, 131–137 (2015)
9. Chemali, E., Kollmeyer, P.J., Preindl, M., Ahmed, R., Emadi, A.: Long short-term memory networks for accurate state-of-charge estimation of Li-ion batteries. IEEE Trans. Industr. Electron. **65**(8), 6730–6739 (2017)
10. Li, C., Xiao, F., Fan, Y.: An approach to state of charge estimation of lithium-ion batteries based on recurrent neural networks with gated recurrent unit. Energies **12**(9), 1592 (2019)
11. Song, X., Yang, F., Wang, D., Tsui, K.-L.: Combined CNN-LSTM network for state-of-charge estimation of lithium-ion batteries. IEEE Access. **7**, 88894–88902 (2019)
12. Vidal, C., Kollmeyer, P., Naguib, M., Malysz, P., Gross, O., Emadi, A.: Robust xEV Battery State-of-Charge Estimator Design Using a Feedforward Deep Neural Network. SAE Technical Paper (2020)
13. Yan, B., et al.: An improved method for the fitting and prediction of the number of covid-19 confirmed cases based on LSTM. Cmc-Computers Materials & Continua (2020)
14. Cheng, H., Xie, Z., Shi, Y., Xiong, N.: Multi-step data prediction in wireless sensor networks based on one-dimensional CNN and bidirectional LSTM. IEEE Access. **7**, 117883–117896 (2019)
15. Koprinska, I., Wu, D., Wang, Z.: Convolutional neural networks for energy time series forecasting. In: 2018 International Joint Conference on Neural Networks (IJCNN), pp. 1–8. IEEE (2018)
16. Pei, S., et al.: 3DACN: 3D augmented convolutional network for time series data. Inf. Sci. **513**, 17–29 (2020)
17. Lara-Benítez, P., Carranza-García, M., Luna-Romera, J.M., Riquelme, J.C.: Temporal convolutional networks applied to energy-related time series forecasting. Appl. Sci. **10**(7), 2322 (2020)
18. Zhao, W., Gao, Y., Ji, T., Wan, X., Ye, F., Bai, G.: Deep temporal convolutional networks for short-term traffic flow forecasting. IEEE Access. **7**, 114496–114507 (2019)
19. Bai, S., Kolter, J.Z., Koltun, V.: An empirical evaluation of generic convolutional and recurrent networks for sequence modeling. arXiv preprint arXiv:1803.01271 (2018)
20. Kollmeyer, P.: Panasonic 18650PF Li-ion Battery Data, Mendeley (2018)

Research on Security Methods of Wireless Sensor Networks Based on Internet of Things Technology

Chengjun Yang[1]([⊠]), Minfu Tan[2], and Weihong Huang[3]

[1] Embedded Technology and Artificial Intelligence, Hechi College of Electronic Information Engineering, Hechi University, Hechi 546300, China
05062@hcnu.edu.cn
[2] Big Data Development and Research Center, Guangzhou College of Technology and Business, Guangzhou 528138, China
[3] School of Computer Science and Engineering, Hunan University of Science and Technology, Xiangtan, China

Abstract. Wireless sensor network operation is an important part of the perception layer of the Internet of Things. However, due to its own physical structure and some characteristics of the use environment, wireless sensor network nodes can easily become the main targets of physical attacks and network attacks. These attacks affect the correctness, reliability and real-time performance of the information transmitted by wireless sensor networks. In this article, we first introduce the characteristics of wireless sensor network technology. We further conducts a security analysis on the network intrusion vulnerability in the wireless sensor network environment. Subsequently, we analyze the main solutions to these attacks. Finally, summarize several issues of wireless sensor network security, and look forward to the key research directions in the future.

Keywords: Internet of Things (IOT) · Wireless Sensor Network (WSN) · Wireless Sensor Network Security (WSNS)

1 Introduction

Wireless Sensor Network (WSN) is a new type of information collection and processing technology, which can monitor data in real time and facilitate users to obtain real and reliable data, which has a huge impact on people's lives. These small devices generally exchange data through some low-power, small-range protocols, such as Zigbee, Z-Wave, Thread Bluetooth low energy, IEEE 802.15.4, etc., and then pass various protocols through a smart gateway like the central node performs unified coordination. In recent years, with the influx of a large number of sensor modules into the market, the application of the Internet of Things technology has been greatly popularized, and the WSN use cases of the Internet of Things combination have been widely deployed in the Internet of Vehicles, smart homes, smart cities, healthcare and industrial monitoring applications. The diversity of equipment makes it difficult to unify equipment security standards,

© Springer Nature Switzerland AG 2021
M. Qiu (Ed.): SmartCom 2020, LNCS 12608, pp. 40–49, 2021.
https://doi.org/10.1007/978-3-030-74717-6_5

and it also brings many new challenges to wireless sensor network security issues. Due to the limitations of computing power, storage capacity, battery capacity, installation location, and difficulty in unifying communication protocols, when deploying these wireless sensor nodes, how to perform access control and authentication while taking into account the installation cost, and ensure the sensor information Confidentiality, integrity, freshness, availability, communication security, and non-repudiation are the main issues currently facing [1]. Attackers often use the lack of these security attributes to launch attacks on wireless sensor networks. This article will summarize and comment on these attacks.

According to the different network architectures and specific application ranges, the protocol levels of WSN are not completely the same, and so far, no standard protocol stack has been formed. The most typical situation is that in addition to the five-layer TCP/IP protocol, the protocol stack also includes an energy management platform, a mobile management platform, and a task management platform, as shown in Fig. 1. The physical layer is responsible for providing simple and robust signal modulation and wireless transceiver technology. The data link layer is responsible for medium access control (MAC) and error control. The network layer is responsible for route discovery and maintenance. It is an important protocol layer of WSN and one of the current research hotspots. The transport layer is responsible for providing sensor network data to external networks, such as the Internet. The application layer includes a series of application software based on monitoring tasks. The energy management platform manages how the sensor nodes use energy, and energy saving needs to be considered at each protocol layer. The mobility management platform monitors and registers the movement of sensor nodes, maintains routes to sink nodes, and enables sensor nodes to dynamically track the location of their neighbors. The task management platform balances and schedules monitoring tasks in a given area.

2 Characteristics of Wireless Sensor Networks

Common wireless networks currently include mobile communication networks, wireless local area networks, Bluetooth networks, networks, etc. Compared with these networks, WSN has the following characteristics:

Limited Hardware Resources. Due to the limitations of price, size, and power consumption, nodes have much weaker computing power, storage space, and memory space than ordinary computer functions. This determines that the protocol level cannot be too complicated in the node operating system design.

The Power Supply Capacity is Limited. Nodes are generally powered by batteries, and the battery capacity is generally not very large. Its special application area determines that during use, the battery cannot be charged or replaced. Once the battery energy is used up, this node will lose its function. Therefore, the basic principles of sensor network design must be based on energy conservation.

No Center. There is no strict control center in WSN. All nodes have equal status. It is a peer-to-peer network. Nodes can join or leave the network at any time. The failure of any node will not affect the operation of the entire network. It has strong survivability.

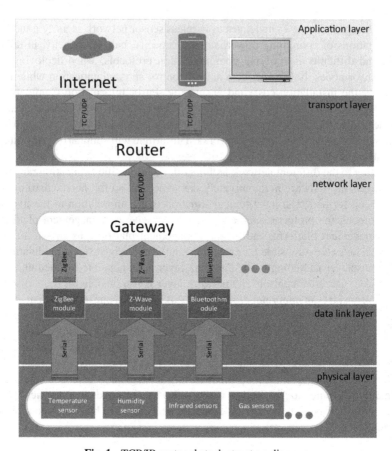

Fig. 1. TCP/IP protocol stack structure diagram

Self-organizing. The layout and deployment of the network do not need to rely on any preset infrastructure. The nodes coordinate their behaviors through layered protocols and distributed algorithms. After the nodes are turned on, they can automatically and quickly form an independent network.

Multi-hop Routing. The communication distance of nodes in the network is limited, generally within a few hundred meters, the node can only communicate directly with its neighbors. If you want to communicate with nodes outside of its radio frequency coverage, you need to route through intermediate nodes. Multi-hop routing in common networks is realized by using gateways and routers, while multi-hop routing in WSN is done by ordinary network nodes without special routing equipment. In this way, each node can be both the initiator and the forwarder of the information.

Dynamic Topology. WSN is a dynamic network, and nodes can move anywhere. A node may exit the network due to battery exhaustion or other failures, or it may be added to the network due to work needs. These will make the topology of the network change at any time, so the network should have the function of dynamic topology organization.

There are a Large Number of Nodes and Densely Distributed. In order to perform monitoring or tracking tasks on an area, there are often thousands of sensor nodes deployed in the area. The distribution of sensor nodes is very dense, and the high connectivity between nodes is used to ensure the fault tolerance and survivability of the system.

3 Several Commonly Used Attack Methods for Wireless Networks

There are various attack methods against sensor networks. According to different system vulnerabilities, hackers will develop different attack plans. These attack plans often require a combination of multiple attack methods. This section will review several representative attacks, listed in total. There are 10 main attack methods and the harm they cause, as shown in Table 1.

Tampering. Directly modify the program that does not set up secure access control, user authentication, or sensors installed in unsafe locations to obtain a legal node control. Use this node to launch malicious code injection, Dos, and man-in-the-middle attacks.

Malicious Code Injection. Attackers can use wireless sensor communication protocol vulnerabilities to implement malicious code injection, such as pairing a virtual Bluetooth device with a smart gateway, and the attacker can obtain system sensitive information and special permissions through the injected malicious code.

Fake Node Injection. The attack method of fake node injection is mainly aimed at the sensor network deployed in the distributed IoT architecture. The attacker inserts a fake node between two legitimate nodes to achieve the purpose of controlling the communication data flow between them. Due to the resource constraints and heterogeneous characteristics of sensor devices, traditional authentication and key management solutions are difficult to implement.

Side Channel Attack. Attackers use some of the weaknesses of sensor equipment such as simple protocol and obvious installation location to analyze the encryption key by detecting side channel information such as power consumption, time consumption, or electromagnetic radiation. With the help of these keys, some confidential data can be encrypted/decrypted.

False Routing Information. By deceiving, tampering or retransmitting routing information, attackers can create routing loops, cause or resist network transmission, extend or shorten the source path, form false error messages, and ultimately achieve the purpose of segmenting the network and increasing end-to-end delay.

Wormhole Attack. Wormhole attacks usually involve two or more malicious nodes working together to launch an attack. Two malicious nodes in different locations will detour the received information and pass it to another malicious node through a private communication channel., Although the two malicious nodes are far apart, there is only one step between the two malicious nodes. In this way, the number of hops through the malicious node will have a great chance of being shorter than the number of hops of the normal path, thereby increasing the chance of obtaining the right of way [3].

Sinkhole Attack. The sewage pool attack makes the leaky node have special attraction to the surrounding nodes in the routing algorithm, and induces almost all the data flow in the area to pass through the leaky node, forming a metaphorical sewage pool centered on the attacker. The cesspool attack changed the direction of network data transmission and disrupted the network load balance. It also provides convenience for other attack methods (such as selective forwarding).

Hello Flood Attack. In many protocols, a node needs to broadcast a Hello message to declare that it is a neighbor of other nodes, and the node that receives the Hello message will assume that it is within the normal wireless transmission range of the sender. A laptop attacker [4] (an attacker with a more powerful device than ordinary sensor nodes) broadcasts routing or other information by using a large enough transmit power to make each node in the network think that the attacker is theirs Neighbors. In fact, the node is far away from the malicious node, and the data packet transmitted with ordinary transmission power may not reach the destination at all, which causes the network to be in a state of chaos and affects the availability of the system. The attacker performing the Hello flooding attack does not need to construct a legitimate data stream. It only needs to rebroadcast the overheard data packet with enough power, so that every node in the network can receive it. Routing protocols that rely on local information exchange between neighboring nodes for topology maintenance and flow control are also vulnerable to this attack.

Table 1. 12 main attack methods and the harm caused by them

Attack name	Harm caused	Possible next attack
Tampering	Special permissions are obtained	Malicious code injection, DOS
Malicious code injection	System sensitive information leakage	DOS
Fake node injection	Control communication data flow	Man in the middle attack
Side channel attack	Data breach	——
False routing information	Split network, network delay	——
Wormhole attack	Disrupt routing	Sinkhole attack
Sinkhole attack	Disrupt network load balance	Selective forwarding
Hello flood attack	Network chaos	——
Selective forwarding	Undermine the integrity of information	——
Sybil attack	Unfair distribution of network resources	——

(Note: —— means not verified)

Selective Forwarding. When hackers use selective forwarding attacks, malicious nodes can simply modify, delete or selectively forward some messages to other nodes in the network [3], which affects the authenticity and integrity of the information to the destination.

Sybil Attack. When hackers launch sybil attacks, malicious nodes claim to have multiple identities (called Sybil nodes) and locate themselves in different locations on the network. This has led to huge unfair distribution of resources.

Therefore, the WSN system faces various attacks. In addition to the above attacks, the secure storage of device data is also very important [6], because attackers or legitimate administrators can easily steal the information.

4 Main Defense Methods of Wireless Sensor Network

For the above 10 attack methods, Table 2 provides the main solutions for each attack method.

Solutions for Tampering and Malicious Code Injection. How to improve the non-repudiation of sensors has become the main research direction for solving tampering and malicious code injection. The Physical Unclonable Functions based Authentication (PUFs) proposed by Aman et al. in 2017 [7] uses microscopic physical changes in sensor integrated circuits to verify the credibility of devices. However, Aman proposed that PUFs are not very reliable; Guin et al. proposed a new verification method SRAM based PUF in 2018, which uses PUF to generate a fixed device ID and stores it in SRAM. The acceptable range reduces the difficulty of verification, that is, it can ensure the non-repudiation range of the equipment to achieve the purpose of improving the reliability of PUFs, but on the other hand, it increases the production cost of the sensor. If deep learning technology is introduced into gateways or routing to perform fuzzy identification of devices, it may be possible to improve the reliability of PUFs without increasing the cost of SRAM.

Fake Node Injection Solution. Porambage et al. proposed an implicit certificate-based PAuthKey identity authentication scheme suitable for distributed sensor network architecture in 2014 [9]. This authentication method can be applied to the link layer and network layer of sensor networks, and provides an end-to-end. An end-to-end identity authentication method. This method can accurately identify illegal nodes, but requires edge network equipment to have higher data processing capabilities.

Side Channel Attack Solution. In 2013, Yao Jianbo et al. proposed a side-channel attack-resistant Elliptic Curve Cryptosystem (ECC) suitable for resource-constrained embedded mobile environments. The system uses fragmented window technology to improve FWNAF (Fractional Width-w NAF) The algorithm's defense against side channel attacks. To deal with this technology, you can also try to use the PUF technology proposed by Aman et al. to make the sensor key more complicated and difficult to obtain from the side channel. However, to solve the problem fundamentally, it is necessary to shield multiple signals, which will undoubtedly increase Sensor installation cost. For example, the sensor is installed in an AESS box that is resistant to electromagnetic side channel attacks [12].

False Routing Information Solution. Documents [13–15] proposed routing and node filtering algorithms based on energy perception. By consuming part of the energy, the transmission distance, the remaining energy of the node, and the key are used as the identity authentication factors. The time-based trust-aware routing protocol (SecTrust-RPL) proposed by Glissa et al. in 2016 [16] also achieves the purpose of suppressing false routing information attacks by sacrificing part of the underlying routing network survival time. Compared with the previous method, This method is more suitable for IoT environments such as smart homes with low energy storage and low requirements for real-time data.

Wormhole Attack and Cesspool Attack Solutions. In 2017, Shukla et al. proposed three machine learning-based IoT intrusion detection systems [17], namely unsupervised learning intrusion detection systems based on K-means clustering, IDS based on decision trees, and a combination of K-means Hybrid two-stage IDS with decision tree learning method, the accuracy of these three methods reaches between 70% and 93%. The disadvantage is that this method requires high system hardware performance and also has false alarms.

Hello Flood Attack Solution. The LEACH defense against Hello flooding scheme proposed by Liu Ai Dong et al. in 2012 [18] adds two functions: two-way authentication of the link and authentication of the sender by the receiver, which can prevent Hello flooding attacks at the expense of some energy. Very good defensive effect. In 2010, Singh et al. proposed a scheme to detect Hello flood attacks based on signal strength and client puzzles [19]. Each node checks the signal strength of the Hello message it receives based on the known radio distance and strength. If it accords with the same known strength, it is classified as "friend", otherwise it is classified as "stranger". The more Hello messages sent by each "stranger" node, the more difficult puzzles need to be solved. In this way, the computing power and energy advantage of the attacker's device can be reduced. In 2020, Atass et al. proposed a method to detect Hello flood attacks based on deep learning [20], and can be combined with meta-heuristic models for transmission path optimization.

Selective Forwarding and Sybil Attack Solutions. In the sensor network architecture with routing as the central node, a more secure routing protocol for Low-Power and Lossy Networks (The Routing Protocol for Low-Power and Lossy Networks, RPL) can be introduced, such as Glissa et al. in 2016 Proposed, time-based trust-aware routing protocol (SecTrust-RPL) [16]. The protocol can use a trust-based mechanism to detect and isolate attacks while optimizing network performance. Pu et al. proposed a monitoring-based method called CMD in 2018 to detect illegal packet loss behavior in RPL networks [21], because the detection object prefers the parent node, which consumes less energy and isolates the network delay. In the distributed sensor network architecture, these two attacks are less harmful, but there is no effective method to isolate malicious nodes for the time being. When cost permits, consider using the CUTE MOTE method proposed by Gomes et al. in 2017 to combine the core microcontroller (MCU) and reconfigurable computing unit (RCU) with the IEEE 802.15.4 radio transceiver., To achieve the purpose of increasing the computing power of a single sensor node, so that a single sensor has the ability to detect and isolate attacks.

Through the above solutions, we can solve the security problem of the WSN system. In addition, we also need to use effective attack detection methods [22] to detect system abnormalities in time and make effective countermeasures.

Table 2. The main solution for each attack

Attack name	Defense
Tampering and malicious code injection	PUFs; SRAM based PUF
Fake node injection	PAuthKey
Side channel attack	Elliptic Curve Cryptosystem Against Side Channel Attacks; PUF
False routing information	Volume-aware routing and node filtering algorithms
Wormhole attack and cesspool attack	Internet of Things Intrusion Detection System Based on Machine Learning
Hello Flood attack	LEACH defense against Hello flood attacks
Selective forwarding and Sybil attack	SecTrust-RPL; CMD

5 Summary and Outlook

As the most intensive and widely deployed part of the Internet of Things system, the wireless sensor network's ability to resist attacks will better reflect the safety and reliability of the entire Internet of Things system. This article lists multiple attack methods that may appear in sensor network technology, and analyzes the current mainstream defense mechanisms against these types of attacks. In the next research work, we will continue to look for security solutions for sensor networks from two directions. One is to solve the defects of sensor equipment itself in accordance with low-cost requirements, and to improve the computing and storage capabilities of sensor nodes. In addition, we will also combine the emerging Internet of Things technology and machine learning technology to deeply study the real-time detection accuracy of wireless sensor network intrusion.

6 Funding Statement

The authors are highly thankful to the Research Project for Young Teachers in Hechi University (NO. 2019XJQN013).

References

1. Liang, W., Xie, S., Long, J., et al.: A double PUF-based RFID identity authentication protocol in service-centric internet of things environments. Inf. Sci. **503**, 129–147 (2019)
2. Wen, H., Ming, X.: Analysis of security routing protocol for wireless sensor networks. Beijing Univ. Posts Telecommun. **29**(S1), 107–111 (2006)
3. Varga, P., Plosz, S., Soos, G.: Security threats and issues in automation IoT. In: Hegedus, C.: 13th International Workshop on Factory Communication Systems (WFCS), pp. 1–6. IEEE (2017)
4. Karlof, C., Wagner, D.: Secure routing in wireless sensor networks: attacks and countermeasures. Ad hoc Netw. **1**(2–3), 293–315 (2003)
5. Andrea, I., Chrysostomou, C.: Internet of things: security vulnerabilities and challenges. In: Hadjichristofi, G.: 2015 IEEE Symposium on Computers and Communication (ISCC), pp. 180–187. IEEE (2015)
6. Liang, W., Fan, Y., Li, K.C., et al.: Secure data storage and recovery in industrial blockchain network environments. IEEE Trans. Ind. Inform. **99**, 1–1 (2020)
7. Aman, M.N., Chua, K.C.: A light-weight mutual authentication protocol for iot systems. In: Sikdar, B.: GLOBECOM 2017–2017 IEEE Global Communications Conference, pp. 1–6. IEEE (2017)
8. Guin, U., Singh, A., Alam, M., Caedo, J., Skjellum, A.: A secure low-cost edge device authentication scheme for the internet of things. In: 31st International Conference on VLSI Design and 17th International Conference on Embedded Systems (VLSID), pp. 85–90 (2018)
9. Guin, U., Singh, A., Alam, M., Canedo, J.: A secure low-cost edge device authentication scheme for the internet of things. In: Skjellum, A. (eds) 2018 31st International Conference on VLSI Design and 2018 17th International Conference on Embedded Systems (VLSID), pp. 85–90. IEEE (2018)
10. Jianbo, Y., Tao, Z.: Elliptic curve cryptography algorithm against side channel attacks. Comput. Appl. Softw. **30**(005), 203–205 (2013)
11. Möller, B.: Improved techniques for fast exponentiation. In: Lee, P.J., Lim, C.H. (eds.) ICISC 2002. LNCS, vol. 2587, pp. 298–312. Springer, Heidelberg (2003). https://doi.org/10.1007/3-540-36552-4_21
12. Xiao-Long, C., Guo-Liang, D., Cui-Xia, W.U., et al.: Design of AES S-box against electromagnetic side-channel attacks. Comput. Eng. **37**(17), 93–95 (2011)
13. Tian, Y., Wang, L.: Routing algorithm for wireless sensor networks by considering residual energy and communication cost. Nanjing Li Gong Daxue Xuebao/J. Nanjing Univ. Sci. Technol. **42**(1), 96–101 (2018)
14. Weiwei, S.: Design and implementation of an improved routing protocol for wireless sensor networks based on the minimum hop count. School of Computer Science and Engineering, Nanjing University of Science and Technology, Nanjing (2016)
15. Lifeng, Z., Yonghua, X.: False data filtering algorithm based on energy-aware routing and node filtering. J. Nanjing Univ. Sci. Technol. (Nat. Sci. Edn.) **4**, 409–415 (2018)
16. Glissa, G., Rachedi, A., Meddeb, A.: A secure routing protocol based on RPL for Internet of Things. In: Global Communications Conference. IEEE (2017)
17. Aidong, L., Zhongwu, L., Jiaojiao, G.: Research on LEACH's defense against HELLO flooding attack. J. Sens. Technol. **25**(003), 402–405 (2012)
18. Singh, V.P., Jain, S., Singhai, J.: Hello flood attack and its countermeasures in wireless sensor networks. Int. J. Comput. Sci. Issues **7**(3), 23 (2010)
19. ATASS, BSSM. Prevention of hello flood attack in IoT using combination of deep learning with improved rider optimization algorithm. Computer Communications (2020)

20. Pu, C., Hajjar, S.: Mitigating forwarding misbehaviors in rpl-based low power and lossy networks. In: 2018 15th IEEE Annual Consumer Communications Networking Conference (CCNC), pp. 1–6 (2018)
21. Gomes, T., Salgado, F., Tavares, A., Cabral, J.: Cute mote, a customizable and trustable end-device for the internet of things. IEEE Sens. J. **17**(20), 6816–6824 (2017)

Research on Multi-channel Pulse Amplitude Analyzer Based on FPGA

Han Deng[1], Chong Wang[2(✉)], Shuang Xie[1], Aishan Mai[1], Weihong Huang[3], and Shiwen Zhang[3]

[1] Big Data Development and Research Center, Guangzhou College of Technology and Business, Guangzhou 528138, China
[2] College of Computer Science and Electronic Engineering, Hunan University, Changsha 410000, China
f2010w0139@hnu.edu.cn
[3] School of Computer Science and Engineering, Hunan University of Science and Technology, Xiangtan 411201, China
shiwenzhang@hnust.edu.cn

Abstract. Multi-channel pulse amplitude analyzer is a key component of energy spectrometer, of which performance determines the measurement accuracy. Therefore, analysis technology is an important research topic in nuclear radiation measurement technology. This paper presents a design method of multi-channel pulse amplitude analyzer based on FPGA technology. In the design, 100 MHz crystal oscillator is used, and the frequency is increased to 200 MHz by digital phase-locked frequency multiplier (PLL). Compared with the traditional 12 MHZ analyzer, this analyzer not only greatly improves the accuracy of the system, but also improves the differential nonlinearity of the system. Through the USB controller composed of FPGA, the design realizes the function of communication with USB2.0 interface. The two-stage constant current source linear discharge method is adopted to improve the channel width equilibrium of ADC and reduce the dead time of the system. Finally, Modelsim and other testing software are used to simulate the designed system and test the real object. The experiment shows that the performance of the system is well.

Keywords: Multi-channel pulse amplitude analyzer · FPGA · VHDL · Phase-locked frequency multiplier

1 Introduction

Intelligent nuclear instruments have become a leading direction in the development of radioactive measuring instruments, which will have a profound impact on the application of nuclear technology, international nuclear security and weapons verification, national defense engineering and scientific testing and other fields [1]. Multi-channel pulse amplitude analyzer is a key part of energy spectrometer. Its performance determines the measurement accuracy of energy spectrometer. Compared with traditional

© Springer Nature Switzerland AG 2021
M. Qiu (Ed.): SmartCom 2020, LNCS 12608, pp. 50–58, 2021.
https://doi.org/10.1007/978-3-030-74717-6_6

analog MCA, digital MCA has the advantages of faster pulse speed and can process digital signals [2].

A complete multi-channel pulse amplitude analyzer should have two major functions:1, Analog to digital conversion and address storage of detected signals, that is, data conversion and storage control function; 2, Display the result directly or sent to the microcomputer for processing through serial port and parallel port, that is, data transmission function. Therefore, the performance of multichannel pulse analyzer is generally analyzed from these two aspects.

The research content of this paper is to design a multi-channel pulse amplitude analyzer with excellent performance. The system uses FPGA chip as the digital logic control core of multi-channel pulse analyzer, and adopts embedded ROM to store data, and makes full use of FPGA internal resources. FPGA is an editable logic device with high performance: it emerged as a semi-custom circuit in the field of special purpose integrated circuits (ASIC). It not only solves the deficiency of custom circuit, but also overcomes the shortcoming of limited logic gate of original editable logic device, greatly improves the overall anti-interference ability, stability and the accuracy of the system, and can make the volume smaller. FPGA provides high performance while maintaining flexibility and programmability. As a result, FPGA is widely used in fields such as communication, high performance computing and machine learning [3, 4].

In terms of data transmission, USB2.0 is adopted in the system to speed up and improve the data processing capacity. Therefore, the multi-channel pulse amplitude analyzer in this design is small, low in power consumption, fast in processing speed, and convenient for outdoor work, which can be widely used in mining, metallurgy, security inspection, industrial analysis and other fields. It is an upgrade of the traditional multi-channel pulse amplitude analyzer. At same time, the data collected by the amplitude analyzer could be stored and processed in the cloud servers [5, 6]. It is also vital to find a highly efficient, user-friendly interface design method for showing information [7].

2 Research Status

At present, the multi-channel pulse amplitude analyzers developed and produced by major instrument companies in the world, such as the United States, Japan and Germany, can be roughly divided into two categories: One is represented by ORTEC [8] and CANBBER, whose design principle uses successive comparison method. It is characterized using digital integrated circuit chips to achieve A/D (analog-digital conversion), the use of digital channel homogenizer to adjust the channel width. Its advantage is that it can satisfy high counting rate, but its disadvantage is that integral nonlinear and differential nonlinear are not ideal. The other is represented by Xilinx Company, whose design principle adopts the linear discharge method [9, 10]. Its characteristic is to use the small scale digital integrated circuit chip to realize the control core. Its advantage is the integral nonlinear, the differential nonlinear good, the disadvantage is only to meet the low counting rate [11]. In addition, Siemens and other companies have also implemented a multi-channel pulse amplitude analyzer with single chip microcomputer AT89C55 as the control core and A/D chip MAX191 as the analog to digital converter device with high speed and low power consumption [12].

The spectral data measured by the multi-channel pulse amplitude analyzer is sent to the host for data analysis and processing through the communication interface. The communication interface includes serial port (RS232), parallel port (EPP), USB port and bus mode (ISA, PCI). At present, many enterprises mainly take PC/104 embedded microcomputer as the control core and adopt bus mode to realize system communication and data transmission. FPGA CHIP EPF10K20 is usually used as the core of logic control, and combined with EPP interface technology, the function of multi-channel pulse amplitude analyzer is realized. In the above communication interface circuit, the performance of USB port is the best, but now it is only 12M, that is, it only meets USB protocol 1.1. How to use P89C51RD2 single-chip microcomputer, A/D converter and USB interface to design a portable multi-channel pulse analyzer and how to use USB2.0 interface in analyzer to achieve data transmission are still at the early stage of research [13]. At same time, voltage assignment satisfying real-time multiprocessor should be paid more attention to get better performance [14].

3 Multi-channel Pulse System Design

The software design of this system is mainly compiled according to the workflow. When the system starts to work, the standard voltage of the channel width and upper and lower thresholds is set first, then the linear gate opens, and the signal begins to be input. The pulse signal charges the capacitor. If the peak is exceeded, close the linear gate, and if the peak is not exceeded, keep the linear gate open. At the same time, judge whether the signal exceeds the threshold value. Start analog-to-digital conversion if it is below the upper and lower thresholds. In the process of discharge, first judge fast and slow discharge. If the capacitor voltage is greater than the slow reference voltage, the fast release and vice versa. The total time of discharge is: $T = T1 + T0$. FPGA counts T with a certain frequency clock, and the count is the track address. The system adds the address of this channel to one, stores it to FPGA, and finally sends data through USB or RS-232.

In order to improve the utilization rate of FPGA and reduce the overall time of software in system synthesis, we adopt the top-down mode to compile the program. In other words, the function of the system is divided into several program submodules, and finally these submodules are integrated through the top-level program. Figure 1 is the division of FPGA functional modules and the control relationship diagram of each module. ① Inside the dotted line is the FPGA chip. In the FPGA, it is divided into three functional modules, namely, the main control module, the data storage module and the data transmission module. ② COMP1 through COMP5 are signals from analog circuits; the BCD code data is the output signal of the BCD converter; SELECT is the combined switch output; CLR is the system reset signal. They are all inputs to the FPGA. ③ XX is the control linear door open/close signal; KF is the control fast release/off signal; ZF is the control self-opening/closing signal; Charge is an on/off signal that controls constant current source discharge; SY is the upper threshold PWM modulation signal, XY is the lower threshold PWN modulation signal. They are the output signals of the FPGA that control the analog circuit according to the state of the input signal. ④ Finally, FPGA outputs the converted state to digital tube display, outputs the converted amplitude data to PC for data processing, and plots the amplitude spectrum.

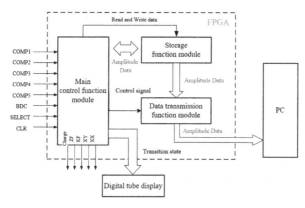

Fig. 1. FGPA functional module structure diagram

The functions of the master control module include the output of analog circuit control signals, the control of memory reading and writing, the control of data transmission, and the control of digital tube data display. The function of the data storage module includes the data in the converted address and the calculated address. The function of the data transfer module is mainly to send data under the control of the controller, including USB or RS-232 is the driver. This module mainly receives the external comparison signal (COMP), sends out the corresponding external control signal, and quantifies the sampled analog signal digitally. The status profile is shown in Fig. 2:

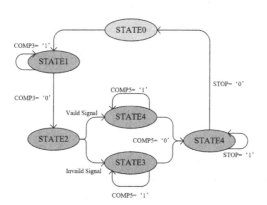

Fig. 2. State flow chart of master control module

STATE0: initial state, all reset;
STATE1: wait for over peak, then enter the STATE2.
STATE2: determine whether the signal is valid, enter the STATE4 if it is valid, or enter the STATE3 if not;
STATE3: empty the capacitor and enter the STATE0, otherwise the state will remain unchanged;

STATE4: slow discharge, internal counter count. When the discharge ends, enter the STATE0, otherwise the state will remain unchanged;

STATE5: check whether the measurement time has arrived. If not, return to the initial state; otherwise, keep the state unchanged.

Figure 3 is the mapping after compiling with VHDL. As can be seen in the picture, we realized the functions of each module of FPGA through compilation. The Hs_majorcontrol module is the master control mapping; the Hs_dif module is the frequency division mapping; the Hs_pwm module is the PWM modulation mapping; the Hs_timecontrol module is the timing function mapping; the Hs_bkramcontrol module is the storage control function mapping; the Hs_rs232_send module is the data transmission function mapping of RS-232 and the Hs_bkram module is the storage function mapping of BLOCK RAM.

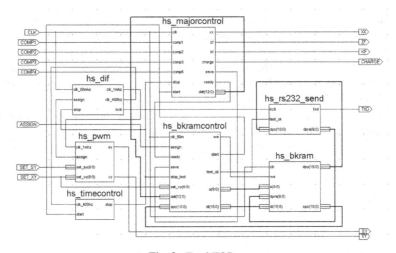

Fig. 3. Total TOP map

4 Experimental Results and Analysis

4.1 Master Function Timing Test

As shown in Fig. 4, ① a moment: start = '1', start measur'ing conversion, xx = '0' linear door open. ② b moment: comp3 = '0', signal over peak, xx = '1', linear door closed. Because at this time, comp1 = '0', comp2 = '1' (invalid signal), after two clock cycles, the discharge door opens, kf = '0' (slow down), zf = '1' from the start, until the charge in the capacitor is discharged (i.e., Comp5 = '0'). ③ c moment: xx = '0' linear gate opens again, and the system begins the second measurement conversion. ④ d moment: comp1 = '1', comp2 = '0', that is, the signal is valid. After two clock cycles, charge = '0', kF = '0', and the slow discharge starts. Meanwhile, the internal counter counts are started. ⑤ e moment: comp5 = '0', save = '1' The data can be stored and output by DAT. The next effective measurement is started after two cycles. ⑥ f moment: start the internal counter count again. ⑦ g moment: after starting the second effective measurement, stop = '1', but the module does not stop working immediately. Instead, it sends out the ready signal after this analog to digital conversion to stop the module action.

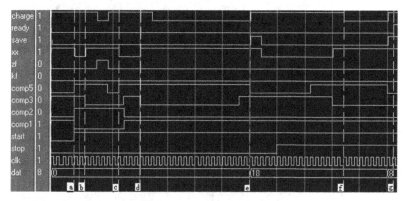

Fig. 4. Analog - digital transformation sequence simulation diagram

4.2 PWM Chopper Amplitude Modulation Test

Before the first data is sent, a set of "10101111" is sent as the starting flag frame, and the PC side begins to receive data. When all the data is sent, a set of "11111111" is sent as the end flag frame, at which time the PC side stops receiving data. In this simulation, 4 groups of valid data were sent, namely "00000010 00000001", "0000000010100101", "1111111111101010101", "0011110011001111". Each set of data is divided into high eight bits and low eight bits respectively. The data is first sent from low eight bits and then sent high eight bits.

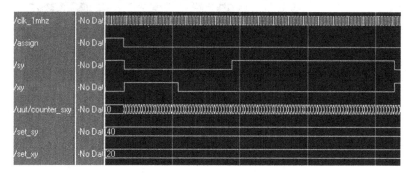

Fig. 5. Global simulation diagram

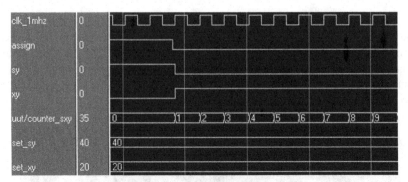

Fig. 6. Set up and down thresholds

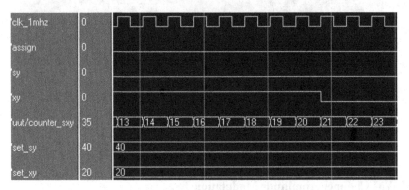

Fig. 7. When the count value exceeds the set threshold, the lower valve level changes

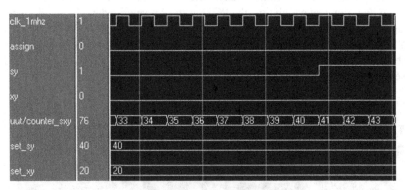

Fig. 8. When the value exceeds the set upper threshold, the upper valve level changes

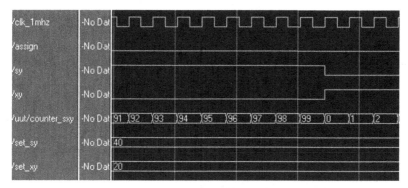

Fig. 9. All reset when counting 100 1 MHz clock frequencies

Timing simulation diagram of PWM chopper output module. Figure 5 is the global simulation diagram, while Fig. 6, 7, 8 and 9 are the local enlarged diagrams.

According to the PWM timing sequence simulation diagram, when the assign signal is high voltage, the counter will not act, and the system will load up or down. When the assign signal is low, turn on the counter and start counting. When the counting value exceeds the upper and lower thresholds, "Sy" and "Xy" will start to change (Fig. 7 and 8), and when 100 1 MHz clock frequencies (namely 10 KHz) are counted, they will all reset (Fig. 9).

5   Conclusion and Prospect

In this paper, we design a multi-channel pulse amplitude analyzer composed of two-stage linear discharge method, which can store 4096 channels of address. This system not only greatly improves the accuracy of the system; And it also improves the differential nonlinearity of the system. The functional modules of the system are tested, and the test results show that the design of the scheme is reasonable and feasible. In the next step, FPGA chip will continue to be used as the digital logic control core of multi-channel pulse analyzer, and the data will be stored in the embedded ROM within the chip, and the internal resources of FPGA will be fully utilized to analyze multi-channel pulse signal.

Acknowledgment. This work is supported by the National Natural Science Foundation of China (No. 61702180) and the Doctoral Scientific Research Foundation of Hunan University of Science and Technology (No. E52083).

References

1. Yang, J.: Improved low power multichannel pulse amplitude analyzer. Open Access Libr. J. **7**(08), 1 (2020)
2. Zeng, W.: The design of digital multi-channel analyzer based on FPGA. Energy Procedia **39**, 428–433 (2013)

3. Murray, K.E., et al.: VTR 8: high-performance CAD and customizable FPGA architecture modelling. ACM Trans. Reconfigurable Technol. Syst. (TRETS) **13**(2), 1–55 (2020)

4. Richardson, J., Fingulin, S., Raghunathan, D., Massie, C., George, A., Lam, H.: Comparative analysis of HPC and accelerator devices: computation, memory, I/O, and power. In: 2010 Fourth International Workshop on High-Performance Reconfigurable Computing Technology and Applications (HPRCTA), p. 10. IEEE (2010

5. Dai, W., Qiu, M., Qiu, L., Chen, L., Wu, A.: Who moved my data? privacy protection in smartphones. IEEE Commun. Mag. **55**(1), 20–25 (2017). https://doi.org/10.1109/MCOM. 2017.1600349CM

6. Gai, K., Qiu, M., Zhao, H.: Security-aware efficient mass distributed storage approach for cloud systems in big data. In: 2016 IEEE 2nd International Conference on Big Data Security on Cloud (BigDataSecurity) (2016)

7. Qiu, M.K., Zhang, K., Huang, M.: An empirical study of web interface design on small display devices. In: IEEE/WIC/ACM International Conference on Web Intelligence (WI 2004). IEEE (2004)

8. Fitzgerald, R., King, L.: Accurate integral counting using multi-channel analyzers. Appl. Radiat. Isot. **159**, 109101 (2020)

9. Esmaeili-sani, V., Moussavi-zarandi, A., Akbar-ashrafi, N., Boghrati, B.: Triangle bipolar pulse shaping and pileup correction based on DSP. Nucl. Instrum. Methods Phys. Res. Sect. A **665**, 11–14 (2011)

10. Abbene, L., Gerardi, G., Del Sordo, S., Raso, G.: Performance of a digital CdTe X-ray spectrometer in low and high counting rate environment. Nucl. Instrum. Methods Phys. Res. Sect. A **621**(1–3), 447–452 (2010)

11. García-Durán, Á., Hernandez-Davila, V.M., Vega-Carrillo, H.R., et al.: Nuclear pulse generator embedded in FPGA. Appl. Radiat. Isot. **147**, 129–135 (2019)

12. Chen, W., Ma, X., Hu, G.: The pulse vacuum sterilization controller based on AT89C55. J. Qingdao Univ. Sci. Technol. (Nat. Sci. Ed.) **3** (2006)

13. Susanto, A.T., Prajitno, P., Wijaya, S.K., et al.: A systematic literature reviews of multichannel analyzer based on FPGA for gamma spectroscopy. In: Journal of Physics: Conference Series, vol. 1528, no. 1, p. 012016. IOP Publishing (2020)

14. Qiu, M., Jia, Z., Xue, C., Shao, Z., Sha, E.H.M.: Voltage assignment with guaranteed probability satisfying timing constraint for real-time multiproceesor DSP. J. VLSI Signal Process. Syst. Signal Image Video Technol. **46**(1), 55–73 (2007)

Towards Smart Building: Exploring of Indoor Microclimate Comfort Level Thermal Processes

Aigerim Altayeva[1], Karlygash Baisholanova[2], Lyailya Tukenova[3], Bayan Abduraimova[4], Marat Nurtas[1], Zharasbek Baishemirov[5], Sanida Yessenbek[6], Bauyrzhan Omarov[2(✉)], and Batyrkhan Omarov[2,7]

[1] International Information Technology University, Almaty, Kazakhstan
[2] Al-Farabi Kazkah National University, Almaty, Kazakhstan
[3] Narxoz University, Almaty, Kazakhstan
[4] L. N. Gumilyov Eurasian National University, Astana, Kazakhstan
[5] Abai Kazakh National Pedagogical University, Almaty, Kazakhstan
[6] Almaty University of Power Engineering and Telecommunications named after G.Daukeyev, Almaty, Kazakhstan
[7] A. Yassawi International Kazakh-Turkish University, Turkistan, Kazakhstan

Abstract. Modern requirements to reduce the consumption of energy resources while maintaining comfortable conditions for people in residential, public and administrative buildings pose the task of developing new approaches to assessing the comfort of the microclimate. Currently used methods for assessing the comfort of the microclimate do not take into account the specific hazards characteristic of non-industrial premises, and for this reason, the introduction of energy-saving measures may lead to a violation of the comfort conditions in the premises of buildings. In this regard, the development of methods and methods to take into account the impact of energy-saving measures on the microclimate is an urgent task.

This research paper is devoted to solving the urgent problem – energy efficiency of buildings. We explore mathematical model of indoor microclimate thermal processes, parameters that affect to indoor microclimate, comfort microclimate serving, and represent simulation results of the developed mathematical model of thermal processes. Also, we explore how to control heating, ventilation and air conditioning equipments considering indoor and outdoor temperature and humidity level, and the problem, how to keep stable indoor comfort temperature and humidity.

Keywords: HVAC system · Indoor microclimate · Thermal process

1 Introduction

Individuals spend most their time in an indoor environment and comfort is one of the most important issues with respect to staying indoors. Environmental quality of an interior environment is directly dependent on its organization and content [1]. Therefore, the task of maintaining a comfortable environment is extremely important for health, good

© Springer Nature Switzerland AG 2021
M. Qiu (Ed.): SmartCom 2020, LNCS 12608, pp. 59–67, 2021.
https://doi.org/10.1007/978-3-030-74717-6_7

spirit, and human activity [2]. Several parameters can be adjusted to achieve comfort in a room—air temperature, humidity, air quality, speed of movement of air throughout the room, oxygen content in the air, ionization of air, and noise level [3–6]. A deviation of the aforementioned parameters could result in the deterioration of the normal state of an individual. This can lead to the disruption of thermal balance as well as result in a negative impact on health and productivity [7].

Energy saving has become an urgent problem for the whole world in recent decades. The solution of this problem is connected not only with improving the environment, but also with ensuring the energy security of individual States [8]. At the same time, for countries with limited reserves of fuel resources, energy security means reducing the economy's dependence on fuel imports, and for resource - producing countries, it consists in ensuring growing domestic demand for energy resources through more efficient use of the energy that is already produced, rather than by increasing fuel production and building new sources of heat and electricity [9].

Since a significant potential for energy saving is the modernization of building enclosing structures, it is most appropriate to introduce energy-saving measures that increase the thermal protection characteristics of the walls, Windows and floors of the building. Insulation and sealing of buildings by applying thermal insulation or replacing individual elements of enclosing structures naturally leads to a reduction in heat losses and, as a result, a decrease in the required amount of heat for heating [8]. However, as a rule, in buildings with natural ventilation, the hygienic conditions of people's stay deteriorate due to changes in the microclimate parameters [10]. Therefore, when solving the problem of increasing the efficiency of the use of fuel and energy resources, it is necessary to take into account the level of comfort of premises.

Thus, a comfortable dwelling can be achieved when the temperature, humidity, and air flow rates are known.

This paper considers mathematical model of the thermal process, indoor microclimate thermal balance, heat balance on indoor microclimate, and temperature control parameters' correlation in the next section. After, in the simulation results section, we demonstration experimental environment and simulation results. Also, the experiment results in this paper were conducted taking into consideration the comfort data of the premises, given based on international European standards and recommendations ISO 7730 [11].

2   Literature Review

When performing the task of designing a smart home, an important aspect is the development of an effective heating, ventilation and air conditioning (HVAC) control system [12], which is aimed at maintaining comfortable microclimatic parameters while ensuring the highest possible level of energy savings.

Over the years, simulations and experiments have developed strategies to find an acceptable balance between residential comfort and low energy consumption. So, some models took into account the air temperature and the availability of clothing, while others focused on factors such as, for example, a draft [13]. As this area has been studied and smart home management systems have developed, there have been needs for more efficient maintenance of comfortable temperature, ventilation, and air conditioning parameters in building design [14–17].

In [18], models of providing comfortable indoor microclimate conditions were considered, taking into account the conditions for minimizing energy costs. as part of the article, it was found that all models can be divided into three classes as "white box" models, "black box" models, and "gray box" models, which are hybrid.

Analysis of the literature on this issue allowed us to form four principles that determine the policy of energy saving:

1. Energy Resources are of great importance both for improving the quality of life of citizens, and for ensuring energy security and independence of the country.
2. Energy resources must have all the characteristics of a commodity, since they can be created, sold, purchased, and otherwise participate in commodity-money relations.
3. In the twenty-first century, non-traditional and renewable energy sources will be actively used.
4. Priority in the selection and implementation of energy-saving measures will be given to solutions that are effective not only in technical or economic terms, but also at the same time contribute to improving the microclimate of premises [15].

Thus, the implementation of energy-saving measures in buildings should be carried out taking into account the comfortable stay of a person in them. Currently, the most common method for determining the comfort of the microclimate is the measurement and evaluation of individual components of the temperature and humidity regime of the room: temperature, mobility, relative humidity, as well as the characteristics of thermal radiation [19]. A significant disadvantage of this approach is the neglect of the mutual influence of microclimate parameters on each other. To improve the accuracy of determining the comfort of the microclimate, it is necessary to develop a comprehensive indicator that takes into account the maximum possible number of parameters and their mutual influence.

3 Mathematical Model

3.1 Heat Balance Equation on Indoor Environment

All Thermal processes of heated space are a series of interconnected heat exchange sub-processes and heat transfer between the elements of the system include: indoor air, building fence, internal content, heaters, ventilation system [20]. These elements interact with each other and with the environment by means of heat and mass transfer. On the fence of air and heat exchange element boundaries is carried out by convection and by radiation. Also, the heat is transferred by convection from the radiators to the internal air space [21]. The heat from the inner to the outer surface of the enclosure is passed through heat transfer. Also on the basis of heat transfer processes are carried out through the facilities of energy partitioning. In addition to thermal processes and should be considered as mass transfer processes due to the need for room ventilation [22–25].

Thus, the heated building can be displayed on a graph $G = \{X, U\}$, where X - the set of vertices, each of which is xk (k = 1, n) corresponds to the internal air space of the k-th building and U - a plurality of edges, each of which corresponds to the heat or material flow between adjacent rooms. We should also highlight the vertex of x0, representing the ambient air.

Each vertex of the graph xk may have thermodynamic parameters of air (Vk, Tk) - the volume and temperature respectively. Uij each edge corresponds to a vector (Hij, cij, ρij, Fij, λii), where, Hi – thickness, cij – specific heat, ρij - density, Fij – surface, λij - fencing the thermal conductivity between the elements of the structure of i and j.

In addition to these system parameters defining the purely thermal processes, you must also enter the parameters responsible for the mass transfer associated with the ventilation of premises and the possible exchange of air masses between the rooms. As this parameter can be viewed obliquely symmetric matrix Gij (i, j = 0, n) - weight air flow between the i-th and j-th element.

Each edge of the structural graph uii symbolizes the heat flow j qii the element xi to xj element. In turn, the heat flow qii determined by the laws of heat transfer, in accordance with which in the fence between the i-th and j-th rooms Tii(x,t) unsteady temperature field is formed, where $x \in [0, Hii]$ linear coordinate perpendicular to the surfaces of the fence. The specified temperature field described by the differential equation of heat conduction

$$\frac{\partial T_{ij}}{\partial t} = a_{ij}\frac{\partial^2 T_{ij}}{\partial x^2} \tag{1}$$

where $a_{ij} = \frac{\lambda_{ij}}{c_{ij}\rho_{ij}}$ – coefficient of thermal conductivity.

At the boundaries of the fence x = 0 and x = Hij temperature satisfies the boundary conditions of the 3rd kind, whereby the convective heat flux from the air to the fence is the heat flow inside the enclosure, i.e.

$$\alpha\left(T_i - T_{ij}(x, t)\right)_{x=0} = -\lambda_{ij}\left(\frac{\partial T_{ij}(x, t)}{\partial x}\right)_{x=0} \tag{2}$$

$$\alpha\left(T_{ij}(x, t) - T_j\right)_{x=H_{ij}} = -\lambda_{ij}\left(\frac{\partial T_{ij}(x, t)}{\partial x}\right)_{x=H_{ij}} \tag{3}$$

where α – the convective heat transfer coefficient.

In turn, the temperature Ti of each structural element can be determined based on the heat balance equation has the following form

$$c_v\rho_B V_i\frac{dT_i}{dt} = \alpha \sum_{\substack{j=0 \\ j\neq i}}^{n} F_{ij}\left(T_{ij}(0, t) - T_i\right)$$

$$+ c_0 \sum_{\substack{j=0 \\ j\neq i}}^{n} G_{ij}\left(T_j - T_i\right) + \alpha F_{ai}(T_{ai} - T_i) + Q_i(t) \tag{4}$$

Where cv, cp - specific isochoric and isobaric heat capacity of air, ρ_a – the density of air, F_{ai} and T_{ai}– the surface area and the temperature of the internal battery, the i-th filling the room, Qi(t) - the heat flux due to supply heat from heaters.

The system of Eqs. (1) and (4) need to be supplemented by a system of differential equations describing the thermal processes in the internal battery, consisting of various production equipment, furniture and other space filling:

$$c_i \frac{dT_{ai}}{dt} = \alpha F_{ai}(T_i - T_{ai}) \tag{5}$$

where c_i – capacity of heat accumulators.

Thus, the mathematical model of thermal processes of building to be heated is a system of n $(n-1)/2$ differential equations in partial derivatives of the form (1) with the boundary conditions (2), (3), and a system of 2n ordinary differential thermal balance Eqs. (4), (5). For the integration of the system (1), (4), (5) it is necessary to set the initial conditions for the temperature T_i (i = 1,n) and for the initial distribution of temperature fields T_{ij} (x, 0), as well as the laws of the heat $Q_i(t)$ and Mass Transfer $G_{ij}(t)$. The result of the integration will function $T_i(t)$ - changes in air temperature.

The above-formulated problem is a problem analysis and provides an answer to the question how will change the room temperature depending on the outdoor environment temperature changes and heat input from heaters of indoor environment. This problem can be solved by numerical methods or finite element based on the asymptotic expansions of the one-dimensional temperature field [26].

Additional systems and technologies are required for heating or cooling in buildings in which a thermal system is not naturally in equilibrium. It is necessary for the net heating energy to be provided by a heating system. The heating system experiences losses when heat is produced from a primary energy source. The total heating energy demand of a building corresponds to the sum of heating energy and heating equipment losses.

4 Simulation Results and Discussion

The microclimate in the workplace is determined by the air temperature, relative humidity, air velocity, and the intensity of radiation from heated surfaces. The optimal parameters of the microclimate is a combination of temperature, relative humidity and air velocity, which in long-term and systematic exposure does not cause variations in the human condition. High air temperature in the premises, while maintaining the other parameters causes fatigue working, body overheating and sweating a lot. Low temperatures can cause local and general cooling of the body and cause colds.

The experiments were conducted at the Artificial Intelligent and Smart City laboratory in Gachon University. In the laboratory space, an area of 3 × 6 m^2 with a height of 2.5 m on one of the outer walls included six work areas separated by translucent partition walls (with a height 1.5 m) with personal computers. Figure 1 illustrates the plan of the laboratory.

Furthermore, as shown by field studies, the humidified air flowed in the duct and over obstacles (e.g., silencers, etc.). There were significant losses due to moisture condensation on solid surfaces. Thus, a power saving humidification process is necessary to determine the minimum amount of normal air and water supply wherein the relative humidity of indoor air was given. For this purpose, a numerical simulation of the spatial distribution of relative humidity in the room was initially conducted through corresponding field experiments.

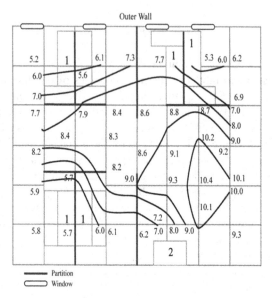

Fig. 1. Experimental dependence of air temperature and the contour temperature to time

The room provided fresh air supply through the three dimension 400×400 mm windows. The experimental studies indicated that at an outside temperature below $T_{outdoor}$ $(-5\ °C)$ during heating and ventilation operations, the air temperature significantly exceeded the critical temperature, and the relative humidity was in the range of 5–10%. The low relative humidity primarily negatively affected the respiratory organs. Furthermore, humid air is a poor conductor of static electricity. This contributed to the accumulation of humid air on the surface, and this exceeded the performance of the electromagnetic field with respect to the remote control at the workplace with PC and led to failure of the electronic equipment. Humidification units were installed to ensure standardized relative humidity values in the installation of central air conditioning. Nevertheless, the generation of vapor or fine water required significant energy costs.

In the study, the results indicated the minimum amount of moisture in the supply air by which it was possible to determine the rated value of relative humidity of internal air. Experiment for the temperature dependence of the room and finding the circuit parameters experiment was conducted. The essence of the experiment was as follows. Temperature sensors were placed on the walls and ceilings in several places, so that you can then get the average integral factor associated with the thermal capacity of walls and ceilings. Simultaneously, the sensors were placed in the air, and the coolant in the external environment. The temperature data were taken at intervals of one minute, and these data can be averaged over the population, which was mentioned earlier.

During two weeks we explored average internal air temperature and comparison with math model calculated data was provided. Figure 2 illustrates the comparison of

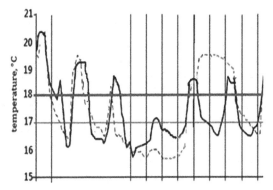

Fig. 2. Average internal air temperature change (dotted line means actual temperature change. full line means math model calculated temperature change)

measured data with the math model calculated data. The foregoing comparison proves adequacy of the mathematical model. The numerical simulation results shown in Fig. 3 include temperature field distribution in the horizontal plane data distance of 1.500 m from the room floor.

5 Conclusion

Thermal processes are building a set of interrelated heat and mass exchange processes arising in different areas of the building. A mathematical model of these processes, based on the classical methods of thermal physics, is of little use to solve the heating process automation tasks. Proposed and justified in the structure of the mathematical model of the thermal building process in the form of a finite-dimensional linear system of differential equations paves the way for effective adaptation of the entire spectrum of modern methods of analysis and synthesis of automatic control systems for heating processes.

References

1. Preethichandra, D.: Design of a smart indoor air quality monitoring wireless sensor network for assisted living. In: Conference Record – IEEE Instrumentation and Measurements Technology conference (2013)
2. Dong, B., Prakash, V., Feng, F., O'Neill, Z.: A review of smart building sensing system for better indoor environment control. Energy Build. **199**, 29–46 (2019)
3. Abraham, S., Li, X.: A cost-effective wireless sensor network system for indoor air quality monitoring applications. Procedia Comput. Sci. **34**, 165–171 (2014)
4. Afram, A., Janabi-Sharifi, F.: Theory and applications of HVAC control systems – a review of model predictive control (MPC). Build. Environ. **72**, 343–355 (2014)
5. Földváry, V., Bukovianska, H., Petráš, D.: Analysis of energy performance and indoor climate conditions of the slovak housing stock before and after its renovation. Energy Procedia **78**, 2184–2189 (2015)

6. Burns, J.A., Cliff, E.M.: On optimal thermal control of an idealized room including hard limits on zone-temperature and a max-control cost term. In: 2013 IEEE 52nd Annual Conference on Decision and Control (CDC) (2013)
7. Merabti, S., Draoui, B., Bounaama, F.: A review of control systems for energy and comfort management in buildings. In: 2016 8th International Conference on Modelling, Identification and Control (ICMIC), pp. 478–486. IEEE (2016)
8. Mien, T.L.: Design of fuzzy-PI decoupling controller for the temperature and humidity process in HVAC system. Int. J. Eng. Res. Technol. (IJERT). **5**(01), 589–594 (2016)
9. Moon, J.W., et al.: Determining optimum control of double skin envelope for indoor thermal environment based on artificial neural network. Energy Build. **69**, 175–183 (2014)
10. Marche, C., Nitti, M., Pilloni, V.: Energy efficiency in smart building: a comfort aware approach based on Social Internet of Things. In: 2017 Global Internet of Things Summit (GIoTS), pp. 1–6. IEEE (2017)
11. ISO/FDIS 7730:2005, International Standard, Ergonomics of the thermal environment—Analytical determination and interpretation of thermal comfort using calculation of the PMV and PPD indices and local thermal comfort criteria (2005)
12. Castilla, M., et al.: Neural network and polynomial approximated thermal comfort models for HVAC systems. Build. Environ. **59**, 107–115 (2013)
13. Saeed, K., Homenda, W., Chaki, R. (eds.): CISIM 2017. LNCS, vol. 10244. Springer, Cham (2017). https://doi.org/10.1007/978-3-319-59105-6
14. Mirinejad, H., Welch, K.C., Spicer, L.: A review of intelligent control techniques in HVAC systems. In: Energytech. IEEE (2012)
15. Song, Y., Wu, S., Yan, Y.Y.: Control strategies for indoor environment quality and energy efficiency—a review. Int. J. Low-Carbon Technol. **10**(3), 305–312 (2013)
16. Altayeva, A., Omarov, B., Suleimenov, Z., Cho, Y.I.: Application of multi-agent control systems in energy-efficient intelligent building. In: Joint 17th World Congress of International Fuzzy Systems Association and 9th International Conference on Soft Computing and Intelligent Systems (2017)
17. Marcello, F., Pilloni, V., Giusto, D.: Sensor-based early activity recognition inside buildings to support energy and comfort management systems. Energies **12**(13), 2631 (2019)
18. Ostadijafari, M., Dubey, A., Liu, Y., Shi, J., Yu, N.: Smart building energy management using nonlinear economic model predictive control. In: 2019 IEEE Power & Energy Society General Meeting (PESGM), pp. 1–5. IEEE (2019)
19. Perera, D.W.U., Pfeiffer, C.F., Skeie, N.-O.: Control of temperature and energy consumption in buildings - a review. Int. J. Energy Environ. **5**(4), 471–484 (2014)
20. Yu, T., Lin, C.: An intelligent wireless sensing and control system to improve indoor air quality: monitoring, prediction, and preaction. Int. J. Distrib. Sensor Netw. **11**, 140978 (2015)
21. Taleghani, M., Tenpierik, M., Kurvers, S.: A review into thermal comfort in buildings. Renew. Sustain. Energy Rev. **26**, 201–215 (2013)
22. Chuan, L., Ukil, A.: Modeling and validation of electrical load profiling in residential buildings in singapore. IEEE Trans. Power Syst. **30**(3), 2800–2809 (2015)
23. Zhang, H., Ukil, A.: Framework for multipoint sensing simulation for energy efficient HVAC operation in buildings. In: 41st IEEE Annual Conference on Industrial Electronics-IECON, Yokohama, Japan (2015)
24. Coelho, V., Cohen, M., Coelho, I., Liu, N., Guimarães, F.: Multi-agent systems applied for energy systems integration: state-of-the-art applications and trends in microgrids. Appl. Energy **187**, 820–832 (2017)

25. Khatibzadeh, A., Besmi, M., Mahabadi, A., Haghifam, M.: Multi-agent-based controller for voltage enhancement in AC/DC hybrid microgrid using energy storages. Energies **10**(2), 169 (2017)
26. Luppe, C., Shabani, A.: Towards reliable intelligent occupancy detection for smart building applications. In: 2017 IEEE 30th Canadian Conference on Electrical and Computer Engineering (CCECE), pp. 1–4. IEEE (2017)

Container Memory Live Migration in Wide Area Network

Yishan Shang[1]([envelope]), Dongfeng Lv[1], Jianlong Liu[2], Hanxia Dong[1], and Tao Ren[3]

[1] Data Information Office of Henan Provincial Military Command Region,
Zhengzhou 450000, Henan, China
[2] China Unicom Corporation, Beijing 100033, China
[3] Hangzhou Innovation Institute of Beihang University, Hangzhou 310051, Zhejiang, China

Abstract. Container memory live migration is an iterative process of memory pages transmission, which is challenging in Wide Area Network (WAN) because the migration often takes too long to finish or even fails due to limited bandwidth. To facilitate container memory live migration in WAN, a key point is to reduce the size of data transmission. In light of this, we leverage popularity values to express the refresh frequency of container memory pages, and propose a new container memory live migration method based on dirty page popularity perception. Our method calculates memory dirty page popularity values and uses these values to perform adaptive memory transmission during migration. We illustrate the proposed method with a detailed discussion on the key algorithm and data structure, and we also implement a prototype system for evaluation. Experimental results reveal that our method improves the container memory live migration efficiency by reducing the total transmitted data with a negligible performance overhead.

Keywords: Container · Memory live migration · Dirty page · Iterative transmission · Popularity

1 Introduction

Cloud computing provides on-demand computing resources, which has been widely used in various fields. In order to achieve load balancing, cost optimization and high availability, cloud computing environments typical span among multiple data centers and has been evolving from public/private clouds to hybrid clouds [17, 20]. Virtual computing environment refers to the basic environment in clouds to distribute computing resources. The most common virtual computing environments are virtual machines (VM) and containers, among which containers are known to be more light-weight and thus becoming the de-facto virtual computing environment.

In cloud computing environment, service reliability is an important performance indicator. To guarantee reliability, virtual computing environment migration has been widely researched. Live migration (also referred to as hot migration) aims to ensuring that services are not affected during the migration process, which has great practical value.

© Springer Nature Switzerland AG 2021
M. Qiu (Ed.): SmartCom 2020, LNCS 12608, pp. 68–78, 2021.
https://doi.org/10.1007/978-3-030-74717-6_8

Container memory could reach GB size and is always frequently refreshed. To conduct container memory live migration, a large amount of data needs to be transferred quickly in a short time from the source to the destination. Although data center network can be efficient, the inter data center network is generally based on WAN, meaning that their bandwidth is very limited. Therefore, during the migration, the transmitted data volume should be reduced as much as possible. Based on this observation, we propose a container memory live migration method based on dirty page popularity. The core of the method is to identify the memory dirty page popularity value, and use this value to optimize the iterative transmission. Our method is capable of optimizing the data transmission during migration, leading to a significant reduction of the bandwidth demand. Therefore, our method can support container memory live migration in WAN. We implement a prototype system and verifies the availability and efficiency.

This paper is organized as follows: The Sect. 1 gives a brief introduction of the problem, the Sect. 2 discusses related works. Section 3, 4 and 5 presents our method, Sect. 6 evaluates the performance of our system. Section 7 concludes this paper.

2 Related Work

Live migration could guarantee the availability of service during migration. There are many researches on VM live migration. Clark et al. propose a pre-copy-based VM memory live migration [1], which has been integrated into KVM [2] and XEN [3]. Pre-copy, also known as iterative copy, divides memory migration into multiple stages. It could significantly reduce the service downtime. However, using the naive iterative copy method for memory live migration requires to transmit all dirty pages in each round. When migrating VMs that perform frequent memory updates, the migration may take a very long time or even fail. MA [4] et al. propose an optimized pre-copy method, which uses bitmaps to mark and distinguish the generated dirty pages, so as to achieve differential iterative transmission of dirty pages and reduce the amount of data transmitted. MR Hines [5] et al. propose a post-copy method for memory live migration. First, the VM is suspended at the source node, and then a minimal set of processor states is copied to the destination node. Finally, the VM is restored at the destination node. The destination processor accesses the memory pages through network from the source node. In the post-copy method, the memory data is only transmitted once, which eliminates the need for iterative copying. However, network-based memory access could lead to serious performance degradation in the VM. Pre-copy and post-copy methods could be combined into hybrid copy methods, which is beyond the scope of this paper.

Containers become the de-facto of cloud computing due to their light weight character. They achieve resources allocation and isolation by using Namespace [6] and Cgroups [7]. Container runs as a group of processes, so container migration is similar with process migration. The naive process migration can be divided into three steps: (1) suspend the process at the source, (2) transmit the process memory space and status to the destination, (3) resume the process at the destination. With the advent of checkpoint restore (CR) [8] technology, process migration has become more efficient. On one hand, the checkpoint operation can dump all the information of the process into a series of files. On the other hand, the process operation can be restored from the dumped files. CRAK [8] and Zap

[11] both support the execution of process CR operations. The former is used as a kernel module of the operating system, while the latter is a user space tool. CRIU [10, 16] supports process CR operation in user space. Based on CRIU, OpenVZ implements the container live migration as an operating system module, which supports the development of subsequent researches [9]. [12] aims at the migration of containers between user devices, Internet data centers and edge devices while adapting to user mobility, however they need to shut down the container. [13] studies container migration in nested virtualization and optimizes the container (memory) migration on the same host by remapping the memory space. However, this method does not support container migration across hosts. [14] proposes a container migration method based on software-defined rules [18, 19], which helps to describe container migration tasks and dependencies more clearly. CloudHopper [15] uses NFS to facilitate data transfer, however, it would lead to long downtime while migrating memory-intensive workloads.

3 Methodology

The entire migration procedure includes four stages, namely the preparation stage, iterative transmission stage, termination stage and recovery stage. The first three stages rely on the coordination between the source and destination, while the recovery stage is conducted purely at the destination. In the preparation stage, the source and destination nodes exchange important parameters such as OS kernel version, network status, etc., to evaluate if the migration condition is satisfied. In the iterative transmission stage, a network channel is established between the source and destination node to facilitate memory transmission. The memory transmission may last for several iterative rounds until the remaining data or the total number of iterations reaches the pre-set threshold (i.e. the termination condition). In the termination stage, the source node performs a complete checkpoint operation, records the transferred memory pages, dumps important information such as process status, associated files, and namespace, and then stops the process at the source node. In the recovery phase, the destination has already received the entire container memory state, so that the container could resume operation directly.

In an active container, the memory update speed could be very fast. Therefore, during migration we need to transmit a large amount of memory dirty pages in a short period of time. However, it is very challenging to achieve this goal, for the reason that the limited bandwidth in WAN would prolong the migration time or even result in the failure of migration. Thus, to support container memory live migration in WAN, we must optimize the dirty pages transmission procedure.

In modern operating systems, the memory usage pattern follows the principle of locality. Based on this observation, we assume that the dirty pages of container memory exhibit different popularity, and those with higher popularity means they are likely to be updated more frequently. We further propose a container memory live migration method through dirty page popularity perception. To be detailed, first, we identify the memory working set, track and record the real-time information of dirty pages. Then we use the dirty page popularity algorithm to calculate a value (i.e., dirty page popularity value). The dirty page popularity value is used to estimate the possibility of the dirty pages being dirty again after the current migration iteration, a higher popularity value means that the page is more possible to be changed in the following time period. Finally, the decision of container memory migration is made based on the popularity value of dirty pages. In every iteration, pages with lower popularity value should be transmitted. Through such a process, the amount of data transmission in the iterative copy process is reduced, as shown in Fig. 1.

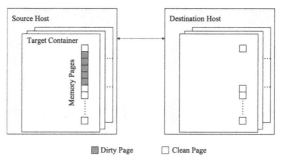

Fig. 1. Container memory migration illustration

4 Key Designs

4.1 Dirty Page Tracking

The container memory live migration depends on an iterative data transmission. During the entire migration, dirty pages of the container's memory needs to be tracked to update the popularity value. However, the memory management mechanism of the operation system makes it difficult to track and predict the address space of a running process. In addition, during the migration, processes may be dynamically created and terminated, and system memory may also be continuously allocated and reclaimed. These factors lead to the difficulty to track dirty pages. KVM uses bitmaps to track dirty pages for pre-copy migration, however this method is not suitable for containers due to its huge overhead.

We aim to optimize container memory live migration through differential transmission of dirty pages. In each round of iterative transmission, memory pages with lower popularity values are transmitted. To calculate the popularity of memory dirty pages, we must solve the following problems: (1) The clear_refs interface provided by Linux kernel could help to track the dirty pages of process memory, however, the update frequency (i.e., the popularity) of dirty pages is still unknown, meaning that it's still impossible to find out the least recently used pages. Hence, we need a new mechanism to calculate the popularity of pages on which basis we could determine what to transmit in each iteration. (2) Although the number of processes in each container is not large, processes may be dynamically created or terminated during migration. Therefore, we should be able to detect the process status in real-time. (3) During live migration, the container memory continues to refresh, so the popularity value of memory dirty pages is evolving. We should be able to updated the popularity value for each iteration.

To trace memory dirty pages and maintain their popularity information efficiently, we design a two-layer data structure which uses a linked list as the first layer and each list node points to one red-black tree as the second layer. This data structure allows us to continuously track the changes in container memory (problem "(2)", and "(3)"). Before each iteration, the dirty page management module calculates a popularity value (problem "(1)") according to memory usage patterns, based on which our method determines which memory page to transfer in each iteration afterwards.

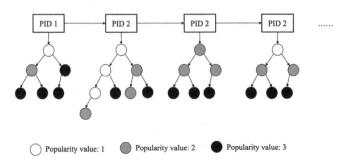

Fig. 2. The two-layer data structure to support the management of memory dirty pages.

The data structure is shown in Fig. 2. "PID x" indicates the container process, "circle" represents memory page of a given process, different colors of "circle" represent different popularity values. At the first layer, the container processes are recorded in the linked list. Each node of the list points to a red-black tree which records the popularity of dirty memory pages of the process. The linked list supports insert, delete and traverse in liner time, therefore it's easy to update the process status in each iteration. We use the red black tree to record the memory information of the process, and the node of the tree represents a specific dirty page unit. In each iteration, the dirty pages are sorted with the help of the red-black tree in order to transfer the less popular dirty pages.

4.2   Dirty Page Popularity-Based Migration Algorithm

We present the dirty page popularity-based migration algorithm in **Algorithm 1**. Three popularity values (i.e., 0, 1 and 2) are defined. In each iteration, our method fetches the information of all the newly generated dirty pages, and uses this information to update the popularity values. All pages with a popularity of 0 and 1 are transmitted in each iteration. A page with popularity value 2 is very likely to be refreshed shortly, therefore it will not be transmitted in the current iteration. The details are as follows.

Algorithm 1 Dirty page popularity-based memory migration algorithm

---

Input: Memory pages: Pages. The Set of Dirty Pages: WorkingSet
Output: The Set of Pages to Transmit: To_Send

---

1. function DirtyPageControl(Pages, WorkingSet)
2. Reset WorkingSet nodes flag to 0
3. for *page* in *Pages* do
4.   if *page* is dirted and *page* in *WorkingSet* then
5.   Get *i*: the index of *page* in *WorkingSet*
6.   if *i.state* == 1 then
7.   set *i.state* = 2
8.   end if
9.   set *i.flag* = 1
10.   else if *page* is dirted and *page* not in *WorkingSet* then
11.   Put *page* into *WorkingSet*
12.   Get *i*: the index of *page* in *WorkingSet*
13.   set *i.state* = 1
14.   set *i.flag* = 1
15.   Add *page* to *To_Send*
16.   else if *page* is not dirted and *page* in *WorkingSet* then
17.   Get *i*: the index of *page* in *WorkingSet*
18.   if *i.state* == 2 then
19.   Add *page* to *To_Send*
20.   end if
21.   Remove *i* from *WorkingSet*
22.   else if *page* is not dirted and *page* not in *WorkingSet* then
23.   Skip
24.   end if
25. end for
26. for *node* in *WorkingSet*:
27.   Remove *node* from *WorkingSet* if *node.flag* equals 0
28
29. return *To_Send*
30. end function

---

(1) If *page* is dirty and is already included in *WorkingSet*, and its popularity value equals 1, then update the node corresponding to *page* in the red-black tree and change its popularity value to 2, and modify the tag bit to indicate that *page* has been checked. If the popularity value equals 2, then do not change popularity value, and modify the tag bit to indicate that the page has been checked.

(2) If *page* is dirty and not in *WorkingSet*, then added it into the *WorkingSet* and set its popularity value to 1. Modify the flag to indicate that the *page* has been checked.

(3) If *page* is not dirty and it is in the *WorkingSet*, when its popularity value is 1, delete the *page* node from the corresponding red-black tree. When the popularity value is 2, delete the *page* node from the corresponding red-black tree and transmit the *page* to the destination.

(4) If *page* is not dirty and *page* does not exist in *WorkingSet*, then skip.

(5) In the process of container memory migration, the dirty page information needs to be updated before each iteration. Dirty page management module (DPM) is used to manage dirty page based on the process number. DPM needs to accurately track the processes activities (i.e., creation and destroy), and the memory activity (i.e., allocation and release). For a process that has been destroyed, it is not necessary to copy its memory page to the destination, so the corresponding data structure can be directly deleted. For newly created process or newly allocated memory, the corresponding memory pages need to be copied to the destination. Therefore, the newly created processes or newly allocated memory page should be included into the dirty page collection (i.e., *WorkingSet*) with a popularity value of 1. If the dirty page exists in the collection, the dirty page popularity value could be changed directly.

For released memory pages, their information will not appear in the output of Ptrace system call. Therefore, the dirty page management module should actively track and record the analyzed memory pages, remove the released memory page from the dirty page collection and delete the corresponding memory pages from the destination in each iteration. In the last iteration, the dumped information generated by checkpoint operation contains the process memory space and the memory binary data, the process could be resumed at the destination using these data.

5 System Modules

We implement the core method in the container migration task manager. As shown in Fig. 3, the container migration manager includes container migration scheduling module, metadata management module, dirty page management module and file management module. Migration manager is responsible for enforcing memory transmit decisions during the migration process, recording dirty page information in the process of multiple rounds of copying, and controlling the iteration process.

The container metadata management module is responsible for obtaining the container's process group number, process relationship, associated files, pipes, sockets, etc. The dirty page management module (DPM) is the core module of the system. We use it to track dirty pages and calculate popularity value. DPM sorts the dirty pages according to the popularity values, and reports them to the container migration scheduling module. The file management module is mainly used to manage the process image files and process information files in the iterative copy process. The scheduling module controls the migration process.

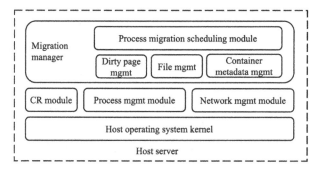

Fig. 3. System architecture

Migration task manager is implemented on top of system tools provided by host operating system, such as Ptrace, CR module, etc. These tools run in user space and therefore could be easily replaced according to different user demands. In addition, we implement the communication module to support authentication and data transmission between source and destination nodes.

6 Evaluation

We evaluate the performance of dirty page popularity-based container memory live migration method (our method). In the experiments, we choose naïve pre-copy method (pre-copy) as a baseline. We have configured two containers running representative workloads, namely Apache server and Linux kernel compilation. The number of iterations, the overall amount of transferred pages and the total migration time are recorded.

Table 1. Experiment environment configuration

Server number	Role	Hardware and software configuration
1	Source Node	Intel Core i5, 8 GB RAM, Ubuntu 16.04 LTS Operation System
2	Destination Node	
3	Request Client	

We set the number of iterations to 30, and record major performance metrics such as the number of memory pages transferred in each iteration. The experimental environment includes three machines, two of which are used as the source node and destination node of migration, and the container to be migrated runs an Apache http server. The third machine is used to generate http requests to the container. During the migration process, HTTP requests were continuously sent to the target Apache container. The configuration of the above servers is shown in Table 1.

The experimental results are shown in Fig. 4. We can see that in our method, the transferred memory pages in each iteration is less than that of pre-copy. Dirty pages

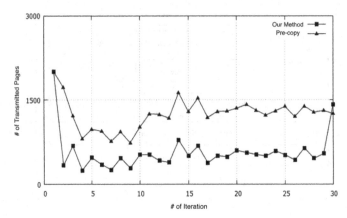

Fig. 4. The amount of transmitted memory pages

will be generated continuously under Apache request, so the number of dirty pages transferred in each iteration keep stable during the migration process.

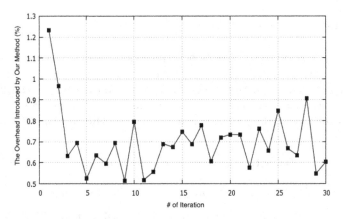

Fig. 5. The overhead of updating working set

Our method needs to update and query the dirty page popularity value before each iteration, which could lead to extra performance overhead. We record the additional performance cost of our method. The results in Fig. 5 indicate that the overhead of updating working set in our method is a very small proportion, which is negligible.

Different workload varies in memory usage patterns, which could finally impact the migration process. To evaluate the migration performance under different workloads, we deploy two containers running a standby Apache server and a Linux kernel compilation task respectively. The memory update frequency of the standby Apache container is significantly lower than that of the Kernel compilation container. Kernel compilation task is known to be computation-intensive so that the corresponding container memory is frequently updated. The results are shown in Table 2, Table 3 and Fig. 6. The two

Table 2. Experimental result of idle apache container

Migration method	# of Iterations	Migration time	Transmitted memory pages
Pre-copy	3	2902209	1166
Our Method	3	2562308	1060

Table 3. Experimental result of kernel compilation container

Migration method	# of Iterations	Migration time	Transmitted memory pages
Pre- Copy	30	129307477	332984
Our Method	6	14473401	35555

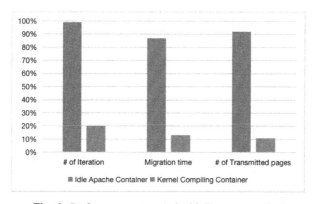

Fig. 6. Performance compared with Pre-copy method

migration methods perform almost the same on the standby Apache container. While on the other hand, for the kernel compilation container, our method outperforms pre- copy with a migration time reduction of 11.19%, and total transferred data volume reduction of 10.68%.

7 Conclusion

Existing container memory live migration methods assume high bandwidth network environment, which fails to comply with multi-data center cloud environment. The limited bandwidth of inter data center network could cause long completion time or even failure of migration. We propose a dirty page popularity-based method, which demands less data transmission during container memory migration. The performance is evaluated through a series of experiments. As future work, we will explore network adaptive methods to further improve the performance of migration.

References

1. Clark, C., Fraser, K., Hand, S., et al.: Live migration of virtual machines. In: Proceedings of the 2nd Conference on Symposium on Networked Systems Design & Implementation-Volume 2. USENIX Association, pp. 273–286 (2005)
2. Kivity, A., Kamay, Y., Laor, D., et al.: KVM: the Linux virtual machine monitor. In: Proceedings of the Linux Symposium, vol. 1, pp. 225–230 (2007)
3. Barham, P., Dragovic, B., Fraser, K., et al.: Xen and the art of virtualization. ACM SIGOPS operating systems review. ACM **37**(5), 164–177 (2003)
4. Ma, F., Liu, F., Liu, Z.: Live virtual machine migration based on improved pre-copy approach. In: 2010 IEEE International Conference on Software Engineering and Service Sciences (ICSESS), pp. 230–233. IEEE (2010)
5. Hines, M.R., Gopalan, K.: Post-copy based live virtual machine migration using adaptive pre-paging and dynamic self-ballooning. In: Proceedings of the 2009 ACM SIGPLAN/SIGOPS International Conference on Virtual Execution Environments, pp. 51–60. ACM (2009)
6. Namespace [EB/OL]. https://en.wikipedia.org/wiki/Linux_namespaces
7. Cgroups [EB/OL]. https://en.wikipedia.org/wiki/Cgroups
8. Zhong, H., Nieh, J.: CRAK: linux checkpoint/restart as a kernel module. Technical Report CUCS-014–01, Department of Computer Science, Columbia University (2001)
9. Hacker, T.J., Romero, F., Nielsen, J.J.: Secure live migration of parallel applications using container-based virtual machines. Int. J. Space-Based Situated Comput. **2**(1), 45–57 (2012)
10. Criu, Inc. Criu [EB/OL]. https://criu.org
11. Osman, S., Subhraveti, D., Gong, S., et al.: The design and implementation of Zap: a system for migrating computing environments. **36**(SI), 361–376 (2002)
12. Elgazar, A., Harras, K.A.: Enabling seamless container migration in edge platforms. In: Workshop Challenged Networks, pp. 1–6 (2019)
13. Sinha, P.K., Doddamani, S., Lu, H., et al.: mWarp: accelerating Intra-host live container migration via memory warping. In: Conference on Computer Communications Workshops, pp. 508–513 (2019)
14. Tao, X., Esposito, F., Sacco, A., et al.: A policy-based architecture for container migration in software defined infrastructures. In: IEEE Conference on Network Softwarization, pp. 198–202 (2019)
15. Benjaponpitak, T., Karakate, M., Sripanidkulchai, K.: Enabling live migration of containerized applications across clouds. In: IEEE INFOCOM 2020 - IEEE Conference on Computer Communications. IEEE (2020)
16. Berg, G., Brattlöf, M., Blanche, A.D., et al.: Evaluating distributed MPI checkpoint and restore using docker containers and CRIU. In: International Conference on Electrical, Communication, Electronics, Instrumentation and Computing (2019)
17. Qiu, M.K., Xue, C., Sha, H.-M.E., et al.: Voltage assignment with guaranteed probability satisfying timing constraint for real-time multiproceesor DSP. J. VLSI Sig. Proc. **46**(1), 55–73 (2007)
18. Dai, W., Qiu, M., Qiu, L., Chen, L., et al.: Who moved my data? privacy protection in smartphones. IEEE Commun. Mag. **55**(1), 20–25 (2017)
19. Qiu, M.K., Zhang, K., Huang, M.L.: An empirical study of web interface design on small display devices. In: IEEE/WIC/ACM International Conference on Web Intelligence. ACM (2004)
20. Gai, K., Qiu, M.K., Zhao, H.: Security-aware efficient mass distributed storage approach for cloud systems in big data. In: IEEE International Conference on Big Data Security on Cloud. IEEE (2016)

*v*RAS: An Efficient Virtualized Resource Allocation Scheme

Jaechun No[1(✉)] and Sung-soon Park[2]

[1] Department of Computer Engineering, Sejong University, Seoul, Korea
jano@sejong.ac.kr
[2] Department of Computer Engineering,
Anyang University and Gluesys Co, Ltd., Anyang, Korea
sspark@gluesys.com

Abstract. In this paper, we present a virtualized resource allocation algorithm (vRAS), which is capable of dynamically predicting VM resource demands based on the real-time snapshot of server resource states. Furthermore, VMs experiencing the heavy resource utilization in the previous time period can be incentivized by obtaining more resources in the next time period, while trying to not aggravate other VMs' application bandwidth running on the same host. We compared our method to other schemes to verify its effectiveness.

Keywords: Resource allocation · *v*RAS · Resource estimation · Burst usage

1 Introduction

Virtualization is becoming an essential component in the cloud environment. It provides the ability of dividing physical server resources into find-grained virtual resources of virtual machines (VMs) to enhance server consolidation and resource utilization [1–3].

The critical aspect of the virtualization is the resource management because the limited number of server resources, such as CPU and memory, should effectively be multiplexed among VMs to achieve the strengths of virtual flexibility, with the help of hypervisor. Although there have been several researches related to resource management [4–8], efficiently allocating VM resources is not an easy task as compared to what would be done in the non-virtualization environment, due to the resource constraint on the host.

In this paper, we propose a resilient VM resource allocation scheme, called *v*RAS (*v*irtualized Resource Allocation Scheme), which is capable of dynamically estimating VM resource demands based on the real-time snapshot of server resource states. Furthermore, VMs experiencing the heavy resource utilization in previous time period can be incentivized by obtaining more resources in the next time period, while trying to not aggravate other VMs' application bandwidth running on the same host.

2 Problem Definition

We address the issue of optimally allocating resources to VMs while taking into account application execution patterns. Among resources, we especially concentrate on assigning CPU and memory, because their allocated sizes directly affect application bandwidths.

© Springer Nature Switzerland AG 2021
M. Qiu (Ed.): SmartCom 2020, LNCS 12608, pp. 79–86, 2021.
https://doi.org/10.1007/978-3-030-74717-6_9

Although I/O storage capacity given to VMs has an effect on application performance, the issue of accelerating I/O bandwidth usually involves I/O related optimizations, such as preloading data or caching, rather than resource allocation policy. There have been several researches related to resource allocation. For example, max-min fairness (MMF) divides resources to maximize minimum allocations among users. This method was adopted to Hadoop in such a way that each slave running on the node is defined to run a fixed number of map and reduce slots (Fixed slot) [7]. Also, in DRF [4], resource allocations are determined by the user's maximum share.

To gain a better insight into those three allocation policies, we evaluated them with the PXZ parallel compression tool using about 7 GB of virtual machine image file. Table 1 demonstrates the performance comparison of three policies. In the evaluation, we captured the resource usage every 1 min. (time unit) and took an average at every 5 min. In the table, we noticed two performance issues:

Resource Shortage: This situation takes place in the case that VMs are not given enough resources after resource reallocation. In DRF and MMF, the performances of VM2 and VM4, respectively, slow down as compared to those of the previous state (from 266 s to 390 s with VM2 in DRF and from 213 s to 542 s with VM4 in MMF).

Wasted Resources: in this case, the large performance difference occurs between VMs because resources allocated some VMs are wasted, while the others suffer from the lack of resources. For example, in all three policies, the performance of VM3 becomes excessively fast, compared to performances of others (92 s of VM3 and 272 s of VM4 with fixed slot, 101 s of VM3 and 542 s of VM4 with MMF, and 950 s of VM3 and 390 s of VM2 with DRF).

3 vRAS

3.1 System Structure

Figure 1 represents an overview of vRAS for competing VMs on the host. The functionality is composed of two main steps and is periodically applied to VMs. The first step is to estimate the resource demands of VMs by keeping track of the resource utilizations accumulated in database since the last allocation functionality was applied to VMs.

The second step is to perform the actual resource partitioning and reallocation among VMs by accounting for the estimated resource sizes in the first step and also by incentivizing in terms of the recent utilization ratios.

Let $t_1, t_2, ..., t_k$ be time units of an allocation period at which unit each VM under consideration accumulates its resource usage to database.

At time unit t_k, vRAS goes through two steps to estimate and partition resources across VMs, by referring to the utilization ratios stored in database. The amount of resources measured at t_k are reallocated to VMs to be used for the next k time units. This procedure is repeatedly executed for every allocation period consisting of k time units.

Table 1. PXZ compression.

		VM1	VM2	VM3	VM4
Previous state	Resources < CPU, MEM >	< 3, 11 >	< 4, 8 >	< 3, 4 >	< 6, 5 >
	Average Resource Usage	< 0.63, 0.24 >	< 0.87, 0.92 >	< 0.36, 0.40 >	< 0.68, 0.93 >
	Performance (sec)	192	266	123	213
	Resource Demands	< 3, 10 >	< 8, 14 >	< 4, 6 >	< 2, 6 >
Fixed Slot	Allocation < CPU, MEM >	< 4, 7 >	< 4, 7 >	< 4, 7 >	< 4, 7 >
	Average Resource Usage	< 0.51, 0.39 >	< 0.87, 0.99 >	< 0.23, 0.25 >	< 0.82, 0.75 >
	Performance (sec)	164	257	92	272
Max-min fairness	Allocation < CPU, MEM >	< 3, 8 >	< 4, 8 >	< 4, 6 >	< 2, 6 >
	Average Resource Usage	< 0.66, 0.34 >	< 0.87, 0.92 >	< 0.23, 0.30>	< 0.99, 0.89 >
	Performance (sec)	192	250	101	542
DRF	Allocation < CPU, MEM >	< 2, 6 >	< 3, 5 >	< 4, 6 >	< 3, 9 >
	Average Resource Usage	< 0.8, 0.46 >	< 0.97, 0.99 >	< 0.23, 0.3 >	< 0.93, 0.58 >
	Performance (sec)	213	390	95	252

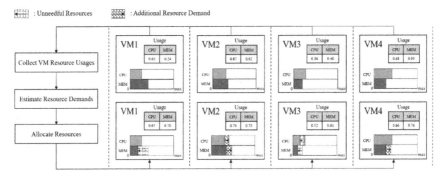

Fig. 1. vRAS Architecture. In case of VM1, CPU and memory sizes were changed from <3,11> to <3,4>.

3.2 vRAS Functionality

To describe vRAS functionality, let V_1, V_2, ..., V_n be VMs executing on a host and C and M be total number of CPUs and total memory sizes of the host, respectively. Also, let S_c and S_m be the maximum allowable usage ratio for CPU and memory that the host can offer to VMs. In other words, if $S_c=1$, then all CPUs on the host can be given to a single VM.

To represent vRAS steps in detail, we use an ordered vector pairs $\{(c_i, m_i)|1 \leq i \leq n\}$ to denote the amount of CPUs and memory sizes assigned to n VMs at time unit t_k of an allocation period. These sizes are used for the next allocation period until the new resource adjustment takes place at t_k of the period. We use the second ordered vector pair to denote CPU and memory usages of VMs at a time unit t_j:

$$\exists t_j (1 \leq j \leq k),$$

$\{(uc_i^{t_j}, um_i^{t_j})|1 \leq i \leq n)\}$ where $0 < uc_i^{t_j}, um_i^{t_j} < 1$.

For V_i with (c_i, m_i) resource sizes, CPU and memory usages are changed to (uc_i^{t1}, um_i^{t1}), (uc_i^{t2}, um_i^{t2}),..., (uc_i^{tk}, um_i^{tk}) during k time units while being accumulated in database. Based on the usages, the resource sizes of V_i are adjusted at time unit t_k and then used until the next allocation period reaches at t_k.

3.3 Estimation of Resource Requirements

The primary objective of the first step of vRAS is to measure the desirable resource sizes of VMs for the next allocation period. This goal is accomplished by facilitating the resource usages stored in database during the current period, while those sizes to be estimated are forced not to exceed the maximum allowable sizes on the host.

Let λ_i^c and λ_i^m and be CPU and memory sizes to be estimated, respectively, for V_i. Then, λ_i^c and λ_i^m can be computed as follows:

$$\lambda_i^c = \{(\text{ceiling}\{(\left(\sum_{1 \leq j \leq k} uc_i^j / k\right) \times c_i)/S_c\} \tag{1}$$

$$\lambda_i^m = \text{ceiling}\{(\left(\sum_{1 \leq j \leq k} um_i^j / k\right) \times m_i)/S_m\} \tag{2}$$

To verify both Eqs. (1) and (2), let x and y be CPU and memory sizes, respectively, to be adjusted at k^{th} time unit of the current allocation period. If $x = 0$ or $y = 0$, then no change takes place in resource sizes and thereby the same amount of resources would be allocated to V_i for the next period.

First of all, computing λ_i^c and λ_i^m needs to keep track of CPU and memory usages for k time units while maintaining the appropriate resource ratio at the host. Therefore, λ_i^c and λ_i^m satisfy the followings:

$$(\varphi \times c_i)/(c_i + x) \leq S_c, \text{ where } \varphi = \sum_{1 \leq j \leq k} uc_i^j / k \tag{3}$$

$$(\theta \times m_i)/(m_i + y) \leq S_m, \text{ where } \theta = \sum_{1 \leq j \leq k} um_i^j / k \tag{4}$$

In Eqs. (3) and (4), $\phi \times c_i$ and $\theta \times m_i$ reflect how much V_i utilized the allocated CPUs and memory on average for the current period. Since $c_i + x$ and $m_i + y$ implies the desirable CPU and memory sizes in terms of resource usages, those values correspond to λ_i^c and λ_i^m. As a result, Eqs. (3) and (4) can be rewritten to Eqs. (1) and (2).

3.4 Determination of Resource Allocation Sizes

The second step of vRAS takes λ_i^c and λ_i^m calculated in the first step as a basis and further weights those values by taking into account the burst usages of a specific time unit. In the case that applications with high resource demands are assigned to VM, the resource utilization of the VM goes upwards as compared to that of other VMs.

Although more resources should be reallocated to the VM to achieve high application performance, such a resource reassignment should not be harmful to other VMs running on the same host. In the second step, vRAS attempts to incentivize the VM exposing high resource requirements, at the same time doing so little adversely affect other VMs. vRAS first detects the time unit revealing the maximum usage difference between two consecutive units and takes it as the starting point to compute the resource weight.

Let a and b be two time units where CPU and memory usages are sharply increased:

$$a \ s.t. \ \max\{|uc_i^{t_a} - uc_i^{t_{a-1}}|, where \ 2 \leq a \leq k\} \tag{5}$$

$$b \ s.t. \ \max\{|uc_i^{t_b} - uc_i^{t_{b-1}}|, where \ 2 \leq b \leq k\} \tag{6}$$

With a and b, resource weights, w_i^c for CPU and w_i^m for memory, are calculated as follows.

$$w_i^c = (\sum_{a \leq j \leq k} uc_i^{t_j})/(k - a + 1) \tag{7}$$

$$w_i^m = (\sum_{b \leq j \leq k} uc_i^{t_j})/(k - b + 1) \tag{8}$$

It implies that weights are measured by leveraging the resource usages of the selected time duration generating the boot utilization. Next, the actual resource allocation sizes, π_i^c for CPU and π_i^m for memory, are computed in algorithm 1. Because steps for computing π_i^c and π_i^m are the same, algorithm 1 describes the steps for computing only π_i^c.

In algorithm 1, x_i implies the amount of resources that can be allocated to a VM. Also, π_i^c is determined only when x_i is no less than λ_i^c, while recomputing x_i for the remaining VMs. If there exist VMs left unallocated, then the estimated CPU size, λ_i^c, is reduced (currently the reduced ratio is set to 0.9) and recalculated for the remaining VMs unallocated.

Algorithm 1: Determination of CPU allocation sizes

$W = \Sigma_{1 \leq i \leq n} w_i^c$;

for each V_i $x_i = C \times (w_i^c / W)$ **end for**

while (there exist VMs not allocated yet) **do**

    for each V_i

        if ($x_i \geq \lambda_i^c$)

            $\pi_i^c = \lambda_i^c$; $x_i = 0$;

            $X = \Sigma_{1 \leq k \leq n} x_k$; $\Pi = \Sigma_{1 \leq n \leq k} \pi_k^c$;

            for each V_i $u_i = (C - \Pi) \times (x_i / X)$ **end for**

        end if

    end for

    for each V_i $\lambda_i^c = \lambda_i^c \times$ *reduced ratio* **end for**

end while

4   Performance Evaluation

We compared the performance of *v*RAS with that of three methods referred at Table 1. Table 2 represents the experimental platform for the performance measurement.

Table 2. Experimental platform

CPU	Intel(R) Xeon(R) CPU E5–2620 v3 2EA
Memory	Samsung PC4-17000R 8GB(2Rx8) 4EA
Host OS	CentOS Linux release 7.2
Hypervisor	KVM, QEMU 1.5.3
VM OS	CentOS Linux release 7.2
Tools	Parallel PXZ 4.999.9
	Filebench 1.4.9.1
	stress-ng-0.07.20

Table 3. Resource distribution and PXZ performance in vRAS (current state)

$<S_c, , S_m>$		VM1	VM2	VM3	VM4
$<0.9,0.9>$	Allocation <CPU, MEM(GB)>	<3, 3>	<4, 9>	<2, 2 >	<5, 6>
	Performance (sec)	190	254	147	233
$<0.75,0.75>$	Allocation <CPU, MEM(GB)>	<3, 4>	<5, 10>	<2, 3>	<6, 7>
	Performance (sec)	190	226	155	202
$<0.6, 0.6>$	Allocation <CPU, MEM(GB)>	<3, 5>	<5, 12>	<2, 3>	<5, 8>
	Performance (sec)	186	221	145	204

Table 3 shows the times for allocating resources in vRAS, while running PXZ. The previous state is the same as that in Table 1, with S_c and S_m (maximum allowable usage ratio for CPU and memory) being varied as described.

First of all, we observed vRAS to see if each VM can get enough resources in the current state to produce better execution time than that in the previous state. With S_c and S_m of 0.75 both, VM2 showing the highest usages gets 5 CPU and 10 GB memory and its execution time is enhanced from 266 s to 226 s. On the other hand, as can be seen in Table 1, the execution time of VM2 in DRF was increased from 266 s to 390 s. Based on this observation, we can notice that vRAS gives more resources to the VM generating high resource usages, in order not to cause resource shortage.

Table 3 also demonstrates that vRAS does not waste resource by not assigning excessive resources to VMs producing little resource usages. For instance, VM3 generating the least average usages at the previous state gets the smallest amount of resources at the current stage, whereas DRF in Table 1 gives 4CPU and 6 GB memory to VM3 and 3CPU and 5GB memory to VM2 denoting highest usages. The same vRAS allocation behavior can be observed even if S_c and S_m both are changed to 0.6 and 0.9.

Figure 2 denotes the usage difference of four allocation methods between the real observed usage and the desirable target usage to execute applications on VMs. In this case, the desirable usage is set to 0.75. To measure the usage difference, we used RMSD (root mean square deviation) that implies the average distance between each real usage measured by four allocation methods and the target usage. Figure 2 shows that vRAS generates the least usage difference as compared to the others, meaning that VMs retaining resources through vRAS demonstrate the usage rates much closer to the target value than the others.

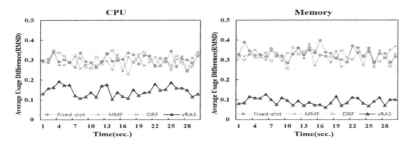

Fig. 2. Usage difference (S_c, $S_m = 0.75$)

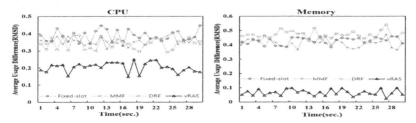

Fig. 3. Usage difference (S_c, $S_m = 0.9$)

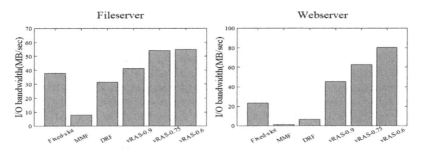

Fig. 4. I/O bandwidth

Also, such a low usage difference implies that the problem of either resource shortage or waste is alleviated in vRAS, compared to that of other methods. The same observation can be seen in Fig. 3, where the desirable target usage is set to 0.9.

In Fig. 4, we compared νRAS I/O bandwidth to the others, by using Filebench that produces two kinds of workloads: fileserver and webserver. In νRAS, S_c and S_m are both changed to 0.9, 0.75 and 0.6, respectively. In this figure, we chose the lowest I/O bandwidth among VMs for the comparison. We can notice that the minimum I/O bandwidth of νRAS is higher than that of others, meaning that in three other methods, there exists a VM that received less resources than it actually needs to produce better bandwidth. Figure 4 shows that such a problem is mitigated in νRAS.

5   Conclusion

In this paper, we present a VM resource allocation scheme, called νRAS. It consists of two steps: first step is to estimate the desirable resource sizes of VMs for the next allocation period and second step is to incentivize VMs demonstrating the burst resource usages during specific time units. We compared νRAS with three other methods including fixed slot, MMF and DRF and found out that the problem of resource shortage and waste obstructing bandwidth speedup is alleviated in νRAS.

Acknowledgements. This research was supported by the National Research Foundation of Korea (NRF) grant funded by the Korea government (2019R1F1A1057082). Also, this work was supported by Institute of Information & communications Technology Planning & Evaluation (IITP) grant funded by the Korea government (MSIT) (No. 2020–0-00104, Development of low-latency storage module for I/O intensive edge data processing). We would like to thank Youngwoo Kim for his contribution to this paper.

References

1. Har'El, N., Gordon, A., Landau, A.: Efficient and scalable pravirtual I/O system. In: USENIX Annual Technical Conference, pp. 231–241 (2013)
2. Razavi, K., Kielmann, T.: Scalable Virtual Machine Deployment Using VM Image Caches, In: Proc. SC'13 on High Performance Computing, Networking, Storage and Analysis, Denver, USA (2013)
3. Ji, X., e t al.: Automatic, application-aware I/O forwarding resource allocation. In: 17th USENIX Conference on File and Storage Technologies, Boston, MA, USA (2019)
4. Ghodsi, A., Zaharia, M., Hindman, B., Konwinski, A., Shenker S., Stoica, I.: Dominant resource fairness: fair allocation of multiple resource types. In: Proceedings of the 8th USENIX Conference on Networked Systems Design and Implementation, pp. 323–336 (2011)
5. Lam, V., Radhakrishnan, S., Pan, R., Vahdat, A., Varghese, G.: NetShare and Stochastic Net-Share: predictable bandwidth allocation for data centers. In: ACM SIGCOMM Computer Communication Review, vol. 42, no. 3 pp. 5–11(2011)
6. Kim, H., Jo, H., Lee, J.: XHive: efficient cooperative caching for virtual machines. IEEE Trans. Comput. **60**(1), 106–119 (2011)
7. Hindman, B., et al.: Mesos: a platform for fine-grained resource sharing in the data Center. In: Proceedings of NSDI' 11, Boston, MA, USA, pp. 295–308 (2011)
8. Funaro, L., Ben-Yehuda, O., Schuster, A.: Stochastic resource allocation. In: Proceedings of the 15 ACM SIGPLAN/SIGOPS International Conference on Virtual Execution Environments, pp. 122–136 (2019)

HOOD: High-Order Orthogonal Decomposition for Tensors

Weitao Tang, Xiang Liu, Huyunting Huang, Ziyang Tang, Tonglin Zhang[✉],
and Baijian Yang[✉]

Purdue University, West Lafayette, USA
{tang384,xiang35,huan1182,tang385,tlzhang,byang}@purdue.edu

Abstract. Tensor decompositions are becoming increasingly important in processing images and videos. Previous methods, such as ANDECOMP/PARAFAC decomposition (CPD), Tucker decomposition (TKD), or tensor train decomposition (TTD), treat individual modes (or coordinates) equally. Their results do not contain a natural and hierarchical connection between a given tensor and its lower-order slices (e.g., a video and its frames). To overcome the practical limitation of existing tensor decomposition methods, we propose an innovative High-Order Orthogonal Decomposition (HOOD) for arbitrary order tensors. HOOD decomposes a given tensor using orthogonal linear combinations of its lower-order slices. Each orthogonal linear combination will be further decomposed. In the end, it decomposes the given tensor into orthogonal rank-one tensors. For object detection and recognition tasks in high-resolution videos, HOOD demonstrated great advantages. It is about 100 times faster than CPD with similar accuracy detection and recognition results. It also demonstrated better accuracy than TKD with similar time overhead. HOOD can also be used to improve the explainability because the resulting eigenimages visually reveal the most important common properties of the videos and images, which is a unique feature that CPD, TKD, and TTD do not have.

Keywords: Tensor decomposition · Object detection · Video processing

1 Introduction

Tensors are extensions of matrices. They are important in processing images and videos in machine learning or deep learning. Due to recent advances in computer technology, image and video data are collected every data in a wide range applications, leading to the need of more computationally efficient and more interpretable tensor decomposition methods in artificial intelligence [3,5]. Previous tensor decompositions are mostly carried out by CANDECOMP/PARAFAC decomposition (CPD) [2], Tucker decomposition (TKD) [17], or tensor train decomposition (TTD) [15]. These methods, however, do not provide a natural and hierarchical connection between a given tensor and its lower-order slices

© Springer Nature Switzerland AG 2021
M. Qiu (Ed.): SmartCom 2020, LNCS 12608, pp. 87–96, 2021.
https://doi.org/10.1007/978-3-030-74717-6_10

(e.g., for the relationship between a video and its frames). The proposed High-Order Orthogonal Decomposition (HOOD) can address this issue.

All CPD, TKD, and TTD can be treated as extensions of singular value decomposition (SVD) for matrices. They are identical for second-order tensors (i.e., matrices) but different for high-order tensors (i.e., orders ≥ 3). CPD decomposes a given tensor into the sum of rank-one tensors, each of which can be expressed as an outer product of vectors. TKD decomposes a given tensor into a core tensor multiple by orthogonal matrices along with their modes. As it requires the core tensor be orthogonal, a property called all orthogonality is satisfied. TTD decomposes a given tensor into two matrices and a few third-order tensors, such that the given tensor can be expressed by an inner train product between the components. In all of those, the roles of individual modes (or coordinates) are treated equivalently. They cannot be used to identify the connections between lower-order slices and the given tensor.

In certain types of high order data, such as a video containing multiple frames, the relationship between modes is important [8,13]. However, this is not contained in CPD, TKD, and TTD. In particular, suppose that a color video is expressed by a fourth-order tensor, where the first and second modes represent the pixel location, and the third and fourth modes represent the RBG channels and frames. Treating the four modes equivalently is inappropriate since the role of the fourth mode differs from the roles of the first and the second modes. It is better to decompose the tensor along its fourth mode first and the rest three modes next. This motivates us to propose HOOD.

Arbitrary tensor can be obtained by adding individual modes sequentially, and HOOD utilizes this idea reversely. It sequentially decomposes an arbitrary tensor into orthogonal combinations of lower-order tensors. We proposed a generalized SVD (GSVD) for tensors to implement HOOD. GSVD decomposes any tensor into the sum of orthogonal rank-one tensors. In HOOD, a low-rank decomposition can be obtained by simply removing rank-one tensors with lower weight values analogous to principal component analysis (PCA) for matrices. This is not viable in CPD, TTD, and TKD. Note that CPD is orthogonal and any decomposition based on it must be computed by an iterative algorithm. The previous result cannot be used to construct a new decomposition if another rank is used [19]. Because all orthogonality is too rigorous, the previous result in TKD cannot be used either. Similar issues also appear in TTD as it requires the two matrices be orthogonal. The numerical algorithms in CPD, TKD, TTD need to be carried out multiple times when multiple candidate ranks are desired, which is not an issue in HOOD.

Although HOOD can be used on arbitrary order tensors, for ease of presentation, we investigated it based on videos, which are fourth order tensors. Our experiments show that HOOD is computationally efficient and need less memory than CPD. HOOD saves 99% computing time compared to CPD in general. When used in conjunction with object detection and recognition in high-resolution videos, HOOD ties CPD, and overwhelms TKD and CPD in terms of detection performance. Based on a few eigenimages introduced by HOOD, we can also visually investigate the contents of the video (or a set of images). This cannot be achieved by any of CPD, TKD, or TKD. Due to its computational efficiency and well structures, HOOD is powerful for long videos.

The rest of the paper is organized as follows. We briefly review related work in Sect. 2, followed by an introduction of preliminaries in Sect. 3. Section 4 introduces our method. Section 5 describes the experiments and discusses the results. We conclude the paper in Sect. 6.

2 Related Work

Numerous tensor decomposition methods and their applications have been proposed in the literature [11]. Among those, the three most popular methods are CPD, TKD, and TTD. For instance, [16] proposes a statistical model based on CPD with applications to medical image data; [18] combines CPD with a sketching method and applies it in topic modeling; [4] uses CPD to process images in deep learning; [13] studies CPD in image detection and recognition problems; and [15] use TTD to do color image denoising. Meanwhile, TKD models have been investigated in all-orthogonal tensor decomposition for large tensors [14], randomized High-Order SVD (HOSVD) [7], smoothed TKD [9], and multilinear SVD (MSVD) [6]. Computational difficulties have been found in both CPD and TKD for large tensors, which may be overcome by memory-efficient algorithms for CPD [1] or TKD [12], respectively. Because rank-one tensors in CPD and TTD are not orthogonal but all orthogonality in TKD is too rigorous, it is important to have an orthogonal tensor decomposition method without all orthogonality [10], but such a method has not been successfully developed yet.

3 Preliminaries

A dth-order tensor $\mathcal{X} \in \mathbb{R}^{n_1 \times \cdots \times n_d}$ is a d-way array. For any $i \in \{1, \ldots, d\}$ and $j \in \{1, \ldots, n_i\}$, the jth reduced one slice of $\mathcal{X}$ along the ith mode, denoted by $\mathcal{X}_{\{(i,j)\}}$, is $(d-1)$th-order tensor composed by setting its ith mode at j. For any distinct $i_1, i_2 \in \{1, \cdots, d\}$, $j_1 \in \{1, \ldots, n_{i_1}\}$, and $j_2 \in \{1, \ldots, n_{i_2}\}$, the (j_1, j_2)th reduced two slice of $\mathcal{X}$ along the (i_1, i_2)th modes, denoted by $\mathcal{X}_{\{(i_1,j_1),(i_2,j_2)\}}$, is $(d-2)$th-order tensor composed by setting its i_1th and i_2th modes at j_1 and j_2, respectively. Similarly, we can define the reduced k slices for any $k \leq d$. The inner product between $\mathcal{X}, \mathcal{Y} \in \mathbb{R}^{n_1 \times \cdots \times n_d}$, denoted by $\langle \mathcal{X}, \mathcal{Y} \rangle$, is the sum of their pairwise elements. We say $\mathcal{X}$ and $\mathcal{Y}$ are orthogonal, denoted by $\mathcal{X} \perp \mathcal{Y}$ if $\langle \mathcal{X}, \mathcal{Y} \rangle = 0$. The Frobenius norm of $\mathcal{X}$ is $\|\mathcal{X}\| = \langle \mathcal{X}, \mathcal{X} \rangle^{1/2}$. The outer product between dth-order tensor $\mathcal{X} \in \mathbb{R}^{m_1 \times \cdots \times m_d}$ and eth-order tensor $\mathcal{Y} \in \mathbb{R}^{n_1 \times \cdots \times n_e}$, denoted by $\mathcal{X} \circ \mathcal{Y} \in \mathbb{R}^{m_1 \times \cdots \times m_d \times n_1 \times \cdots \times n_e}$, is $(d+e)$th order tensor with elements given by their combinations. A dth-order tensor is said rank-one if it can be expressed by an outer product of d vectors. CPD decomposes $\mathcal{X} \in \mathbb{R}^{n_1 \times \cdots \times n_d}$ as $\mathcal{X} \approx \sum_{r=1}^{R} \mathcal{X}_r$, where $\mathcal{X}_r = \boldsymbol{x}_{r1} \circ \cdots \circ \boldsymbol{x}_{rd}$ are rank-one tensors, $\boldsymbol{x}_{rj}$ for $j \in \{1, \cdots, d\}$ and $r \in \{1, \ldots, R\}$ are called factors, and R is the CP-rank, such that $\|\mathcal{X} - \sum_{r=1}^{R} \mathcal{X}_r\|^2$ is minimized. CPD does not require the factor matrices (columns given by x_{r_j} for each individual j) be orthogonal. TKD is decomposes it as $\mathcal{X} \approx \sum_{r_1=1}^{R_1} \cdots \sum_{r_d=1}^{R_d} g_{r_1 \ldots r_d} \mathcal{X}_{r_1 \cdots r_d}$, where $g_{r_1 \ldots r_d}$ is the $(r_1, \ldots, r_d)$th entry of core tensor $\mathcal{G} \in \mathbb{R}^{R_1 \times \cdots \times R_d}$, $(R_1, \ldots, R_d)$ is called TK-rank, $\mathcal{X}_{r_1 \cdots r_d} = x_{r_1 1} \circ \cdots \circ x_{r_d d}$ are rank-one tensors, and $x_{r_j j}$ with $j \in \{1, \ldots, d\}$ and $r_j \in \{1, \ldots, R_j\}$ are called factors, such that the Frobenious between difference is minimized. TKD requires that core tensor $\mathcal{G}$ be all orthogonal and the factor matrices be orthogonal. Thus, a property called all orthogonality is satisfied. TTD decomposes it as $\mathcal{X}_{i_1 \cdots i_d} \approx \sum_{r_1=2}^{R_1} \cdots \sum_{r_{d-1}=1}^{R_{d-1}} g_{1,r_1 i_1} g_{2,r_1,i_2,r_2} \cdots g_{d-1,r_{d-2} i_{d-1} r_{d-1}} g_{d-1,i_d}$, where $g_{1,r_1 i_1}$ and g_{d,i_d,r_d} are (r_1, i_1) and (i_d, r_d) entries of core matrices $\mathbf{G}_1$ and $\mathbf{G}_d$, respectively, and $g_{j,r_{j-1} i_j r_j}$ is the (r_{j-1}, i_j, r_j)th entry of core tensor $\mathcal{G}_j$ for $j \in \{2, \ldots, d-1\}$.

4 Method

HOOD is sequential and hierarchical. It decomposes an arbitrary-order tensor based on a sequence of the modes. Without the loss of generality, assume that the sequence is from the last model to the first mode. The first hierarchy of HOOD is to decompose $\mathcal{X} \in \mathbb{R}^{n_1 \times \cdots \times n_d}$ as

$$\mathcal{X} = \sum_{i_d=1}^{n_d} \alpha_{i_d} \mathcal{A}_{i_d} \circ a_{i_d}, \tag{1}$$

where $\alpha_{i_d} \geq 0$ are real numbers, $a_{i_d} \in \mathbb{R}^{n_d}$ are unit vectors, and $\mathcal{A}_{i_d} \in \mathbb{R}^{n_1 \times \cdots \times n_{d-1}}$ are unit $(d-1)$th-order tensors. The second hierarchy to decompose $\mathcal{A}_{i_d}$ individually, such that it can be expressed as

$$\mathcal{A}_{i_d} = \sum_{i_{d-1}=1}^{n_{d-1}} \beta_{i_{d-1} i_d} \mathcal{A}_{i_{d-1} i_d} \circ a_{i_{d-1} i_d}, \tag{2}$$

where $\alpha_{i_{d-1} i_d}, a_{i_{d-1} i_d} \in \mathbb{R}^{n_{d-1}}$ are unit vectors, and $\mathcal{A}_{i_{d-1} i_d} \in \mathbb{R}^{n_1 \times \cdots \times n_{d-2}}$ are unit $(d-2)$th-order tensors. Combining (1) and (2), we have

$$\mathcal{X} = \sum_{i_{d-1}=1}^{n_{d-1}} \sum_{i_d=1}^{n_d} \alpha_{i_{d-1} i_d} \mathcal{A}_{i_{d-1} i_d} \circ a_{i_{d-1} i_d} \circ a_{i_d}, \tag{3}$$

where $\alpha_{i_{d-1} i_d} = \alpha_{i_d} \beta_{i_{d-1} i_d}$. In the end, HOOD decomposes $\mathcal{X}$ along $\mathcal{S} = (d, d-1, \ldots, 2)$, a selected sequence of the modes, as

$$\mathcal{X} = \sum_{i_2=1}^{n_2} \cdots \sum_{i_d=1}^{n_d} \mathcal{X}_{i_2 \cdots i_d} = \sum_{i_2=1}^{n_2} \cdots \sum_{i_d=1}^{n_d} \alpha_{i_2 \cdots i_d} \circ a_{i_2 \cdots i_d} \circ \cdots \circ a_{i_d}, \tag{4}$$

where $\alpha_{i_2 \cdots i_d}$ are called the weights and $a_{i_{d-k+1} \cdots i_d}$ are called factors in the kth hierarchy. The number of nonzero weights in (4), denoted by $\text{rank}_\mathcal{S}(\mathcal{X})$, is called the *HOOD rank* of $\mathcal{X}$ along $\mathcal{S}$.

We next propose the generalized singular value decomposition (GSVD) for tensors to implement HOOD. We treat it as an extension of SVD for matrices. The key is the derivation of (1). The final decomposition given by (4) can be obtained by an order-decent algorithm. In particular, we define matrix $\mathbf{Q} = (q_{jj'})_{n_d \times n_d} \in \mathbb{R}^{n_d \times n_d}$ with $q_{jj'} = \langle \mathcal{X}_{(d,j)}, \mathcal{X}_{(d,j')} \rangle$, such that

$$\mathbf{Q} = \mathbf{ADA}^\top, \tag{5}$$

where $\mathbf{A} = (\boldsymbol{a}_1, \ldots, \boldsymbol{a}_{n_d})$ is an orthogonal matrix for eigenvectors and $\mathbf{D} = \mathrm{diag}(d_1, \ldots, d_{n_d})$ is a diagonal matrix for eigenvalues. Let $\alpha_{i_d} = d_{i_d}^{1/2}$ for all $i \in \{1, \ldots, n_d\}$ and

$$\mathcal{A}_{i_d} = \frac{1}{\alpha_{i_d}} \sum_{j=1}^{n_d} \mathcal{X}_{(d,j)} a_{j i_d} \tag{6}$$

where $a_{j i_d}$ is the jth component of $\boldsymbol{a}_{i_d}$. We obtain (1). By the implementation of the same method to individual $\mathcal{A}_{i_{d-1}}$, we can obtain (2) and (3). In the end, we obtain (4). Then, we propose the following order-decent algorithm.

Algorithm 1. Order-Decent Algorithm for HOOD

1: **Input:** Tensor $\mathcal{X} \in \mathbb{R}^{n_1 \times \cdots \times n_d}$ and $\mathcal{S}$ for a sequence of modes.
2: **Output:** weights and factors in the kth hierarchy for all $k \in \{1, \ldots, d-1\}$ in (4)
3: Re-order $\mathcal{S}$ such that $\mathcal{S} = (d, d-1, \ldots, 2)$
4: Implement (5) and (6) and obtain $\boldsymbol{a}_{i_d}$ and $\mathcal{A}_{i_d}$ for all $i_d \in \{1, \ldots, n_d\}$
5: Assume that $\mathcal{A}_{i_{d-k+1} \cdots i_d}$ has been derived. Define $\mathbf{Q}_{i_{d-k+2} \cdots i_d} = (q_{ij})_{n_{d-k+1} \times n_{d-k+1}}$ similarly as $\mathbf{Q}$ in the first hierarchy.
6: Implement a method similar to (5) and (6) for the derivation of the weights and factors in the kth hierarchy
7: Re-order $\mathcal{S}$ and the results back
8: Return

Theorem 1. *Rank-one tensors $\mathcal{X}_{i_2 \cdots i_d}$ given by GSVD are orthogonal.*

Proof. Because $\mathbf{Q}$ is symmetric, $\boldsymbol{a}_{i_d} \perp \boldsymbol{a}_{i'_d}$ for any distinct i_d and i'_d in (5) by the property of eigen decomposition for symmetric matrices. By the same method, we conclude that $\boldsymbol{a}_{i_{d-1} i_d} \perp \boldsymbol{a}_{i'_{d-1} i_d}$ for any distinct i_{d-1} and i'_{d-1}. Further, we have that $\boldsymbol{a}_{i_{d-k} i_{d-k+1} \cdots i_d} \perp \boldsymbol{a}_{i'_{d-1} i_{d-k+1} \cdots i_d}$ for any distinct i_{d-k} and i'_{d-k} when $k \in \mathcal{S}$. We draw the conclusion by the property of inner products of outer products of vectors [10]. $\qquad\square$

Corollary 1. *If $\tilde{\mathcal{X}}$ is an approximate HOOD of $\mathcal{X}$ along $\mathcal{S}$ with $\mathrm{rank}_\mathcal{S}(\tilde{\mathcal{X}}) \le m$, then $\|\tilde{\mathcal{X}} - \mathcal{X}\|^2$ is minimized by choosing rank-one tensors with the largest m weights on the right-hand side of (4).*

Proof. We can prove the conclusion by the same method the proof for the approximation given by traditional PCA for matrices. $\qquad\square$

We are particularly interested in the case when $\mathcal{X} \in \mathbb{R}^{n_1 \times n_2 \times n_3 \times n_4}$ is used in video representations, where n_1 and n_2 represent the pixels, n_3 (i.e., $n_3 = 3$) represents RGB channels, and n_4 represents the frames. The j_4th frame of the video is $\mathcal{X}_{(4,j_4)}$. It is the j_4th reduced one slice of $\mathcal{X}$ along its fourth mode. By (6), we have

$$\mathcal{A}_{i_4} = \frac{1}{\alpha_{i_4}} \sum_{j=1}^{n_4} \mathcal{X}_{(4,j)} a_{ji_4}. \tag{7}$$

The size of $\mathcal{A}_{i_4}$ is identical to the size of frames. It is a linear combination of the frames contained by the video. Its role is identical to the role of the left singular (or eigen) vectors in traditional SVD for matrices. Therefore, $\mathcal{A}_{i_4}$ can be interpreted as the i_4th singular (or eigen) tensor in HOOD. We call $\mathcal{A}_{i_4}$ for all $i_4 \in \{1, \ldots, n_4\}$ given by (7) eigen images in HOOD for videos.

5   Experiments

We performed three experiments to evaluate HOOD. The first experiment compared the computing time of HOOD against that of CPD, TKD, and TTD. The second experiment investigated the low rank decomposition quality of HOOD, CPD, TKD, and TTD by relative total variation, Fréchet Inception Distance (FID), and a bench mark object detection algorithm YOLO v3. The third experiment studied the eigen images of HOOD, introduced in the last paragraph of Sect. 4. All the experiments were carried out with different number of rank-one tensors. For CPD, the employed number of rank-one tensors is the exact number of rank-one tensors used in the low rank decomposition. For TKD and TTD, the employed number of rank-one tensors is the product of the ranks in the low rank decomposition. In first and second experiments, the CPD and TKD were performed with a converge tolerance of $1e-3$ and a maximum of 50 iterations.

All the experiments were conducted on a workstation with Intel i7-8700K CPU at 3.70 GHz, 32 GB DDR4 RAM, 1T SSD, and Nvidia 2080 GPU. Two videos were employed in experiment 1 and experiment 2. Both had 300 frames with resolution 1920×1080. Selected frames of the two videos are presented in Fig. 1. Another video was employed for experiment 3 to investigate eigen images, its selected frames were presented in Fig. 3.

Experiment I. We performed HOOD, CPD, TKD, and TTD on both two videos with number of rank-one tensors equals to $1, 2, 4, \ldots, 512, 1024$, to compare their computation time. We found that experiments on two videos show similar trends and only present the result of Video 1.

Figure 2(a) presents the computing time of HOOD, CPD, TKD, and TTD with different options of numbers of rank-one tensors based on Video 1. When the number of rank-one tensor equals one, all the four algorithms have a relatively low computing time. As the number of rank-one tensors increases, the computing time of CPD increases drastically, but that of HOOD, TKD, and TTD does not. This experiment shows that HOOD is as computing efficient as TKD and TTD.

Fig. 1. Illustration of Video 1 and 2

Our results also show that HOOD saved as much as 99% computational time and consumed much less memory in size comparing to that of CPD.

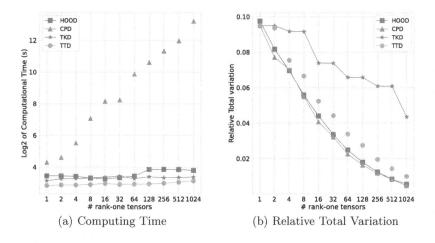

(a) Computing Time   (b) Relative Total Variation

Fig. 2. Computing time and relative total variation

Experiment II. We performed low rank decomposition of the two videos by HOOD, CPD, TKD, and TTD with different numbers of rank-one tensors, and compared the decomposition quality in terms of relative total variation, FID, and performance of YOLO v3 on decomposed videos. Since the trend of relative total variation versus number of rank-one tensors is similar in the decomposition of Video 1 and 2, only the result of Video 1 is presented.

Figure 2(b) presents relative total variation of HOOD, CPD, TKD, and TTD with number of rank-one tensors equals to $1, 2, 4, ..., 512, 1024$. Lower relative total variation represents better decomposition quality. The relative total variation of all four algorithms decreases significantly with increasing number of

(a) Dataset overview   (b) Eigen images

Fig. 3. HK dataset and the first three eigen images

rank-one tensors. HOOD and CPD show the similar relative total variation, which is lower than the relative total variation of TTD and TKD. TKD has the highest relative total variation and worst decomposition quality among all four algorithms.

FID is originally developed for image generation evaluation for visual quality of generated fake images. The lower the FID scores, the better the generative models. We used FID for visual quality of the decomposed videos against the original videos. We fed the low rank decomposed videos and original videos frame-by-frame to the pre-trained inception model and the extracted feature maps.

Table 1 presents the FID of all four algorithms on Video 1 and 2 when number of rank-one tensor equals to 1024. CPD and HOOD show significant lower FID, which implies better decomposition quality, than TKD and TDD in both Video 1 and 2. In Video 1, CPD presents slightly better decomposition quality than HOOD, while in video 2, HOOD presents slightly better decomposition quality than CPD. TTD shows better decomposition quality than TKD in both Video 1 and 2, but is no match for HOOD and CPD.

Table 2 presents detection of performance of YOLO V3 on Video 1 and 2 decomposed by all four algorithms with number of rank-one tensors equals to $128, 256, 512$, and 1024. None object is detected in videos decomposed by TKD, indicating that the performance of TKD is much worse than the other three methods. For HOOD, CPD, and TTD, the detection performance increases with increasing number of rank-one tensors. HOOD shows better detection performance than CPD in Video 1 for all number of rank-one tensors and Video 2 for number of rank-one tensors equals to 256 and 512. TTD shows better detection performance than HOOD and CPD in both two videos when number of rank-one tensors equals to 128. It is bypassed by HOOD and CPD when number of

Table 1. FID of Video 1 and 2. (*# rank-one tensors* = 1024)

Video	HOOD	CPD	TKD	TTD
Video 1	273.449	264.619	325.538	312.889
Video 2	241.58	267.877	379.602	338.502

Table 2. F1 score and AP of Video 1 and 2. It shows the detection performance when the number of rank-one tensors are 128, 256, 512, and 1024, respectively. # objects denotes the number of true positives.

# r1 tensors	128				256				512				1024			
Method	HOOD	CPD	TKD	TTD	HOOD	CPD	TKD	TTD	HOOD	CPD	TKD	TTD	HOOD	CPD	TKD	TTD
# objects in v1	49	41	0	112	128	105	0	104	227	209	0	217	342	331	0	263
F1 score in v1	0.149	0.128	0	0.31	0.346	0.29	0	0.295	0.538	0.505	0	0.518	0.707	0.7	0	0.598
AP (Person) in v1	0.074	0.05	0	0.164	0.194	0.159	0	0.168	0.369	0.335	0	.354	0.563	0.548	0	0.438
# objects in v2	189	265	0	495	555	494	0	437	797	778	0	749	1130	1182	0	944
F1 score in v2	0.117	0.161	0	0.280	0.309	0.28	0	0.253	0.413	0.407	0	0.392	0.539	0.558	0	0.471
AP (Person) in v2	0.059	0.087	0	0.161	0.182	0.161	0	0.144	0.262	0.256	0	0.244	0.372	0.389	0	0.308

rank-one tensors increase to 256, 512, and 1024. Despite slightly variations in different measurement metrics, experiment 2 shows that HOOD has a similar decomposition quality to CPD, and superior to TKD and TTD.

Experiment III. Experiment 3 presents selected frames of Video 3 and their corresponded eigen images. For each row, those from left to right are the original frame, first eigen images, second eigen images, and third eigen images. It is obvious that first eigen image well-preserved most information of the original image. Meanwhile, the rest eigen images preserved the finer information like edges and outlines.

Our experiments on different videos show that HOOD posses high computational efficiency like TKD and TTD, and also high decomposition quality like CPD. Furthermore, the eigen images in HOOD reserved high level of information of the original images, which cannot be achieved by CPD, TKD, and TTD.

6 Conclusion

HOOD is a new tensor decomposition method. It decomposes a given tensor into the sum of orthogonal rank-one tensors, where orthogonality is usually violated in CPD. Our experiment shows that the computation of HOOD is fast and the result is accurate. It also shows that HOOD can be efficiently applied to high-resolution images and videos when the sizes of given tensors are very large. Due to its nice properties, we expect that HOOD can be used to study real-world big data problems for videos when the frames for a video cannot be loaded simultaneously to memory. To fulfill these tasks, we need to understand how to use eigenimages to learn large videos, especially in deep learning, such that the result is more explainable than methods without tensor decompositions. This is left to future research.

References

1. Baskaran, M., et al.: Memory-efficient parallel tensor decompositions. In: 2017 IEEE High Performance Extreme Computing Conference (HPEC), pp. 1–7. IEEE (2017)
2. Carroll, J.D., Chang, J.J.: Analysis of individual differences in multidimensional scaling via an n-way generalization of "eckart-young" decomposition. Psychometrika **35**(3), 283–319 (1970)
3. Chen, M., Zhang, Y., Qiu, M., Guizani, N., Hao, Y.: Spha: smart personal health advisor based on deep analytics. IEEE Commun. Mag. **56**(3), 164–169 (2018)
4. Cho, S., Jun, T.J., Kang, M.: Applying tensor decomposition to image for robustness against adversarial attack. arXiv preprint arXiv:2002.12913 (2020)
5. Dai, W., Qiu, L., Wu, A., Qiu, M.: Cloud infrastructure resource allocation for big data applications. IEEE Trans. Big Data **4**(3), 313–324 (2016)
6. De Lathauwer, L., De Moor, B., Vandewalle, J.: A multilinear singular value decomposition. SIAM J. Matrix Anal. Appl. **21**(4), 1253–1278 (2000)
7. Drineas, P., Mahoney, M.W.: A randomized algorithm for a tensor-based generalization of the singular value decomposition. Linear Algebra Appl. **420**(2–3), 553–571 (2007)
8. Huang, H., Liu, X., Zhang, T., Yang, B.: Regression PCA for moving objects separation. In: The 2020 IEEE Global Communications Conference (GLOBECOM 2020), Accepted. IEEE (2020)
9. Imaizumi, M., Hayashi, K.: Tensor decomposition with smoothness. In: Proceedings of the 34th International Conference on Machine Learning, vol. 70, pp. 1597–1606, JMLR. org (2017)
10. Kolda, T.G.: Orthogonal tensor decompositions. SIAM J. Matrix Anal. Appl. **23**(1), 243–255 (2001)
11. Kolda, T.G., Bader, B.W.: Tensor decompositions and applications. SIAM Rev. **51**(3), 455–500 (2009)
12. Kolda, T.G., Sun, J.: Scalable tensor decompositions for multi-aspect data mining. In: Eighth IEEE International Conference on Data Mining, pp. 363–372 (2008)
13. Liu, X., Huang, H., Tang, W., Zhang, T., Yang, B.: Low-rank sparse tensor approximations for large high-resolution videos. In: 19th IEEE International Conference on Machine Learning and Applications (ICMLA 2020), Accepted. IEEE (2020)
14. Malik, O.A., Becker, S.: Low-rank tucker decomposition of large tensors using tensorsketch. In: Advances in Neural Information Processing Systems, pp. 10096–10106 (2018)
15. Oseledets, I.V.: Tensor-train decomposition. SIAM J. Sci. Comput. **33**(5), 2295–2317 (2011)
16. Tang, X., Bi, X., Qu, A.: Individualized multilayer tensor learning with an application in imaging analysis. J. Am. Stat. Assoc. pp. 1–26 (2019)
17. Tucker, L.R.: Some mathematical notes on three-mode factor analysis. Psychometrika **31**(3), 279–311 (1966)
18. Wang, Y., Tung, H.Y., Smola, A.J., Anandkumar, A.: Fast and guaranteed tensor decomposition via sketching. In: Advances in Neural Information Processing Systems, pp. 991–999 (2015)
19. Zhang, T.: CP decomposition and weighted clique problem. Stat. Probab. Lett. **161**, 108723 (2020)

A GMM-Based Anomaly IP Detection Model from Security Logs

Feng Zhou[1(✉)] and Hua Qu[2]

[1] China Software Testing Center, Beijing, China
[2] Xi'an Jiaotong University, Xi'an, China

Abstract. The intrusion prevention system (IPS) is a widely used security system which generates logs for the attacks blocked by it for management personnel to review and conduct further processing. However, most of the entries in the actual IPS logs are not attack entries, which makes it impossible for us to obtain the attacker's IP through simple log analysis. The traditional log analysis methods rely on the administrator to manually analyze the log text. So it is necessary to use anomaly detection methods for analysis. The majority of existing log data-based automatic detection methods for anomalies cannot get an satisfying result while ensuring computational requirements and the interpretability of the model. This paper chose the Gaussian Mixture Model (GMM) to detect abnormal IP on the log dataset. The GMM method provides better detection results while ensuring relatively low computational requirements, and maintains the interpretability of the model. Experiments show that the ability of GMM method to detect abnormal IP is strong and the GMM is a suitable log data-based automatic detection method for detecting abnormal IP.

Keywords: IPS logs · Abnormal IP detection · The GMM method

1 Introduction

The intrusion prevention system (IPS) is a widely used security system which can block most security attacks and maintain the security of the protected system. The IPS system generates logs for the attacks blocked by it for management personnel to review and conduct further processing. Log data is an important data source for anomaly detection. However, most of the entries in the actual IPS logs are not attack entries, which makes it impossible for us to obtain the attacker's IP through simple log analysis. Therefore, it is necessary to conduct further analysis and detection of IPS logs.

The content and types of events recorded in the log tend to be stable under normal conditions. Except for some highly concealed APT attacks, most of the attacks are not instantaneous and have a fixed pattern. The log data will produce an abnormal pattern when recording malicious behavior, which provides the possibility to detect abnormal IP from the log data.

© Springer Nature Switzerland AG 2021
M. Qiu (Ed.): SmartCom 2020, LNCS 12608, pp. 97–105, 2021.
https://doi.org/10.1007/978-3-030-74717-6_11

The traditional log analysis methods rely on the administrator to manually analyze the log text. This process requires a lot of labor costs, and requires system administrators to understand the network environment, master the system architecture, and follow up in real time to understand possible attack methods and even attackers in reality. This is not realistic in today's rapidly developing network environment.

In addition, in order to avoid being discovered by security devices or administrators, logs generated by attacks are getting closer and closer to logs generated by normal access behavior. In addition, due to the extremely large number and different types of network access in actual application scenarios, IPS log data files are often extremely large and cannot be manually analyzed and processed. So it is necessary to use anomaly detection methods for analysis [6–10].

The existing log data-based automatic detection methods for anomalies can be divided into two categories:

- Label-based supervised learning methods: such as decision trees, LR, SVM, etc.;
- Unsupervised learning methods: such as PCA, Clustering and so on. Supervised learning is very effective in detecting known malicious behaviors or abnormal states, but it relies on prior knowledge and cannot detect unknown attacks. Unsupervised methods can be used to detect unknown anomalies, but the accuracy of most methods needs to be improved.

Comparing the principles, advantages and disadvantages of several methods, we chose the Gaussian mixture model to detect abnormal IP on the log data set "*eventdata0701_0707*". Specifically, we define a feature vector for the daily log generated by each IP and divide the data set into two parts, training and testing. We train the hybrid model on the training set and classify it on the test set.

2   Related Work and Comparision of Existing Methods

For log-based detection tasks, there are some mature detection methods.

1) **Principal component analysis**
   The principal component analysis method learns the low-dimensional representation of the extracted feature vectors, and projects the original data from the original space to the principal component space, and then reconstructs the projection to the original space. If only the first principal component is used for projection and reconstruction, for most data, the error after reconstruction is small; but for abnormal points, the error after reconstruction is still relatively large.
2) **Single-class support vector machine**
   The single-class support vector machine has only two categories of normal and abnormal, and the normal data is far more than the abnormal data. Therefore, the single-class support vector machine can be used to classify the extracted features and complete anomaly detection. Its mathematical essence

is to find a hyperplane to divide the positive examples in the sample, so as to realize detection through this hyperplane. However, the huge computing requirements make this method difficult to adapt to larger data sets.

3) **Text analysis**

Logs are a special type of text, so some studies introduce the concept of the longest common subsequence in order to reduce the number of matching patterns obtained in the calculation process, and realize the simple classification of logs [1]. Further, we can determine the correspondence between log types and characters based on character matching, so as to achieve detection [2]. This type of text-based detection method implements anomaly detection by analyzing the text in detail, but the generalization ability of this method is limited due to the need for a well-defined text processing method.

4) **Methods based on learning**

With the development of neural network technology, detection methods based on learning are gradually being proposed. LSTM [3], graph embedding [4] and other methods directly perform anomaly detection on the results of log vectorization through neural network training. The application of these related models has brought a greater improvement in the accuracy of detection. However, the limitations of the model also make the detection effect extremely dependent on the processing of features. Compared with the traditional method, the computational overhead consumed has increased by orders of magnitude, and the interpretability of the model is poor.

All in all, the Gaussian Mixture Model method provides better detection results while ensuring relatively low computational requirements, and maintains the interpretability of the model. Furthermore, the feature engineering realizes the extraction of data characteristics that need attention, and it is relatively easy to generalize to logs from different sources.

3   Detection Model

We use the Gaussian Mixture Model (GMM) for detection. GMM is one of the most commonly used statistical methods. It uses the maximum likelihood estimation method to estimate the mean and variance of the Gaussian distribution [5], and combines multiple Gaussian distributions to express the extracted feature vectors and find outliers.

3.1   Algorithm Background Knowledge Introduction

3.1.1   Gaussian Distribution and Mixture Model

Gaussian distribution, or normal distribution, is the most common distribution of many variables in conventional statistical data. However, a single Gaussian distribution cannot accurately describe the distribution of multi-dimensional variables, and the feature vectors extracted from the data tend to have relatively large dimensions. Therefore, it is necessary to use the mixed Gaussian distribution to describe the distribution of multi-dimensional variables, i.e.

$$P(x|\theta) = \frac{1}{(2\pi)^{\frac{D}{2}}|\sum|^{\frac{1}{2}}} exp(-\frac{(x-\mu)^T \sum^{-1}(x-\mu)}{2}) \tag{1}$$

where μ is the expectation, σ is the standard deviation, $\sum$ is the covariance, and D is the data dimension.

Based on this idea, the Gaussian mixture model can be regarded as a model composed of several Gaussian distributions, that is, the superposition of several of the above formulas. Gaussian distribution gives the model good fitting performance and calculation performance to real data. Specifically, the parameters of this mixed model are:

$$\theta = (\tilde{\mu}_k, \tilde{\sigma}_k, \tilde{\alpha}_k) \tag{2}$$

It can be understood as the expectation, variance and proportion of each sub-model in the mixed model.

3.1.2   Likelihood Function

We use a priori assumption: each data is independent and identically distributed. Thus the likelihood function can be represented by the product of probability density functions. In practical applications, the logarithmic function is used to calculate, that is, the likelihood function of the Gaussian mixture distribution is obtained:

$$logL(\theta) = \sum_{j=1}^{N} log(\sum_{k=1}^{K} \alpha_k \phi(x|\theta_k)) \tag{3}$$

For Gaussian distribution, the most commonly used learning algorithm is maximum likelihood estimation of parameters. But we don't know a priori which sub-distribution the data belongs to, so we need to solve it by iterative method.

3.1.3   EM Algorithm

The most commonly used algorithm is the EM algorithm in the face of likelihood estimation of a mixture of Gaussian distributions. This is an iterative algorithm. Each iteration contains the following two main steps:

1) **E-step:** Use the current estimated value to find the expectation;
2) **M-step:** Calculate the maximum estimate based on the current expectation to obtain the next iteration Parameters.

Repeat the above steps until convergence and get the result of maximum likelihood estimation. Initialize several different parameters for iteration, and take the best result as the final result.

Fig. 1. The whole process of testing the data set

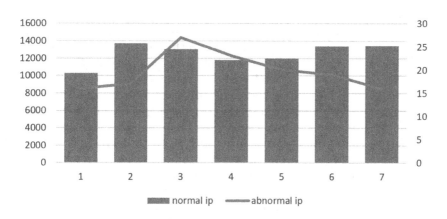

Fig. 2. Types and quantity of IP everyday

Table 1. Some statistics of the data set

Item		Quantity
Log entries	Total entries	5519339
	Normal entries	5475231
	Abnormal entries	44108
IP	IP	64736
	Preset abnormal IP	493
	The actual abnormal IP	79

3.2   Detection Model

The whole process of testing the data set is shown in the Fig. 1.

In addition to the models and algorithms described above, it also includes the following key steps:

1) **Feature Definition**

   First, defining a feature vector per day for each source IP. This feature vector contains all the access information of this source IP on this day. Since the log fields such as dst_ip, dst_port, event_name, url, and content in the statistical field objects have many different values, we count the number of frequent and infrequent occurrences of the values of these fields. The time period of log generation times is divided into 5 time periods according to all_day, 12 am–6 am, 6 am–12 pm, 12 pm–6 pm.

   According to the comparison range of whether the statistical field value frequently occurs, the specificity of the log is divided into src_ip (only compared with the logs of the same time period with the source ip of src_ip) and all (compared with all logs of the same time period) for statistics. For specificity as src_ip, according to the current statistical field such as dst_ip, whether the proportion of all logs whose source ip is src_ip and the same time is less than 0.05, the popularity of the log is divided into unpop and common for statistics; for specificity as all, according to the current statistical field such as whether the number of occurrences of dst_ip in all logs at the same time is greater than the average number of occurrences of dst_ip, the popularity of the log is divided into unpop and common for statistics.m, and 6 pm–12 am.

   By analyzing the content field of the log, there will be an encoded attack code in the url or body_data field of the abnormal log. The characters of these attack codes are encoded in hexadecimal or base64. Their URLs are longer, and there are a lot of numbers and uppercase letters, % and &. However, these characters rarely appear in large numbers in normal access logs. So we count the url length, uppercase letter frequency, number frequency, special character %, & frequency in the content field.

   We use the above popularity, specificity, objects, and times as a Cartesian product, such as unpop_src_ip_d

   st_ip_all_day (representing the number of dst_ip that rarely appears in the current src_ip all-day log).

   Through classification and statistics log information, we extracted a total of 100-dimensional feature vector, plus the five-dimensional feature vector of url length, uppercase letter frequency, number frequency, special character %, & frequency statistics, to form a 105-dimensional feature vector.

2) **Data Processing**

   First, we sort the logs according to the generation time. By analyzing the logs, we can see that the log content is composed of key-value pairs, such as SerialNum = 0113211803309994. After that, we use regular expressions to extract log key-value pairs, count the values of statistical items composed of the Cartesian product of popularity, specificity, objects, and times in the

Table 2. The confusion matrix obtained by detection

Predictive value	Actual value	
	Positive	Negative
Positive	TP:24	FP:6507
Negative	FN:1	TN:30625

experimental method. We also count the length of the URL in the content field of the log and the occurrence frequency of numbers, uppercase letters, %, & in the URL. Finally, the above two are combined into a 105-dimensional feature vector per src_ip per day.

We use the feature vectors of the first four days as the training set, and the latter three days as the test set. Normalize the feature vector in the data set, use the Gaussian mixture model GMM in sklearn to train on the training set to obtain the trained model, and then classify the source IP on the test set. The abnormal IP is 1 and the normal IP is 0.

4 Test Results

4.1 Data Set

The data set is the logs generated by the security device within 7 d. Each log contains 46 fields, but most of the fields have no detection significance. Some statistics of the data set are shown in the Table 1. Plotting the types and numbers of IPs that appear in the daily log. Paying attention to the great difference between the coordinate axes in the Fig. 2. We can intuitively see the high bias of the data, and the number of abnormal IPs and the total number of IPs per day. The distribution is irrelevant.

In this data set, the proportion of abnormalities is relatively low, with abnormal entries accounting for about 0.80%, and abnormal IP accounting for about 0.12%, which is significantly lower than that in conventional anomaly detection tasks.

4.2 Test Results

4.2.1 Confusion Matrix
The confusion matrix obtained by detection is as shown in Table 2.

4.2.2 Index
The index calculation of the test results is shown in Table 3.

4.2.3 ROC Curve
The ROC curve obtained by the test is shown in the Fig. 3.

Table 3. Index score

Index	Value
Accuracy	0.825
Precision	0.004
Recall	0.960
Specificity	0.825
Error	0.175
F1-measure	0.007
AUC	0.004

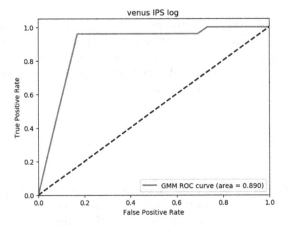

Fig. 3. The ROC curve by the test

5   Conclusion

We apply the Gaussian Mixture Model to abnormal IP detection. We evaluate the performance of GMM method on the log data set "eventdata0701_0707" through comprehensive experiments and the results are promising. Besides, the feature engineering used in our model realizes the extraction of data character- istics that need attention, and it is relatively easy to generalize to logs from different sources.

References

1. Zhao, Y., Wang, X., Xiao, H., Chi, X.: Improvement of log pattern extracting algo- rithm using text similarity. In: International Parallel and Distributed Processing Symposium Workshops, Vancouver Canada, pp. 507–514. IEEE (2018)
2. Xu, K.Y., Gong, X.R., Cheng, M.C.: Audit log association rule mining based on improved apriori algorithm. Comput. Appl. **36**(7), 1847–1851 (2016)

3. Tuor, A., Baerwolf, R., Knowles, N., et al.: Recurrent neural network language models for open vocabulary event-level cyber anomaly detection (2017)
4. Liu, F., Wen, Y., Zhang, D., Jiang, X., Meng, D.: Log2vec: a heterogeneous graph embedding based approach for detecting cyber threats within enterprise. In: ACM SIGSAC Conference (2019)
5. W. Contributors. Maximum Likelihood Estimation. Internet (2015)
6. Gai, K., Qiu, M., Zhao, H.: Security-aware efficient mass distributed storage approach for cloud systems in big data. In: 2016 IEEE 2nd International Conference on Big Data Security on Cloud (2016)
7. Qiu, M., Jia, Z., Xue, C., Shao, Z., Sha, E.H.M.: Voltage assignment with guaranteed probability satisfying timing constraint for real-time multiproceesor DSP. J. VLSI Sig. Process. Syst. Sig. Image Video Technol. **46**, 55–73 (2007)
8. Zhang, Q., Huang, T., Zhu, Y., Qiu, M.: A case study of sensor data collection and analysis in smart city: provenance in smart food supply chain. Int. J. Distrib. Sens. Netw. **9**(11), 382132 (2013)
9. Chen, M., Zhang, Y., Qiu, M., Guizani, N., Hao, Y.: SPHA Smart personal health advisor based on deep analytics. IEEE Commun. Mag. **56**(3), 164–169 (2018)
10. Zhu, M., et al.: Public vehicles for future urban transportation. IEEE Trans. Intell. Transp. Syst. **17**(12), 3344–3353 (2016)

Intelligent Textbooks Based on Image Recognition Technology

Xin Zhang[1]([⊠]), Yuwen Guo[1]([⊠]), Tao Wen[1], and Hong Guo[1,2]

[1] College of Computer Science and Technology, Wuhan University of Science and Technology, Wuhan, China
[2] Hubei Province Key Laboratory of Intelligent Information Processing and Real-Time Industrial System, Wuhan, China

Abstract. In recent years, with the development of science and technology, the education mode has gradually changed from offline teaching to the combination of offline education and online education. Compared with the traditional offline education, the significant advantages of online education are low cost, rich content and flexible use. In this paper, an intelligent textbook based on image recognition technology is designed, which combines online courses with physical textbooks to realize the collaboration of two learning methods. The APP can not only realize the video playback function and other basic functions of the online education platform, but also adopt the image recognition technology, so that users can scan the picture through the APP to watch the explanation video or 3D model projected onto the real picture by AR way. This paper uses SSM framework and OpenCV to build the background, and combined with Objective-C to build an online course APP running on IOS platform. In general, the app not only designs and implements the basic online course learning process, but also realizes the accurate recognition of images by combining the image matching technology based on OpenCV and the perceptual hash algorithm.

Keywords: SSM · OpenCV · Image matching · iOS · AR

1 Introduction

With the rapid development of the Internet, the education mode changes have been taken place. The continuous innovation of technical means makes the development of online education more vigorous. The traditional education mainly depends on the offline way. This mode will be restricted by time, space and other factors in the process of education information transmission, resulting in the imbalance of educational resources, which is the problem that the country has always attached great importance to. Obviously, only relying on offline teaching can not solve this problem well. In addition, traditional education also has higher education cost, so teachers can not teach students according to their aptitude.

However, online education has solved some of the shortcomings of offline education to some extent. First of all, it reduces the cost. For online education, teachers only need

M. Qiu (Ed.): SmartCom 2020, LNCS 12608, pp. 106–115, 2021.
https://doi.org/10.1007/978-3-030-74717-6_12

a computer or even a mobile phone to teach, and students only need a device that can be connected to the Internet to learn. Secondly, it breaks through the regional restrictions and makes the high-quality educational resources reasonably allocated, so that every student can get high-quality teaching resources. In addition, online education makes students more flexible in the arrangement of study time and place. It is these advantages that make it of great significance to develop a simple and efficient online education platform with abundant learning resources [1].

The rest of this paper is organized as follows: the second part mainly describes the general idea of this paper, it describes technologies related to online education and image recognition, and briefly describes the chapter arrangement of this paper. The third part mainly introduces the overall framework of app and the specific description of image recognition technology. The fourth part introduces the experimental platform and results, finally is the summary.

2 Related Work

E-learning, is a method of content dissemination and rapid learning through the application of information technology and Internet technology. The "E" of E-learning means electronic, efficient, exploratory, extended, easy to use and enhanced [2]. By 2020, the infrastructure level of online education will be significantly improved, and modern information technologies such as the Internet, big data and artificial intelligence will be more widely applied in the field of education [3].

Image recognition is an important part of the field of artificial intelligence, which refers to a way to distinguish between targets and objects of different patterns by object recognition of images, that is, to identify an entity through a computer rather than the human eye. In its development process, it can be roughly divided into three stages: character recognition, digital image processing and recognition and object recognition. This technology of simulating human eye function by computer has a very wide range of applications, and plays an important role in the fields of remote sensing image recognition, military criminal investigation, biomedicine and machine vision. It can be said that image recognition technology has brought great convenience to human production and life.

Nowadays, the number of online education platforms is increasing day by day, and users' acceptance of online education is getting higher and higher. There are not only open course platforms with enterprises as the main body, but also MOOC platforms relying on the world's top universities [4]. Domestic e.g. Netease open classes, Chinese University MOOCS, Tencent classes, etc., and foreign ones such as edX, coursera, udacity, etc. all have a large number of users, and the number of users is increasing day by day. However, the function design of these online education platforms only stay at the level of video playback, and there are still great improvements in user interaction. If the image recognition technology is combined with online education to optimize the user experience and platform work efficiency, it will be a great change in the field of online education.

3 Architecture

3.1 SSM Framework

SSM framework (Spring + Spring MVC + MyBatis), spring MVC is a lightweight web framework based on Java that implements the request driven type of Web MVC design pattern [5]. It decouples the responsibilities of web layer by using the idea of MVC architecture pattern [6] (see Fig. 1).

SSM is composed of spring framework, spring MVC and mybatis. Spring framework is a lightweight java development framework, which can only be done by EJB in the past. Now, spring JavaBean technology can be used to implement SSM. Spring MVC uses MVC design method to decouple the program into multiple disjoint parts, which is convenient for development and customization. Mybatis is a Java based persistence layer framework, and almost all JDBC code is eliminated. The interface and Java plain old Java objects are mapped to the database through simple XML annotation [7].

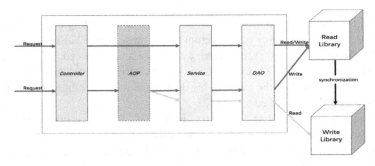

Fig. 1 SSM framework flow chart

3.2 Perceptual Hash Algorithm

Perceptual hash algorithm is the general term of a class of hash algorithms, which describes a class with comparable hash functions. Its function is to generate a unique (not unique) "fingerprint" string through the algorithm, so as to judge the similarity of images by comparing the fingerprint information of different images [8, 9]. The closer the results are, the higher the similarity of images is. Perceptual hashing algorithms can be roughly divided into three categories: aHash, pHash and dHash [10].

After the 64 bit hash value of the image is obtained by the perceptual hash algorithm, the similarity of the two images is quantified by hamming distance. The larger the Hamming distance is, the smaller the image similarity is, and the smaller the Hamming distance is, the greater the image similarity is.

Hamming distance represents the number of bits corresponding to two strings of the same length. Generally, d (M, N) represents the Hamming distance between two strings M, N [11].

aHash Algorithm. aHash algorithm is fast and low precision. The treatment process is divided into the following six steps:

1. Zoom image: in order to reduce the amount of image information while retaining the image structure, the image is uniformly scaled to 8 * 8 (64 pixels).
2. Convert to grayscale image [12]: convert the zoomed image to 256 level grayscale image. In this paper, we use floating point algorithm, where R = red, G = green, B = blue:

$$Gray = R0.3 + G0.59 + B0.11 \tag{1}$$

3. Calculate the average value: calculate the average value of all pixels after gray processing.
4. Compare pixel gray value: traverse through each pixel of gray image, greater than the average value is marked as 1, less than 0.
5. Construct hash value: combine 64 bits to generate hash value.
6. Compare fingerprint: calculate the fingerprint of two pictures and calculate Hamming distance.

pHash Algorithm. The pHash algorithm is highly accurate and slow, and uses DCT (discrete cosine transformation [13]) to pull down the frequency [14]. The process is divided into the following eight steps:

1. Zoom out: To simplify the calculation of the DCT, scale the picture uniformly to 32 × 32.
2. Simplify the color: convert the image to 256 grayscale image.
3. Calculate DCT: decompose the image into frequency aggregation and ladder shape. The full name of DCT is discrete cosine transform, which is mainly used to compress data or image. It can transform spatial signal to frequency domain, and has good performance of decorrelation. The principle of DCT is as follows:

One dimensional DCT transform:

$$F(u) = c(u) \sum_{i=0}^{N-1} f(i) \, cos\left[\frac{(i+0.5)\pi}{N}u\right] \tag{2}$$

$$c(u) = \begin{cases} \sqrt{\frac{1}{N}}, u = 0 \\ \sqrt{\frac{2}{N}}, u \neq 0 \end{cases} \tag{3}$$

Where f(i) is the original signal, F(u) is the coefficient after the DCT transformation, N is the number of points of the original signal, and c(u) is the compensation factor, making the DCT transformation matrix an orthosecting matrix.

Positive transformation formula for 2D DCT transformation:

$$F(u, v) = c(u)c(v) \sum_{i=0}^{N-1} \sum_{j=0}^{N-1} f(i,j) cos\left[\frac{(i+0.5)\pi}{N}u\right] cos\left[\frac{(j+0.5)}{N}v\right] \tag{4}$$

$$c(u) = \begin{cases} \sqrt{\frac{1}{N}}, u = 0 \\ \sqrt{\frac{2}{N}}, u \neq 0 \end{cases} \tag{5}$$

4. Zoom out DCT: Keep the 32-by-32 matrix of the DCT calculation on tche upper-left 8-by-8 matrix, which represents the lowest frequency in the picture.
5. Average calculation: Calculate the DCT mean.
6. Further reduce the DCT: Compared to the DCT matrix of 8×8, set to 1 for the DCT mean and 0 for less than the DCT mean.
7. Construct hash value: combine 64 bits to generate hash value.
8. Compare fingerprints: Calculate the fingerprints of the two pictures, calculate the distance between Heming.

dHash Algorithm. Compared with pHash, dHash is faster. Compared with aHash, dHash works better with almost the same efficiency. It is based on gradient. It is divided into the following six steps:

1. Zoom out of the picture: Shrink to a size of 9×8 (72 pixels).
2. Convert to a grayscale map: Transforms a scaled picture into a 256-order grayscale map.
3. Calculate the difference value: calculate the difference value between adjacent pixels, so that there are 8 different differences between 9 pixels in each row, a total of 64 differences in 8 lines.
4. Compare the difference value: if the color intensity of the previous pixel is greater than that of the second pixel, the difference value is set to 1; if it is not greater than the second pixel, it is set to 0.
5. Construct hash value: combine 64 bits to generate hash value.
6. Compare fingerprint: calculate the fingerprint of two pictures and calculate Hamming distance.

3.3 OpenCV's Template Matching Algorithm

OpenCV is a cross platform computer vision and machine learning software library based on BSD license [15]. OpenCV mainly provides a set of common infrastructure for the development of computer vision programs and enhances the perception ability of machines in commercial products [16].

Template matching is one of the many applications of OpenCV, that is to find a small area matching the given sub image in the whole image area. Here, a template image (given sub image) and an image to be detected (original image) are needed. On the image to be detected, the matching degree between the template image and the overlapped sub image is calculated from left to right and from top to bottom. The greater the matching degree is, the greater the possibility of the two images being the same [17].

The template matching algorithm provided by OpenCV has the following three types of situations [18]:

The first kind uses square difference to match, the best match is 0, the worse the matching, the bigger the matching value. There are square difference matching (CV_TM_SQDIFF) and standard square difference matching (CV_TM_SQDIFF_NORMED).

The second type uses the multiplication operation between template and image, so the larger number indicates the higher matching degree, and 0 represents the worst case of matching. There are correlation matching (CV_TM_CCORR) and standard correlation matching (CV_TM_CCORR_NORMED).

In the third category, the relative value of template to its mean value is matched with the correlation value of image to its mean value. 1 represents perfect matching, − 1 represents poor matching, and 0 indicates nó correlation (random sequence). There are two kinds of matching: correlation coefficient matching (CV_TM_CCOEFF) and standard correlation coefficient matching (CV_TM_CCOEFF_NORMED).

OpenCV provides the following feature detection methods:

1. SURF: the full name of SURF is "Speeded Up Robust Feature", which has the characteristics of scale invariant and high computational efficiency.
2. SIFT: SURF algorithm is an accelerated version of SIFT algorithm. SIFT (Scale Feature Transform) is a scale invariant feature detection method. Although the speed is slower than SURF algorithm, the features measured by SIFT algorithm are more accurate in space and scale [19].
3. ORB: ORB is the abbreviation of ORiented Brief, which is an improved version of brief algorithm. ORB algorithm is 100 times faster than SIFT algorithm and 10 times faster than SURF algorithm. In the field of computer vision, there is a saying that ORB algorithm has the best comprehensive performance than other feature extraction algorithms in various evaluation. However, it has the disadvantages of no rotation invariance, noise sensitivity and scale invariance [20].
4. Fast: Fast (Features from Accelerated Segment Test). This operator is specially used for fast detection of interest points. Only a few pixels can be compared to determine whether they are key points.
5. Harris corner: when searching for valuable feature points in the image, using corner is a feasible method [21]. Corner is a local feature that is easy to locate in the image, and it exists in many artificial objects (such as corners produced by walls, doors, windows, tables, etc.). The value of a corner is that it is the junction of two edge lines, a two-dimensional feature that can be accurately located (even at sub-pixel level). Harris feature detection is a classical method for corner detection.

SURF Algorithm. SURF (Speed up Robust Features) is a robust local feature detection algorithm. It was first proposed by Herbert Bay et al. In 2006 and improved in 2008 [22]. SURF algorithm is inspired by SIFT algorithm, which has the characteristics of high repeatability detector, good distinguishable descriptor, strong robustness and high operation speed. The reason for the high efficiency of SURF algorithm lies in its proper simplification and approximation on the premise of its correctness. On the other hand, it uses the concept of integral image many times.

The SURF algorithm consists of the following stages:

The first part: feature point detection.

- Feature point detection based on the Hessian matrix [23]
- Scale spatial represents.
- Feature point positioning.

The second part: feature point description.

- Distribution of directional angles.
- Feature point descriptors based on Haar waves.

4   Implementation and Results

4.1   System Analysis

Smart Textbook APP based on image recognition technology uses development tools such as the MySQL database, IntelliJ IDEA, and Xcode. The back end runs on Tomcat on alibaba cloud ubuntu servers, and the APP runs on iOS systems (see Fig. 2).

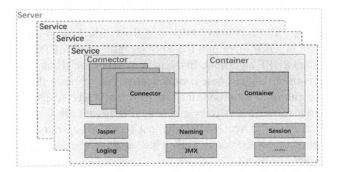

Fig. 2 The structure of tomact

Smart Textbook APP based on image recognition technology is a B/S architecture-based application. The B/S architecture refers to the Browser/Server (browser/server) structure, where the client runs the software in a browser manner and maintains only one server. The B/S architecture has the advantages of not having to install clients running directly in a web browser, controlling multi-client access directly on an Internet network, and not having to install clients [24]. So for app flexibility, the entire application is developed using a B/S architecture.

The intelligent textbook app based on image recognition technology can be divided into eight parts: course recommendation module, account management function module, user browsing history management module, user collection management module, course list calendar module, course details module, system announcement module, image scanning recognition and playback module.

4.2 Implementation of iOS Client

The image scanning and recognition playing module is the core function of the system. After the scanning is successful, the multimedia resources are projected to the real world in the form of AR.

By logging in to the APP, you enter the course recommendation interface, the top right corner of the recommendation page is a quick entrance to the picture scan (see Fig. 3).

Click on the course item to go to the course details page. The upper part of the page has a quick entry to scan the picture (see Fig. 4).

Fig. 3 Course recommendation interface

Fig. 4 Lesson chapter section page

The personal information column consists of five parts: personal information display, personal information modification, course browsing record, course collection and system notice (see Fig. 5).

Finally, scan the picture recognition course and play the video's scan page and scan effect map with AR (see Fig. 6).

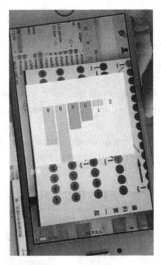

Fig. 5 Personal information page   **Fig. 6** Scan and AR play page

5 Conclusion

The main work and innovations in this article are as follows:

1. Image recognition technology is used, and by combining paper books with online education, it is beneficial to the popularization of online education.
2. The use of AR technology combined with online course playback, not limited to the simple video playback, video and other multimedia resources projected in the real world, this kind of "touch by hand" course is not available in the existing online learning platform.
3. The project basically completed the design and implementation of the service side and the client, and the six core functions of curriculum recommendation, course presentation, course review, course inquiry, course collection, course history and course scanning all met the design requirements as expected.

Reference

1. Lei, A., Bo, Z.: Design and implementation of online learning platform. Comput. Program. Skills Maint. **2020**(03), 17–19 (2020)
2. Chen, Y., Yan, G.: Online education, cool learning coming. Xinmin Wkly. **2014**(13), 58–62 (2014)
3. Heng, L.: Research on online education ecosystem and its evolution path. China Distance Educ. **2017**(01), 62–70 (2017)
4. Anderson, V., Gifford, J., Wildman, J.: An evaluation of social learning and learner outcomes in a massive open online course (MOOC): a healthcare sector case study. Hum. Resour. Develop. Int. **23**(3), 208–237 (2020). https://doi.org/10.1080/13678868.2020.1721982

5. Jin, L.: Development of online shopping platform based on SSM. Comput. Knowl. Technol. **16**(11), 281–237 (2020)
6. Qiu, M.K., Zhang, K., Huang, M.: An empirical study of web interface design on small display devices. In: IEEE/WIC/ACM International Conference on Web Intelligence, WI'04, pp. 29–35 (2004)
7. Li, Y.Z., Gao, S., Pan, J., Guo, B.F., Xie, P.F.: Research and application of template engine for web back-end based on MyBatis-Plus. Procedia Comput. Sci. **166**, 206–212 (2020)
8. Wang, Y., Liu, D., Hou, M.: Code cloning based on image similarity detection. Comput. Appl. **39**(07), 2074–2080 (2019)
9. Qin, C., Chen, X., Dong, J., Zhang, X.: Perceptual image hashing with selective sampling for salient structure features. Displays **45**, 26–37 (2016)
10. Gai, K., Qiu, M., Zhao, H., Sun, X.: Resource management in sustainable cyber-physical systems using heterogeneous cloud computing. IEEE Trans. Sustain. Comput. **3**(2), 60–72 (2018)
11. Taheri, R., Ghahramani, M., Javidan, R., Shojafar, M., Pooranian, Z., Conti, M.: Similarity-based Android malware detection using Hamming distance of static binary features. Fut. Gener. Comput. Syst. **105**, 230–247 (2020)
12. Gu, M., Su, B., Wang, M., Wang, Z.: Overview of color image graying algorithm. Comput. Appl. Res. **36**(05), 1286–1292 (2019)
13. Sunesh, Rama Kishore, R.: A novel and efficient blind image watermarking in transform domain. Procedia Comput. Sci. **167**, 1505–1514 (2020)
14. Shim, H.: PHash: a memory-efficient, high-performance key-value store for large-scale data-intensive applications. J. Syst. Softw. **123**, 33–44 (2017)
15. Dai, W., Qiu, L., Ana, W., Qiu, M.: Cloud infrastructure resource allocation for big data applications. IEEE Trans. Big Data **4**(3), 313–324 (2018)
16. Gong, Y., Guo, Q., Yu, C.: Design of OpenCV face detection module under Android system. Electron. Des. Eng. **20**(20), 52–54 (2012)
17. Feng, X.: Marching in step: The importance of matching model complexity to data availability in terrestrial biosphere models. Global Change Biol. **26**(6), 3190–3192 (2020)
18. Wang, C., Li, Q.: Image multi-objective template matching algorithm based on OpenCV. Electron. Technol. Softw. Eng. **2018**(05), 57–59 (2018)
19. Wang, B., Tian, R.: Judgement of critical state of water film rupture on corrugated plate wall based on SIFT feature selection algorithm and SVM classification method. Nucl. Eng. Des. **347**, 132–139 (2019)
20. Yang, H., Li, H., Chen, K., Li, J., Wang, X.: Image feature point extraction and matching method based on improved ORB algorithm. Acta graphica Sinica **2020**(05), 1–7 (2020)
21. Haggui, O., Tadonki, C., Lacassagne, L., Sayadi, F., Ouni, B.: Harris corner detection on a NUMA manycore. Fut. Gener. Comput. Syst. **88**, 442–452 (2018)
22. Shi, L., Xie, X., Qiao, Y.: Research on face feature detection technology based on surf algorithm and OpenCV. Comput. Digit. Eng. **38**(02), 124–126 (2010)
23. Chen, P., Peng, Y., Wang, S.: The Hessian matrix of Lagrange function. Linear Algebra Appl. **531**, 537–546 (2017)
24. Bai, X.: Design of repeater network management based on B/S architecture. Mod. Electron. Technol. **37**(01), 57–65 (2014)

Visual Analytics for Basketball Star Card Market—Under the Background of the NBA Restart

Jinming Ma[1(✉)] and Shenghui Cheng[2]

[1] The Chinese University of Hong Kong, Shenzhen, China
220022086@link.cuhk.edu.cn
[2] Shenzhen Research Instutite of Big Data, Shenzhen, China
chengshenghui@cuhk.edu.cn

Abstract. This paper is to analyze the fluctuation of the basketball cards market and to find the price change of the basketball cards during the past three months under the background of the NBA Restart. Besides, the paper makes a comparison between two online trade markets respectively-eBay and Card Hobby. Mixed methods are employed to gather specific data to answer the research questions. Data scraping is used to gain data from the trade platforms. Inferential statistical analyses and data visualization are conducted to examine whether the price and the quantity of the demand and supply of the basketball star cards are influenced by the NBA Restart. Also, the paper tries to dig out the time series data to understand the behaviors of both the suppliers and consumers in the market. The paper finds that the external incidents have big effects on the price change of the basketball star cards. Both supply and demand sides are sensitive to the external incidents. The price of a basketball star card depends on the price fundamentals of the entire star card market and the value of the card itself and the external impact of the star card itself factor.

Keywords: Basketball star cards · EBay · Card Hobby · Price fluctuation · NBA restart · Economic behavior

1 Introduction

For basketball fans and collectors, they should have heard of NBA star cards. Taking the well-developed United States as an example. In the United States, star cards have a history of 120 years. Star cards are divided into regular cards, special cards, limited cards, jersey cards, patch cards, refraction cards, rookie cards, etc. Star cards are issued according to products, themes, and are closely related to leagues and clubs. The issued series are positioned at high, middle, and low grades. The circulation limitation, appearance, the potential of the star, and the story behind it are all factors that affect the value of a star card, and the development of a player is a fundamental factor that affects the valuation of a star card. At present, the star card has evolved into an important category of sports collections. The paper tries to solve the following questions:

© Springer Nature Switzerland AG 2021
M. Qiu (Ed.): SmartCom 2020, LNCS 12608, pp. 116–126, 2021.
https://doi.org/10.1007/978-3-030-74717-6_13

a) The connection between the eBay market and the Card Hobby market:
b) Summary of the characteristics of the star card auction market;
c) Analyze the price fluctuation of the star card from the behavior and psychology of speculators and investors;
d) Try to explain the determinants of star card price: value, supply, demand, potential, and auction mechanism, etc.

This article is trying to describe the price trend and transaction information of a specific basketball star card in the past three months. I tried to use historical data and in-depth analysis combined with background knowledge and professional knowledge to analyze the multi-party behavior behind the historical data to restore how the market price trend graph was formed.

2 Data Description

2.1 Sample Selection

Due to the wide variety of cards on the star card, we choose LeBron James Prizm 19–20 base in this article in order to simplify the analysis. The reasons for choosing this card are: the market transaction volume is relatively large, and the generated transaction data is relatively large, and, there are more gimmicks, and there are many entry points for analysis. For example, the impact of the NBA restart and the Lakers' championship on the price of this card is a factor worth analyzing.

Since the research in this article is an independent problem, there is no second-hand data that can be directly used and borrowed. The card trading market is relatively scattered, and the possibility of obtaining a large amount of official data at one time is very small. Therefore, this article selects a specific card for analysis. Learn more about the behavioral habits of the card trading market.

Since there is no directly available secondary data (eBay does not directly publish their transaction data, and both eBay and Card Hobby will only publish transaction data for the past three months), I plan to crawl the data by myself. It is obviously more reliable than using the official second-hand data directly, but for the propositions established in this article, this path is obviously not feasible. Therefore, the preparation of data crawling and data cleaning are particularly important.

2.2 Data Crawling

The object of data crawling in this article is the transaction data of star cards in eBay and Card Hobby market.

This is the historical transaction record on the eBay trading platform. For the mother page of each card record, we will crawl the necessary data, which are auction title, bid price, bid number, bid sale time, star card source. In order to simplify the analysis, this article only analyzes the star cards sold in the auction, and in order to unify the product standards, we need to use eBay's screening function to unify certain standards, which can reduce the burden of our subsequent data cleaning. According to the characteristics

Fig. 1. Historical transaction record on the eBay trading platform

and characteristics of the star card, we establish the standard, the star card we select is: *Panini LeBron James Prizm 19–20 base#129 A single sale card (not Lot), no PSA, no refraction, and no flaws.*

2.3 Data Cleaning and Reorganization

The crawled data needs to be further cleaned up for further analysis. After Data cleaning and reorganization. The eBay market data includes the title of the listed auction, the auction sale date, the source of the card, the number of bids, and the bid price. After processing the data, we retained 551 sets of transaction data and performed simple processing of other data, including converting the Date to time index. In the same way, we perform the same crawling and processing on the Card Hobby market data: get the following data, a total of 109 sets of transaction data. The data of the Card Hobby market includes the same information as eBay had.

3 Empirical Analysis

3.1 Price Distribution

First, let's observe the overall price distribution of the two markets: we can clearly find that the eBay market has experienced greater price fluctuations in the past three months (Fig. 2).

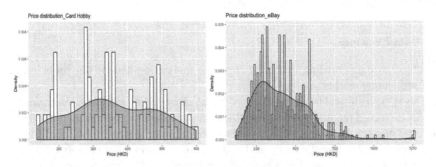

Fig. 2. The price distribution of the eBay and Card Hobby market

3.2   Transaction Volume of Star Cards at Different Times

Before analyzing the price fluctuations of the star card, we want to know more about some time-related characteristics and trading habits of the star card market. So, we conducted statistics on the time distribution of auction transactions on eBay and Card Hobby Market, respectively, according to the daily transaction time distribution, monthly transaction distribution and week-based transaction distribution. First of all, for the analysis of the eBay market, we found that we have already emphasized that the volume of transactions in eBay is larger, bringing together investors and enthusiasts from all over the world. The eBay market adopts PDT time. We can find that the eBay market's daily trading hours are mainly concentrated in the daytime, especially around 6 and 7 at night; from a monthly perspective, the card analyzed in this article has the largest transaction volume in August, accounting for 1/2 of the total transaction volume. The transaction volume in September was only 1/2 of that in August, which we should be sensitive to and an important factor in our analysis of price fluctuations. Finally, we analyze it from the perspective of the week. Sunday is the day with the largest transaction volume, which is also in line with expectations. People spend their holidays at home and have more time to surf the Internet (Figs. 3 and 4).

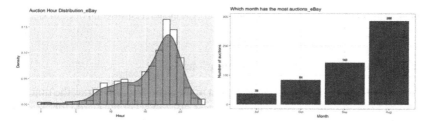

Fig. 3. The number of auctions every hour distribution and its histogram of eBay

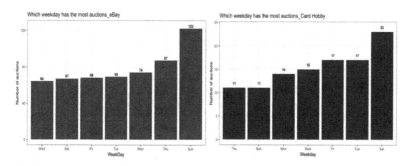

Fig. 4. The histogram of the number of auctions every weekday of eBay and Card Hobby

Let's examine the same characteristics of the Card Hobby market. We can find some similarities by comparing it with the eBay market. The transaction volume in August was still the largest, accounting for nearly 1/2. The transaction volume in September continued to drop by approximately 1/2 of August; Saturday has replaced Sunday as the day with the largest transaction volume in the domestic market; the biggest difference is in the distribution of daily trading hours, almost all domestic transactions are concentrated between 8 pm and 10 pm At this time, this time is usually the time consumers spend playing mobile phones and shopping malls after a busy day of work and study. This also reflects that the Card Hobby market is more concentrated (from China) than eBay's target audience. It is especially obvious (Fig. 5).

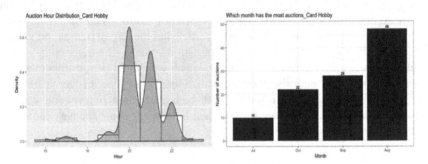

Fig. 5. The number of auctions every hour distribution and its histogram of Card Hobby

3.3   Price Trends and Reasons for Fluctuations

After the above-mentioned preparation, we cut to the topic. First, we draw the time series price chart of the eBay market and the time series price chart of the Card Hobby market. This chart is the starting point and basis of our analysis (Fig. 6).

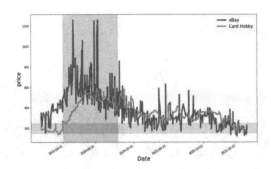

Fig. 6. The price changes graph in the past three months of eBay (linear interpolation)

According to the above figure, we can simply see that card prices continued to rise after the NBA restart, reaching a peak in early August, and subsequent card prices beginning to decline, and by mid-October, the price basically returned to the level of the end of July. In addition, the price fluctuations in the Card Hobby market are basically consistent with eBay, and the price fluctuation trend is similar. From the graph, Card Hobby is a passive response to eBay, and there will be some delays in the response to eBay price fluctuations.

As there are more buyers and sellers on eBay, more buyers participating in the auction, and more speculations, the price volatility of the eBay market is also greater than that of the domestic market. These are in line with our expectations and Economic theory. As the world's largest online secondary trading market for star cards, eBay brings together collectors, enthusiasts, investors, and speculators from different regions of the world. There are a large number of buyers and sellers, and they attract more attention. Auctions There are more participants; as the largest domestic player card earphone trading market, the volume of Card Hobby is about 1/5 of eBay's.

There are many types of cards, and Card Hobby is more in line with what we call the "two-eight distribution" (there will be more demands for popular cards). We can almost make an assumption that all the cards circulated in the Card Hobby market come from eBay. Through communication with relevant collectors, self-expression and collection of relevant information, there are many middlemen in the two markets in China who are responsible. Buy star cards from the eBay market in batches and sell them to buyers in China or list them on the card shopping market. The price of the Card Hobby market is often compared with eBay. This view can also be seen from the title of the listed auction on the card. Many cards listed for auction carry words such as "eBay rose." The analysis of historical data is in line with our assumptions and economic intuition. The price changes in the Card Hobby market closely follow the eBay market, but there are a reaction time and a certain lag (due to information costs, transaction costs, etc.). Due to the difference in volume, the price volatility of the eBay market will be greater, and the price volatility will be greater at the same time. Then the biggest remaining question is how to explain the price fluctuations of this card in the past three months: Before answering this complicated question, let's take the eBay market as an example to further analyze the price fluctuations in the past three months. I then did some descriptive statistical analysis.

First of all, the box chart of the standard card we analyzed on the eBay market is drawn on a weekly basis. According to the box chart, we can easily see the fluctuation trend of the card price. It is worth noting that this box chart can provide us with some things that the price trend chart above cannot tell us. For box charts, the black dots below have a special name-"singular value" or "outlier". There are many outliers Economics and finance are often dismissed in research, but considering the special circumstances of this article, we should further think about the actual meaning of these black dots in the Fig. 1 tend to think of these black dots as auction behaviors that deviate from mainstream prices (Fig. 7).

Next, I used some analysis techniques in the time series to plot the average price per day and average price per hour. It can be seen from the figure that the conclusion is basically in line with our above-mentioned changes. The color changes in the following two pictures can more intuitively reflect the price changes of this star card, and we can find that the eBay star card auction time is not as full as the Card Hobby. Concentrated in the evening, because it is a global market, the auction time is relatively larger (Fig. 8).

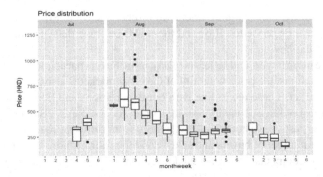

Fig. 7. The price boxplot per week in the past three months of eBay

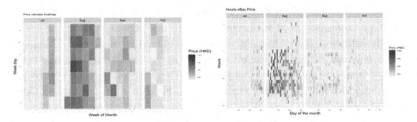

Fig. 8. The price distribution graph per day and per hour in the past three months of eBay

In order to better promote the LeBron James 2019–20 Panini Prizm Base, the price of this card fluctuated violently during these three months. The price of this card has experienced skyrocketing, and then the price has gradually dropped back to the price level three months ago. His story, we first thought to further pay attention to the time period when prices fluctuate sharply, we divide it into an up period and a down period. The rising period is '2020-7-25': '2020-8-15'; the falling period is '2020-8-15': '2020-9-15'. We draw the price trend charts of these two sub-periods separately:

3.4 Price Rising Period

Based on our foreshadowing and data and analysis above, when the NBA announced the restart, it has re-attracted the attention of basketball fans, star card enthusiasts and collectors from all over the world, and the market demand has increased; at the same time, card dealers and speculators have also realized the timing of this restart began to increase card prices through "hype" and other methods, and the supply and demand sides were working at the same time. In addition, after the restart, the Lakers defeated the city rivals Los Angeles Clippers. On this basis, people's expectations for James' card will become higher. At this time, we can continue to analyze deeply. On the demand side, who will provide the demand for this card? A reliable speculation is that there are ordinary star card collectors and star card speculators. From the point of view of ordinary consumers, enthusiasts or collectors, their purpose of buying this card is to own him, so as to obtain the gift of this card. Their own satisfaction, when consumers expect the

Lakers to win the championship, they will expect that this card will bring them more satisfaction in the future, and at the same time they will need to pay a higher price, so their behavior is to buy such "assets in advance". "In order to avoid paying higher prices later, in general, the fundamental reason for the increase in demand is that these consumers believe that the price of the card will continue to rise due to the impact of external factors; and from the perspective of speculators Said that they do not consider the satisfaction that the card brings to them, but only consider whether the price of the card will continue to rise, for speculators there are only four words: buy low and sell high, so their behavior logic is the same. The price is expected to rise, so it will increase the demand at this time. The supply-side analysis may be more complicated. The star cards in the star card trading and auction market are completely determined by the supply side, that is, there is no process of asking prices. The cards are put entirely by suppliers, and the demand side does not have a process of asking for supply prices. Therefore, from this perspective, the card merchants who put the star cards on the auction market expect the price to fall. More rigorously, cash out in time to avoid risks, even if they expect the price to increase. The star card market is relatively affected by external factors such as player performance. At this time, the Lakers are calling for the championship and the market is hot. Price, we can find through the previous analysis that the trading volume of the star card in August reached 1/2 of the total trading volume, and high trading volume and high price appeared at the same time.

To sum up, we can explain the reasons for the rising period as external factors (NBA restart and Lakers title call) increase the demand for star cards and consumer expectations for prices, and speculation on the supply side (singularity) promotes price increases. High transaction volume and high prices mean that the impact on the demand side is greater at this time (Fig. 9).

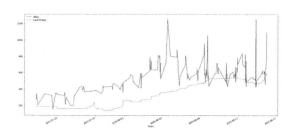

Fig. 9. The price changes graph in the price rising period of eBay

3.5 Price Down Period

As the price of the star card reached its peak in mid-August, the price began to decline gradually, until the end of October returned to the level of three months ago, the price fell as much as three times, usually if in the stock market or other financial the sharp decline in prices in the asset portfolio is mostly due to some kind of negative external shocks, such as the outbreak of the new crown virus in the United States, which caused the US stock market to melt. Such an analysis does not conform to our example in this article.

Then, I think of housing prices and the US subprime mortgage crisis in 2008, the so-called bubble. How to measure whether there is a bubble in this card? If we look at the price trend in the past three months, at the peak of August, the price of this card does have a big bubble. If we start from this assumption, how to understand the bubble burst in mid-August. The most fundamental reason for the bubble burst is that people no longer believe that the price of this card will continue to rise. The high price in mid-August has deviated from the actual value of this card.

First of all, this card is just a Base card from the perspective of a professional star card. Besides, this card has nothing special. Driven by multiple external factors, this card reached its price peak in mid-August and the bubble burst. The card itself is not scarce, a circular trap appeared, once the price started to fall, it could not be stopped, causing the price to return to the price three months ago. Another reason is that the market overdrafted the increase in demand that was originally triggered after the Lakers actually won the championship in October when the Lakers' call for the championship was the highest. Due to the increasing probability of the Lakers winning the championship, the market's early victory over the Lakers may lead to a huge price change response, so that after the Lakers really won the championship, there was no additional demand. These additional demands were consumed by external factors and market hype long before the championship. Maintaining high prices requires a high demand. After the high transaction volume and high price in August, the proportion of speculators in the market is increasing, and the demand for real collectors is decreasing. The price has lost the support of the demand side, the auction price drops, and the chain reaction of the decline in transaction volume occurs (Figs. 10 and 11).

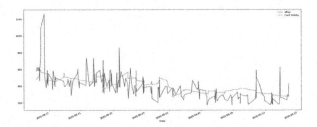

Fig. 10. The price changes graph in the price down period of eBay

The above picture shows that I have selected several different popular star cards and compared their prices three months ago and the current prices. For readers, there is no need to know the specific value of these cards and the players they refer to. Drawing this picture is just to convey the influencing factors of card price fluctuations in the star card market. As we said, there are two factors that affect market fundamentals and specific cards themselves. As shown in the above picture, most of the star cards have experienced different degrees of price decline during this period, but there are still some star cards whose prices have risen or remained stable. This further verifies the particularity of the price fluctuation of this card we selected in this article.

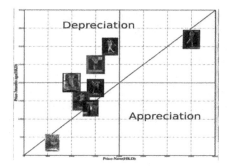

Fig. 11. The basketball star cards price comparison graph (3 months ago and present)

4 Conclusion

Through conversations with professional star card collectors, it is understood that the price of star cards will increase when the NBA starts in the past, and when the championship is produced every year, the price of the star cards of the players of the champion team will rise to a certain extent; and the difference between this time and the past is that after the Lakers won the championship, the price of the star card did not rise. The market seems to have advanced this time point, and we have already conducted a large analysis of the reasons for this situation.

The price of a star card depends on the price fundamentals of the entire star card market and the value of the card itself (such as whether it has a signature) and the external impact of the star card itself factor.

The analysis in this article has many shortcomings and defects, such as data completeness, rigor, and too many assumptions and subjective judgments in the analysis. But what stories can tell us behind the data, everyone has different explanations, and it is happy to use data to tell the story.

Acknowledgement. This work is supported by Shenzhen Research Institute of Big Data.

References

1. Marusak, J.: Charlotte Hornets, 2 p. Herald-Sun, The Durham, NC, 21 February 2019 (2019)
2. Gryc, W.E.: Revenue in first-price auctions with a buy-out price and risk-averse bidders. J. Econ. **129**(2), 103–142 (2019). https://doi.org/10.1007/s00712-019-00674-8
3. Beckett Basketball **31**(3), 6–11 (2020). 7p
4. Beckett Sports Card Mon. **37**(10), 58–69 (2020). 12p
5. Gai, K., Qiu, M., Zhao, H.: Security-aware efficient mass distributed storage approach for cloud systems in big data. In: 2016 IEEE 2nd International Conference on Big Data Security on Cloud (2016)
6. Qiu, M., Jia, Z., Xue, C., Shao, Z., Sha, E.H.M.: Voltage assignment with guaranteed probability satisfying timing constraint for real-time multiproceesor DSPM. J. VLSI Sig. Process. Syst. Sig. Image Video **46**(1), 55–73 (2007)

7. Zhang, Q., Huang, T., Zhu, Y., Qiu, M.: A case study of sensor data collection and analysis in smart city: provenance in smart food supply chain. Int. J. Distrib. Sens. Netw. **9**(11), 382132 (2013). https://doi.org/10.1155/2013/382132

8. Humphreys, B.R., Johnson, C.: The Effect of Superstar Players on Game Attendance: Evidence from the NBA, Working Paper, No. 17-16 (2017)

9. Berri, D.J., Schmidt, M.B.: On the road with the national basketball association's superstar externality. J. Sports Econ. **7**(4), 347–358 (2006)

10. Balafoutas, L., Chowdhury, S.M., Plessner, H.: Applications of sports data to study decision making. J. Econ. Psychol. **75**, 102153 (2019). https://doi.org/10.1016/j.joep.2019.02.009

An Effective Seafile Dockerfile for Raspberry Pi to Make Docker YAML Files for Treehouses

Hiroyuki Terauchi[1] and Naixue Xiong[2(✉)]

[1] Department of Mathematics and Computer Science, Northeastern State University Tulsa,
Tulsa, USA
terauchi@nsuok.edu

[2] Department of Mathematics and Computer Science, Northeastern State University, Tahlequah,
OK 74464, USA

Abstract. Seafile is an open-source, cross-platform, self-hosting file share/sync application. It is installed and configured Seafile via Treehouses, which provides useful functions that host web services on the Tor network. There is no maintainable Docker container of Seafile for the wide range of Raspberry Pis, including RPI0, 3, and 4. Therefore, a custom Seafile Docker container had to be made for the wide range of Raspberry Pis allowing users to install and configure Seafile through Treehouses command. Since the Seafile Dockerfile is old, there were several issues that needed to be resolved for the newer Seafile to work in the Docker container. This paper will explain the process of revising the old Dockerfile and making the YAML file for the Seafile container. The almost all issues pertaining to the revision of the Seafile Dockerfile occurred because the old Dockerfile and configuration script cannot deal with the dependencies and configuration settings of the newer version of Seafile. When the YAML file was made for Treehouses, it required a few of tweaking the Seafile configuration to make the file exchange function work. In addition, the analyze of the troubleshooting during the project suggested that the most crucial technique to revise an old Dockerfile is to be aware of the version of the software.

Keywords: Docker · Docker Compose · Seafile · Raspberry Pi · Raspberian · Treehouses · ARM

1 Introduction

With the rise of 5G and the development of intelligent devices, emerging applications such as the Internet of Things (IoT), the Internet of Vehicles, and the Smart City have gradually become tidal current. On the one hand, in view of the construction of Smart City and the deployment of IoT, many literatures have carried out research on intelligent model [1, 2], Wireless Sensor Network (WSN) and other technologies [3–6], such as increasing the energy efficiency of wireless sensors [7, 8] and improving the fault tolerance of WSN [9]. On the other hand, massive data and complex computing put forward more requirements for cloud computing, cloud storage system and post cloud computing (including fog computing and edge computing, see [10]). For example,

M. Qiu (Ed.): SmartCom 2020, LNCS 12608, pp. 127–135, 2021.
https://doi.org/10.1007/978-3-030-74717-6_14

privacy-preserving [11–13], access control in practical use [14], secure data replication [15].

Treehouses is a Raspberian-based software that allows users to share their different learning platforms via Raspberry Pi [16]. Treehouses provides useful functions which host web services on local and the Tor network. Treehouses hosts each web service on a Docker container, so the web services (which are on the Treehouses system) must be Dockerized, making them easy to manage. Moreover, Treehouses manages containers by the function: services.sh. Therefore, users can download and configure various web services only by executing services.sh function through Treehouses CLI. Treehouses supports a wide range of Raspberry Pi, including RPI4, RPI3, and RPI0. Users do not need to worry about the difference of the architecture in installing and configuring the web services. This paper will explain the process of Dockerization of Seafile for Treehouses.

Seafile is an open-source, cross-platform, self-hostable file share/sync application. Seafile was developed and is distributed by Seafile.Ltd [17], and is composed of the Seafile server, Seahub (Web Interface), and Ccnet server. The Seafile system can use Nginx or Apache as a proxy server [18]. The Seafile.Ltd provides tar.gz files including the Seafile server, Seahub, Ccnet server, and an installer script for Raspberry Pi on their GitHub page [19]. Seafile Ltd. Provides two Seafile: Seafile for x86_x64 and ARM. Although Seafile Ltd. Provides an official Dockerfile of the Seafile system for x86_x64, one is not privided for ARM. Even though Seafile.Ltd does not provide an official Docker image of Seafile for Raspberry Pi, there are several unofficial Dockerfiles of Seafile for Raspberry Pi on Docker Hub and GitHub. Dominik Maier, a developer of Seafile, uoloads his Dockerfile of Seafile for Raspberry Pi to GitHub. However, his Docker image is not compatible with Raspberry Pi 0 because the base image of his Seafile image is Ubuntu [20]. Ubuntu can be used on Raspberry Pi 3 and Raspberry Pi 4 because the architecture of Raspberry Pi 3 is ARMv7, and the architecture of Raspberry Pi 4 is ARMv8 [21]. However, the Ubuntu image does not run on ARMv6, and Raspberry Pi 0 has ARMv6 architecture [22]. Yuri Teixeira has also shared a Docker image of Seafile for Raspberry Pi called rpi-seafile [23]. His Docker image can run on Raspberry Pis, including Raspberry Pi 0. However, his Dockerfile is completely out of date because its base image is resin/raspbian:wheezy. The raspbian:wheezy cannot download any files through APT (Advanced Package Tool). Therefore, it is not practical to use his image. However, although Yuri Teixeria's Docker image is outdated, it does work on all target devices of Treehouses. Therefore, the most practical solution was to alter his Dockerfile. In order to deploy the Seafile Docker image through Treehouses, the two following requirements must be satisfied:

1. The web services must run on all our supported Raspberry Pi devices (Raspberry Pi 0, 3, and 4).
2. The web services must be maintainable.

In this paper, the main technical contributions are about making automatic installation and configuration of the Seafile on Raspberry Pi by Treehouses and satisfying the aforementioned requirements. In addition, common troubleshooting is explained. In order to accomplish this goal, two main tasks are required: build the Docker image of the

Seafile for Raspberry Pi, and automate the initiation of the Seafile container. The process of making a Docker image can be classified into two parts: building a Docker image from a Dockerfile, and starting a Docker container from the created Docker image. The process of building the Docker image will be discussed in Sect. 3, and the process of starting the Docker container in Sect. 4. In Sect. 5, the implementation of the Docker container on Treehouses will be described. In Sect. 6, the problems that were encountered during the revision process of the old Seafile Dockerfile will be analyzed, as well as a description of techniques obtained through the debugging process. Finally, a plan to further improve this project will be demonstrated. The process could be useful for developers for Raspberry Pi based micro-cloud system as a reference.

2 Background Technology

2.1 Treehouses

Treehouses works as a controller for web applications hosted on the Tor network in Raspberry Pi. It uses Docker to manage each web application. Users of Treehouses are able to easily install, configure, run, and manage the web application on the Tor network because the applications are Dockerized. The contributors to Treehouses solve all complicated dependencies problems and configuration issues. Users are not even required to learn how to use Docker and manage Tor port mapping because Treehouses abstracts Docker commands and Tor port management through services.sh [24].

2.2 Services.Sh

services.sh is a bash script which abstracts Docker commands and Tor port management for users to manage web applications. Services.sh mainly wraps Docker Compose commands. Users can host web application on the Tor network with two commands of services.sh: services [app name] install and services [app name] up. The contributors of Treehouses define the YAML file for web applications [21] (see Algorithm 1).

3 Docker Image Building

The Yuri Teixeira's Dockerfile has the following base image: hypriot/rpi-python. The image hypriot/rpi-python also has the base image: resin/rpi-raspbian:wheezy [22]. Unfortunately, Raspbian Wheezy is not supported by Raspbian developers anymore; there are no Wheezy on Raspbian Archive [23]. Therefore, it was necessary to replace the base image of hypriot/rpi-python from resin/rpi-raspbian:wheezy to resin/rpi-raspbian:stretch. Resin/rpi-raspbian:stretch works on any Raspberry Pis as the Docker base image [22]. After that, the new Seafile Dockerfile was made from the original Python image. The new rpi/python:stretch has the alias name treehouses/rpi-python [24]. This name was used in the new Dockerfile, and altered the process building the Seafile image.

After replacing the base image from hypriot/rpi-python to treehouses/rpi-python, the three following issues happened: 1) The image building process stopped due to an error. 2) The container did not start. 3) The container did not start on Raspberry

Pi 0. The first error happened because the following command could not be executed: RUN localedef -i en_US -c -f UTF-8 -A /usr/share/locale/locale.alias en_US.UTF-8. Docker leaves caches in each image building process. It was necessary to enter into the cache container where the image building stoped. It was then discovered that local.alias was not in usr/share/locale but in /etc. Therefore, localedef -i en_US -c -f UTF-8 -A /usr/share/locale/locale.alias en_US.UTF-8 was changed to localedef -i en_US -c -f UTF-8 -A /etc./locale.alias en_US.UTF-8.

The second problem happened because the latest Seafile was installed. At first, it was nearly impossible to figure out why the container would not start. Therefore, the execution of entrypoint.sh, which automatically configured and started Seafile in the container, was halted and manual configuration inside the container was attempted. In this debugging process, it was discovered that certifi and idna were missing. The latest Seafile image requires certifi and idna on the Python package. Therefore, the following command was added: RUN pip install certifi idna. When Yuri Teixeira built the Docker image, the aforementioned Python packages were not required to configure Seafile. However, Yuri Teixeira's Dockerfile installed the latest Seafie image from Seafile's GitHub account, so certifi and idna were required.

The third problem also occurred because the latest Seafile was installed. At the time of the publication, the latest Seafile is version 7, but the officially compiled Seafile version 7 does not work on Raspberry Pi 0 (The source page shows that Seafile should be compiled on armv6 devices if it supports armv6 devices. However, the official Seafile 7.0.3 was compiled on Raspberry Pi 3 B+, so the compiled Seafile missed dependencies to run on Raspberry Pi 0. The latest Seafile is 7.0.5, it can be assumed that the same cause induces this issue.) [28]. Therefore, it was decided to change the Dockerfile to install Seafile version 6.3.4. The base image was replaced from resin/rpi-raspbian:wheezy to resin/rpi-raspbian:stretch.

4   Starting the Seafile Container

Configuring and initiating Seafile requires using the following command: entrypoint.sh. Yuri Teixeira uploaded entrypoint.sh on his GitHub account. The entrypoint.sh automatically configures SQLite, Seafile, and Seahub. The script requires four environmental variables. Users must provide these variables through the Docker command or YAML file. The entrypoint.sh works perfectly for p;der versions of Seafile, but tweaking is required for Seafile version 6.3.4 to satisfy the requirement of Treehouses. Seafile version 6.3.0 and the above variables allow users to manage Gunicorn through gunicorn.conf [30]. Without using a proxy server, such as Apache or Nginx, it is necessary to change a variable bind in gunicorn.conf from 127.0.0.1 to 0.0.0.0 [31]. The default setting of gunicorn.conf is 127.0.0.1 due to security issues [32]. Before starting Seahub, gunicorn.conf needed to be altered by entering the following command into entrypoint.sh:

sed -i s/127.0.0.1/0.0.0.0/ {Seafile path}/gunicorn.conf.

This image can start the fine container. However, the container is unfortunately impeded by the restrictions of Treehouses. In default, Seafile uses port 8082 for Seafile and port 8000 for Seahub. Other containers in Treehouses already occupy port 8000 and 8082. Therefore, it was decided to map port 8085 with port 8000 and port 8086 with

8082. The host and clients can access Seahub UI through port 8085. Port 8086 is used to upload and download files. Users can access Seahub UI from their Browser through port 8086. It seemed to be fine that the web browser could conduct file transfer, too. However, Seafile will not function until the users manually have connected port 8086 to conduct a file transfer through the Seafile server. Docker connects the port on the host side, 8086 and the port on the Seafile container side, 8082. However, the Seahub does not know Docker maps the port 8086 on the host side and the port 8082 on the Seafile container, so Seahub lets the web browser access to port 8082 not port 8086. Therefore, file transfer never occurs (Fig. 1).

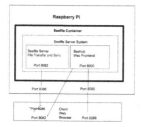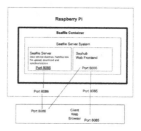

Fig. 1. Left: Visualize the problem, Right: Solution

There are two ways to fix this problem: 1) Use a proxy server; 2) Change the port number of Seafile from 8082 to 8086. In order to change the port number for Seafile, seafile.conf must be altered. Seafile.conf has a variable port, and by default 8082 is assigned. The change from port number 8082 to 8086 allows Seafile to listen to port 8086. The change must happen before the Seafile starts. Therefore, a new sed command was needed to be added in entrypoint.sh before Seafile starting.

5   Automate the Initiation of a Seafile Container

Treehouses has a function, services.sh, which manages several Docker containers from Treehouses CLI. The Seafile container is also managed by services.sh. Services.sh uses Docker Compose to initiate a Seafile container with dependencies, Therefore it was necessary to create a Bash script to install-seafile.sh, which generates docker-compose.yml. The following code is the YAML file for Seafile (see Algorithm 8).

Algorithm 8   Seafile.yaml

```
1:  version: "3"
2:    services:
3:      seafile:
4:        container_name: seafile
5:        image: treehouses/rpi-seafile
6:        ports:
7:        - "8085:8000"
8:        - "8086:8086"
9:        environment:
10:        - SEAFILE_NAME=Seafile
11:        - SEAFILE_ADDRESS=$(treehouses tor)
12:        - SEAFILE_ADMIN=example@seafile.com
13:        - SEAFILE_ADMIN_PW=seacret
14:        volumes:
15:        - /home/data/seafile:/seafile
```

This YAML file is used to make the Seafile container via Docker-compose. When the YAML file is executed, treehouses/rpi-seafile is used to make the Seafile container. If this image is not in the local environment, the Docker engine downloads it from Docker Hub. After that, Docker makes the two port-mapping to the Seafile container. 8085:8000 is used for Seahub and 8086:8086 is used for Seafile. The four environment variables are used for the configuration. SEAFILE_NAME is used to name the Seahub. SEAFILE_ADMIN and SEAFILE_ADMIN_PW are used to enter the Seahub web interface as an administrator. SEAFILE_ADDRESS is used to tell where the Seafile system is hosted. Users can access the Seahub through the URL with port 8085. Because Treehouses primarily hosts its web services on the Tor network, Treehouses tells Seahub which Onion Address is used. Treehouses has a Tor network relating function: treehouses tor. This function returns the currently configured Onion Address if the Raspberry Pi has already started the Tor system. SEAFILE_ADDRESS takes the Onion Address which treehouses tor returns, allowing the Seahub to use the Onion Address to configure the Seafile system. The configuration data and files are stored into /seafile on the container side. Therefore, despite a new Seafile container being spawned, if the container maps /home/data/seafile with /seafile, the container still holds the previous configuration and files.

6 Debugging Technique

Several problems were faced while revising the Seafile Docker image. Even though the change of the code itself is very short and seems to be trivial, many hours were spent to root out the cause of the problems. This allowed two critical debugging techniques to be discovered. First, if the software version is different, the software installation and configuration might change. It is vital to be aware of the different versions of the software. Knowing this could have eliminated a huge number of hours debugging the Dockerfile.

Second, even though issues seem to be impossible to solve, the constant investigation eventually sheds light on the issues.

The difference in software versions induced three main problems:

- The image building process stopped due to an error.
- The container did not start.
- The container did not start on Raspberry Pi 0.

Three reasons caused the aforementioned problems:

- The newer version of Seafile requires certifi and idna.
- gunicorn.conf was changed.
- The Seafile version 7 was not compiled by not Raspberry Pi but Raspberry Pi 3.

The first problem was relatively easy to solve because the configuration program sent error messages, and it was a simple matter of logging into the Seafile container and executing the configuration script manually. The Docker interactive function was very beneficial to discovering the configuration problem. The installation and configuration manual even explicitly describe the requirement. Therefore, it is crucial to know which version of the software that is to be installed and read the manual carefully.

The second and third problem are harder to solve. The basic knowledge relating to Gunicorn or the general application server would lead to solve the second problem easier. Moreover, the third problem had been impossible to solve if I had not known that Yuri Teixeira's Seafile Container had worked on Raspberry Pi 0 beforehand. However, these problems were solved because I was repeatedly asking questions, making hypotheses, conducting experiments, and observing results. Through this simple process, I was able to ask the right questions, find the right information. In order to solve unknown problems, it is very pivotal to conduct experiments, analyze the results, and research continuously.

7 Conclusion

Treehouses services provides a very convenient experience to users via the automatic installation and configuration of web applications. The technical foundations of the treehouses services are Docker and Docker Compose. It is especially important to make Docker images that are compatible with ARM devices in order to run the applications on Raspberry Pi. Throughout the Docker building process, it is critical to be aware of which version of the software to be installed. For future work, Seafile will be hosted not only on the Tor network, but also on general hosting services.

References

1. Yang, Y., Xiong, N., Chong, N.Y., Défago, X.: A decentralized and adaptive flocking algorithm for autonomous mobile robots. In: 2008 3rd International Conference on Grid and Pervasive Computing - Workshops, Kunming, pp. 262–268 (2008)
2. Li, J., Xiong, N., Park, J.H., Liu, C., Shihua, M.A., Cho, S.E.: Intelligent model design of cluster supply chain with horizontal cooperation. J. Intell. Manuf. **23**(4), 917–931 (2012)

3. Wan, Z., Xiong, N., Ghani, N., Vasilakos, A.V., Zhou, L.: Adaptive unequal protection for wireless video transmission over IEEE 802.11 e networks. Multimedia Tools Appl. **72**(1), 541–571 (2014)
4. Shu, L., Zhang, Y., Yu, Z., Yang, L.T., Hauswirth, M., Xiong, N.: Context-aware cross-layer optimized video streaming in wireless multimedia sensor networks. J. Supercomput. **54**(1), 94–121 (2010)
5. Zeng, Y., Sreenan, C.J., Xiong, N., Yang, L.T., Park, J.H.: Connectivity and coverage maintenance in wireless sensor networks. J. Supercomput. **52**(1), 23–46 (2010)
6. Long, F., Xiong, N., Vasilakos, A.V., Yang, L.T., Sun, F.: A sustainable heuristic QoS routing algorithm for pervasive multi-layered satellite wireless networks. Wirel. Netw. **16**(6), 1657–1673 (2010)
7. Lin, C., Xiong, N., Park, J.H., Kim, T.: Dynamic power management in new architecture of wireless sensor networks. Int. J. Commun. Syst. **22**(6), 671–693 (2009)
8. Guo, W., Xiong, N., Vasilakos, A.V., Chen, G., Yu, C.: Distributed k–connected fault–tolerant topology control algorithms with PSO in future autonomic sensor systems. Int. J. Sens. Netw. **12**(1), 53–62 (2012)
9. Lin, C., He, Y., Xiong, N.: An energy-efficient dynamic power management in wireless sensor networks. In: 2006 5th International Symposium on Parallel and Distributed Computing, Timisoara, pp. 148–154 (2006)
10. Zhou, Y., Zhang, D., Xiong, N.: Post-cloud computing paradigms: a survey and comparison. Tsinghua Sci. Technol. **22**(6), 714–732 (2017)
11. Liu, Y., Ma, M., Liu, X., Xiong, N.N., Liu, A., Zhu, Y.: Design and analysis of probing route to defense sink-hole attacks for internet of things security. IEEE Trans. Netw. Sci. Eng. **7**(1), 356–372 (2020)
12. Shahzad, A., et al.: Real time MODBUS transmissions and cryptography security designs and enhancements of protocol sensitive information. Symmetry **7**(3), 1176–1210 (2015)
13. Sang, Y., Shen, H., Tan, Y., Xiong, N.: Efficient protocols for privacy preserving matching against distributed datasets. In: International Conference on Information and Communications Security, pp. 210–227 (2006)
14. Yang, J., et al.: A fingerprint recognition scheme based on assembling invariant moments for cloud computing communications. IEEE Syst. J. **5**(4), 574–583 (2011)
15. Wang, Z., Li, T., Xiong, N., Pan, Y.: A novel dynamic network data replication scheme based on historical access record and proactive deletion. J. Supercomput. **62**(1), 227–250 (2012)
16. Welcome to treehouses. treehouses.github.io. https://awesomeopensource.com/project/treehouses/treehouses.githubiio. Accessed 15 Apr 2020
17. Open Source File Sync and Share Software. Seafile. https://www.seafile.com/en/home/. Accessed 15 Apr 2020
18. Seafile Ltd., Components Overview. Private Seafile. https://download.seafile.com/published/seafile-manual/overview/components.md. Accessed 15 Apr 2020
19. Haiwen: haiwen/seafile. GitHub, 1 April 2020. https://github.com/haiwen/seafile. Accessed 15 Apr 2020
20. Dominik, M.: domenukk/seafile-docker-pi. GitHub. https://github.com/domenukk/seafile-docker-pi. Accessed 15 Apr 2020
21. Murray, B.: RaspberryPi. Ubuntu Wiki, 13 March 2020. https://wiki.ubuntu.com/ARM/RaspberryPi. Accessed 29 April 2020
22. Legacy resin base images. Balena Documentation. https://www.balena.io/docs/reference/base-images/legacy-base-images/. Accessed 29 Apr 2020
23. Yuri Teixeira: yuriteixeira/rpi-seafile. GitHub. https://github.com/yuriteixeira/rpi-seafile. Accessed 15 Apr 2020
24. treehouses: treehouses/cli. GitHub https://github.com/treehouses/cli. Accessed 19 Apr 2020

25. Hypriot: hypriot/rpi-python. GitHub, 22 December 2015. https://github.com/hypriot/rpi-python. Accessed 15 Apr 2020

26. Index of /raspbian/dists/. https://archive.raspbian.org./raspbian/dists/. Accessed 19 Apr 2020

27. treehouses: treehouses/rpi-python. Docker Hub. https://hub.docker.com/r/treehouses/rpi-python. Accessed 19 Apr 2020

28. jobenvil, Sikorra, J.P.: seafile.sh illegal instruction in 7.0.3, Issue #41. · haiwen/seafile-rpi. GitHub. https://github.com/haiwen/seafile-rpi/issues/41. Accessed 15 Apr 2020

29. Hypriot. hypriot/rpi-python. GitHub. https://github.com/hypriot/rpi-python/blob/master/Dockerfile

30. Seafile Ltd.: Deploying Seafile with SQLite. Private Seafile. https://download.seafile.com/published/seafile-manual/deploy/using_sqlite.md. Accessed 15 Apr 2020

31. Steve, Pan, D.: Seafile server 7.0 bind address issues. Seafile Community Forum, 13 June 2019. https://forum.seafile.com/t/seafile-server-7-0-bind-address-issues/9127. Accessed 15 Apr 2020

32. Vmajor and jobenvil: Seafile 7 only listens to 127.0.0.1, Seafile Community Forum, 5 August 2019. https://forum.seafile.com/t/seafile-7-only-listens-to-127-0-0-1/9544/7. Accessed 15 Apr 2020

Design of a Hardware Accelerator
for Zero-Knowledge Proof in Blockchains

B. O. Peng[1], Yongxin Zhu[1,2,3(✉)], Naifeng Jing[1], Xiaoying Zheng[2],
and Yueying Zhou[2]

[1] School of Microelectronics, Shanghai Jiao Tong University, Shanghai 200240, China
[2] Shanghai Advanced Research Institute, Chinese Academy of Science, Shanghai 201210, China
zhuyongxin@sari.ac.cn
[3] University of Chinese Academy of Science, Beijing 100049, China

Abstract. With the popularization and maturity of blockchain technology, more and more industries and projects are gradually trying to combine blockchain technology, including digital currency, Internet of Things, 5G new infrastructure. The most important thing for these applications is to require its safety. These security services are usually provided by cryptographic protocols, and zero-knowledge proof is such a core technology to provide the bottom layer of security services. However, the most widely used protocol named zk-SNARK, involves solving multiple large-scale examples of tasks related to polynomial arithmetic on large prime fields of cryptography and multi-exponentiations on elliptic curve groups. Complicated and huge calculations bring longer prover time, which hinders the implementation of some applications. In this paper, we propose a design of hardware accelerator based on FPGA for zero-knowledge proof. The zk-SNARK engine which is combined of multiple FFT, MAC and ECP units reduces the prover time by 10x and provides the possibility for future blockchain terminals based on mobile devices.

Keywords: Blockchain · Zero-knowledge proof · zk-SNARK · FPGA

1 Introduction

In recent years, the zero-knowledge proof and zero-knowledge succinct non-interactive argument of knowledge (zk-SNARK) have drawn significant attention as a core technology in blockchain security services [1, 2] including the cryptocurrency industry, verifiable computations and IoT because it will provide the privacy that is necessary for data transfer which current security protocols do not provide [3, 4].

Goldwasser [5] first introduced zero-knowledge proof that enable a prover to convince a verifier that a statement is true without revealing anything else.

Blum [6] extended the notion to non-interactive zero-knowledge proofs in the common reference string model.

The commonly used algorithm is Groth16 [7] which introduced the NILP (non-interactive linear proofs) [8]. It is currently used in ZCash [9], Filecoin and Coda. The

© Springer Nature Switzerland AG 2021
M. Qiu (Ed.): SmartCom 2020, LNCS 12608, pp. 136–145, 2021.
https://doi.org/10.1007/978-3-030-74717-6_15

calculation of Groth16 is divided into three parts: setup, prove and verify. In the setup period, Pk/Vk (proof/verification key) are generated. During the prover time, a proof with a given witness/statement is generated. Finally, verify parts verifies whether the proof is correct through Vk. According to the introduction of Groth16, the main calculation of Groth16 algorithm consists of two parts: FFT and Multi-Exponentiations during the prover time. Usually, the prover time is about minutes while the verifier time is about milliseconds. The specific benchmark can be seen in Sect. 5.

Due to the large time costs, zero-knowledge proof technology is difficult to implement in many applications that require immediacy and portability. Therefore, the design proposed in this paper is a customized hardware design for the key steps in calculation to achieve the purpose of reducing time overhead.

Our contributions can be summarized as below:

- We proposed multiple FFT (Fast Fourier Transform) units and large size FFT operation is decomposed into multiple smaller FFT units to perform FFT and iFFT operation [10].
- We proposed multiple MAC (Multiply and Accumulate Circuit) units combined of adders and multipliers which are configured to perform Montgomery multiplication on a finite field [11].
- We proposed multiple ECP (Elliptic Curve Processing) units to reduce the computational overhead due to the calculation of elliptic curve groups [12].
- We implemented the design of this zk-SNARK engine based on FPGA, reduced the prover time by 10x.

The rest of the paper is organized as follows. In the Sect. 2 we laid out the mathematical formulas for Groth16 algorithm and we present the implementation of Groth16 by libsnark [13] using C++. Then we present the implementation of zk-snark engine using Verilog HDL in Sect. 3 In Sect. 4, we compare the prover time between software and hardware implementation. Finally, the conclusion is drawn in Sect. 5.

2 Groth16's Calculation Steps and Software Implementation

2.1 Calculation Steps of Groth16 Algorithm

All the work in this paper is carried out from the perspective of calculation. So, what we focus on is the calculation steps of groth16 algorithm.

QAP (Quadratic arithmetic programs) [7] is the base of groth16 algorithm. In the QAP, we can define the "Relation": $\mathscr{R} = (F, aux, l, \{ui(X), vi(X), wi(X)\}i = 0,1... \boldsymbol{m},$ t(X)). The statement is $(\alpha1,...,\alpha\ell) \in F^\ell$ and the witness is $(\boldsymbol{a\beta}+1,...,\boldsymbol{am}) \in F^{\boldsymbol{m-\ell}}$, in the case of $\alpha0 = 1$, the Eq. (1) is satisfied. The order of t(X) is n.

$$\sum\nolimits_{(i=0)}^{m} \alpha_i u_i(X) \cdot \sum\nolimits_{(i=0)}^{m} \alpha_i v_i(X) = \sum\nolimits_{(i=0)}^{m} \alpha_i v_i(X) + h(X)t(X) \qquad (1)$$

In the process of setup, we randomly select α, β, γ, δ, $\boldsymbol{x}{\leftarrow}F^*$ to generate σ and τ. τ = (α, β, γ, δ, $\boldsymbol{x}$) and

$\sigma = (\alpha, \beta, \gamma, \delta, \{x^i\}_{i=0,1,\ldots,n-1}, \{\frac{\beta u_i(x)+av_i(x)+\omega_i(x)}{\gamma}\}_{i=0,1,\ldots,l}, \{\frac{\beta u_i(x)+av_i(x)+\omega_i(x)}{\delta}\})$ i $= \ell+1,\ldots,m)$.

In the process of prove, we randomly select two parameters r and s, then we calculate $\pi = \prod\sigma = (A, B, C)$. In this part, the Eq. (2), (3) and (4) are satisfied.

$$A=\alpha + \sum_{i=0}^{m} \alpha_i u_i(x) + r\delta \tag{2}$$

$$B=\beta + \sum_{i=0}^{m} \alpha_i v_i(x) + s\delta \tag{3}$$

$$C=\frac{\sum_{i=l+1}^{m} \alpha_i(\beta u_i(x)+av_i(x)+\omega_i(x))+h(x)t(x))}{\delta} + A\,s + rB - rs\delta \tag{4}$$

In the process of verify, we need to calculate whether the Eq. (5) holds.

$$A \cdot B = \alpha \cdot \beta + \frac{\sum_{i=0}^{l}(a_i(\beta u_i(x) + \alpha v_i(x) + w_i(x))}{\delta} \cdot \gamma + C \cdot \delta \tag{5}$$

The calculation steps that we need to pay attention to include the following points. First, according the α, β, γ, δ, $x \leftarrow F^*$, we need to calculate the Lagrange basis to get $u_i(x)$, $v_i(x)$ and $w_i(x)$. Then, we need to calculate t(x) after we calculate xi. The multi-exponentiations involved here are all in the finite field F^*.

In the prove period, we need to do iFFT to get the polynomial coefficients of $u_i(x)$, $v_i(x)$ and $w_i(x)$ and then using FFT to get the value of $u_i(x)$, $v_i(x)$ and $w_i(x)$ in coset, so we can get the value of h(x) in coset according to (1), then through iFFT we can get the polynomial coefficients of h(x). There are 4 times iFFT and 3 times FFT in the prove period which cost large time.

Finally, we can generate the prove according to the $u_i(x)$, $v_i(x)$ and $w_i(x)$, h(x), r and s. The cost of the generation of prove includes 4 times multiply accumulate operations. Once in (2) and (3), twice in (4).

2.2   The Method of Software Implementation

The questions that need zero-knowledge proof are essentially NP (Non-deterministic Polynomial) questions. There is a category of NP question, NPC (NP-complete) questions. All NP questions can be transformed into an NPC problem. As long as one NPC question can be solved in polynomial time, all NP questions can be solved in polynomial time. There are many ways to describe an NPC questions. The way of describing NPC questions is called "language". The "libsnark" library summarizes several description languages and the most widely used are R1CS (Rank-1 Constraint System) and QAP (Quadratic Arithmetic Program). When describing the questions, we generally use R1CS to convert the problem into a polynomial form. When calculating the results, we usually use QAP. The "libsnark" provides us a method to convert R1CS to QAP named "Reduction".

Libsnark provides us enough functions to set a R1CS circuit, calculate the value of parameters and verify the proof, what we need to work is using these functions to build the system. In general, the logic of Groth16's proof generation and verification are shown as follows (Fig. 1).We choose to build a simple Merkel tree to implement the Groth16 algorithm to generate and verify the circuit.

A verification circuit of the Merkel path is constructed to generate and verify the proof. The depth of the Merkle tree is 3 and the calculation of the tree uses the SHA-256 hash function. After we construct Merkle Circuit and generate R1CS, we set the value of each variable and then generate the certificate through function "r1cs_gg_ppzksnark_prover". Knowing Vk, proof and public information(root), we call the interface of r1cs_gg_ppzksnark_verifier_strong_IC to verify.

In the prove period, the prove command needs to provide the 8 leaf nodes of the original 3-layer Merkle tree and specify the path corresponding to the leaf node that needs to be proved.If the verification is passed, it means there is a Merkle path that can generate the root without revealing any specific information about the path and it achieved zero-knowledge proof.

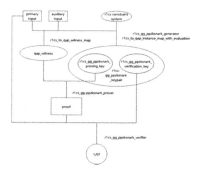

Fig. 1. The logic of Groth16's proof generation and verification

3   Hardware Design and Implementation

After the implementation of groth16 using software, we tried to design a hardware accelerator to optimize the complex calculations in the proof, especially the FFT and Multi-Exponentiations.

In the case of using zk-SNARK protocol for cryptographic verification in blockchain applications, the generation of proof may consume significant computing power and requires a high-performance operating system. Conventionally, the implementation of the calculation requires one or more processors in network server or computer system. The process of generation uses the secret key in the form of a polynomial equation to create the proof without disclosing the secret key. The certification verification verified the validity if the certification without acquiring the knowledge of the secret key. Both operations may require extremely high computing demands on the processor. The zk-SNARK protocol involves solving polynomial arithmetic on large prime number fields of cryptography and multiple large examples of the task of multi-scalar multiplication on elliptic curve groups. In order to meet the demand for computing power, dedicated hardware accelerators such as FPGA can be used to perform at least part of the calculations which are associated with Groth16 algorithm. Therefore, the device can contribute to the most difficult and complex part of the calculation and reduce the calculation time, so that the proof under the zk-SNARK protocol can be performed efficiently, which provides the possibility for the immediacy and portability of future blockchain applications.

In some embodiments, the processor is configured to execute application software that is programmed to respond to the memory instructions stored in storage devices. The processor can communicate via PCI-e bus to couple the instructions and data to the DDR controller. SDRAM is coupled to the zk-SNARK engine to load instructions and data used to run the calculation of the Groth16 algorithm. The zk-SNARK engine (Fig. 5) is configured to execute calculations including finite-field arithmetic operations, such as MAC, FFT, iFFT and ECP. The calculation results are combined to obtain the final results and it can be loaded into the processor which can be remotely accessed on a cloud, a network server, or a node in the blockchain.

Figure 2 presents our calculation steps during the proof. We evaluate the polynomials A, B, and C to calculate the proof. A, B, C and input are data with 256 bits sequence. The calculation in PF can be divided into two parts. The first part is calculation the coefficient of polynomial H. The second part is to use the coefficient of H, QAP evidence and Pk to calculate the proof. The first step is performing the evaluation of A, B, C on the set S according to the following equations $aA[i] = \sum_{k=0}^{m-1} fva[k]csa[i][k]$, $aB[i] = \sum_{k=0}^{m-1} fva[k]csb[i][k]$, $aC[i] = \sum_{k=0}^{m-1} fva[k]csc[i][k]$.

In these equations, parameter I ranges from 0 to 2^{21}, full variable assignment $(fva) = \{$main input, auxiliary input$\}$, constraint system(cs) $= \{csa, csb, csc\}$, the coefficients of polynomials A, B, C are calculated by iFFT.

For the zk-patch, we can calculate it as the following Eq. (6), (7) and (8):

$$Hi = d2 * aAi + d1 * bB[i] \tag{6}$$

$$H0 = H0 - d3 - d1 * d2 \tag{7}$$

$$Hm = Hm + d1 + d2 \tag{8}$$

Among them: $i = 0$ to $m-1$, $m = 2^{21}$, $d1$, $d2$ and $d3$ are 256-bit random numbers. After this step, we need to evaluate the polynomials A, B, C in the set T. We calculate $aA[i] = aA[i] * g^i$, $aA = FFT(aA)$, $aB[i] = aB[i] * g^i$, $aB = FFT(aB)$, $aC[i] = aC[i]*g^i$, $aC = FFT(aC)$, $H_{tmp[i]} = aA[i]*aB[i] - aC[i]$. The set T is the coset of S. According to the above description, the coefficients of polynomial H can be calculated efficiently by FFT and then we calculate the coefficient of $H = iFFT(H\_tmp)$, the coefficient of $H[i] = H[i] + H_{tmp[i]} * g - 1$.

Finally, we calculate the queries based on multiscalar multiplication on the elliptic curve BN256 to obtain the proof. Using the equation $H_{query} = \sum_{i=0}^{m} H[i]P[i]$, we can get the proof. The framework of our zk-SNARK engine is mainly combined with FFT units, MAC units and ECP units, we will introduce these units in the next part.

3.1 FFT Units

The FFT units is configured to perform large-scale FFT operation, for example, a length of 2^{21} points calculation of the FFT algorithm. Figure 3 shows a simplified diagram

Fig. 2. Main calculation steps during the proof

showing the decomposition of the 2^{21} points FFT unit. In order to perform the large-size FFT unit, multiple smaller and more manageable FFT hardware units are used for evaluation. It includes 1024 points FFT unit and 2048 points FFT unit that repeat operations on 2^{21} variable domains. The 1024-point FFT unit uses multiple 4-base FFT units, while the 2048-point FFT unit uses multiple 2-base FFT units and multiple 4-base FFT units. Figure 4 shows a 2-base FFT unit, which is a 2-point FFT unit. In this embodiment, din0 and din1 are input, and dout0 and dout1 are outputs.ω1 is the rotation factor. Figure 5 shows a 4-base FFT unit, which is a 4-point FFT unit. In this embodiment, din0 din1, din2 and din3 are inputs, and dout0, dout1, dout2 and dout3 are outputs. ω1, ω2 and ω3 are the corresponding rotation factor. The calculation flow of the 1024-point FFT unit includes five stages with 256 loop iterations of 4-base FFT unit and the calculation process of the 2048-point FFT unit includes five stages each of which includes 512 loop iterations of 4-base FFT unit and 512 loop iterations of 2-base FFT unit.Assuming that N = 2048, M = 1024 and P = 2^{21}, the FFT or iFFT unit of FPGA can include the following logical steps:

1 Divide variable into 1024 groups, each group has 2^{11} variables.
2 Use 2048-point FFT unit to perform 2048-point FFT on all 1024 groups from step1.
3 Multiply the results from step2 by ω.
4 Use 1024-point FFT unit to perform 1024-point FFT on the 2048 result groups from step3.
5 Calculate the output of the original large FFT and iFFT.

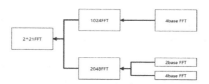

Fig. 3. The decomposition of 2^{21} points FFT

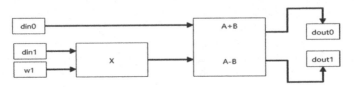

Fig. 4. 2-base FFT unit

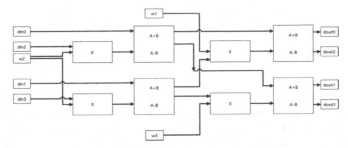

Fig. 5. 4-base FFT unit

3.2 ECP Units

ECP unit is configured to use hardware accelerators to implement fast ECP algorithm: $R_i = \Sigma_{i=0}^{N-1} d_i P_i$, $N = 2^{21}$, d_i is a scalar, R_i and P_i are the point on the curve. When d_i is a large number with 256-bits, the multiplication of the scalar and the elliptic point require high computing power. Therefore, it is advantageous to configure the hardware accelerator to implement a fast ECP algorithm to calculate the addition of $d_i P_i$ in a faster way. The ECP unit includes a sorting engine used to sort the sequence in descending order with the scalar d_i and find the maximum value d_1 and the second maximum value d_2. On the curve, we consider that $d_1 P_1 + d_2 P_2 = d_1 P_1 - d_2 P_1 + d_2 P_1 + d_2 P_2 = (d_1 - d_2)P_1 + d_2(P_1 + P_2) = d_{dif} P_1 + d_2 P_{ad}$. Use the maximum value d_1 and the second maximum value d_2 to calculate d_{dif}, and then we add the corresponding points on the elliptic curve to get P_{ad}. These two pairs of new points (d_{dif}, P_1) and (d_2, P_{ad}) are used to replace the original points.

3.3 MAC Units

In some embodiments, the MAC unit is a finite field computing element configured to perform the data operations. The MAC unit includes multiple MAC blocks. Each MAC block group is configured to perform data operations. As shown in Fig. 6 the MAC unit is a finite calculation element configured to perform multiplication and addition operations, and it can be implemented according to the structure shown in Fig. 7. In the MAC block, DDR commands are coupled to and received by the command buffer. The DDR command is used to control the FSM of the MAC block, which controls the MAC calculation to obtain output data. The DDR data is coupled to the data buffer and the multiplexer to select the data to be multiplied and added using the multiplier and the adder. The multiplier is configured to perform Montgomery multiplication on a finite field with 256×256 bits. If a two-stage MAC is required, the multiplier output (multi-out) of the MAC block is coupled to the multiplier input (multi-in) of the MAC block and the result is coupled to the output buffer.

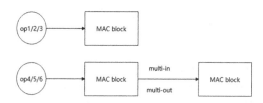

Fig. 6. MAC units

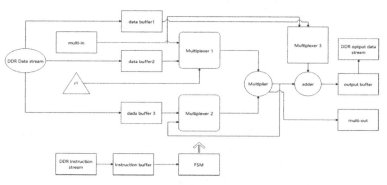

Fig. 7. Data and instruction stream structure in MAC

4   Results and Analysis

For software, we benchmark proof system on the Merkle tree R1CS instance. The benchmarks were obtained using 3.1 GHz Intel Core i5-9400F CPU, in single-threaded mode, using the BN256 curve.The prover spends almost all of its time either doing FFTs or multi-exponentiations. The percentage of FFT execution time is 16% of prover execution time. The Table 1 shows the results of software execution time and hardware execution time in FPGA.

Table 1. Benchmarks of software/hardware implementation.

Proof system	Time		
	Generator time (s)	Prover SW time (s)	Prover HW time (s)
Groth16	70.12	80.01	8

For the hardware implementation, we test the time of calculation in the prove period. The testbench are designed according to the setting and some of the results in the implementation of the software. At a frequency of 200 MHz, a total of 1593164290 clock cycles have been experienced. In other words, the prover time is reduced to about 8 s. It is precisely because of the following points of optimization and strategies that our acceleration We have customized three calculation hardware units for those complex calculation steps.

- In FFT units, the butterfly operation units, the twiddle factor complex multiplication module and the final precision interception module are processed by pipeline.
- In MAC units, the montgomery modular multiplication reduces the number of modulo and simplifies the complexity of division.
- In ECP units, we have reduced the complexity of elliptic curve operations through mathematical approximation.
- Parallel design is adopted between units. The computing tasks that need to be executed sequentially can be calculated in parallel through our scheduling. Results are satisfactory:

5   Conclusion

In this paper, we proposed the hardware implementation of Groth16 algorithm to accelerate zero-knowledge proof calculation in blockchains. By designing the FFT, MAC and ECP units, we solve the calculation of polynomial arithmetic on large prime fields of cryptography and multi-exponentiations on elliptic curve groups in the zero-knowledge proof. Compared with the implementation of software under the Merkle tree R1CS instances, we reduce the prover time by nearly 10x. This hardware accelerator provides the possibility for future blockchain terminals based on mobile devices13. However, our work still needs to be improved because we just consider the time costs and the only evaluation index is prover time and we have not verified the tests on FPGA, so far only basic simulation results have been completed. We just focus on the accelerate of the key calculation steps without considering the overall collaborative design. Our follow-up research will further improve our current work especially in the implementation of software and hardware co-design to make a foundation for related blockchain applications.

Acknowledgment. This work is supported by the National Natural Science Foundation of China (Grant No. 61772331), Natural Science Foundation of China (grant No. U1831118), the Strategic Priority Research Program of Chinese Academy of Sciences (Grant No. XDC02070800), Shanghai Municipal Science and Technology Commission (grant No. 19511131202), Pudong Industry-University-Research Project (grant PKX2019-D02), Independent Deployment Project of Shanghai Advanced Research Institute (grant E0560W1ZZ0). I would like to express my gratitude to all those who helped me during writing this paper. Firstly, thanks to all my team members who gave me much useful advice and encouragement. And gratitude to my parents for their nurturing and support.

References

1. Naganuma, K., Yoshino, M., Inoue, A., Matsuoka, Y., Okazaki, M., Kunihiro, N.: Post-Quantum zk-SNARK for Arithmetic Circuits using QAPs. In: 2020 15th Asia Joint Conference on Information Security (AsiaJCIS), Taipei, Taiwan, pp. 32–39 (2020). doi: https://doi.org/10.1109/AsiaJCIS50894.2020.00017
2. Clerk Maxwell, J.: A Treatise on Electricity and Magnetism, 3rd edn., vol. 2, pp. 68–73. Clarendon, Oxford (1892)
3. Fan, Y., Zhao, G., Lin, X., Sun, X., Zhu, D., Lei, J.: One secure IoT scheme for protection of true nodes. In: Qiu, M. (ed.) SmartCom 2018. LNCS, vol. 11344, pp. 143–152. Springer, Cham (2018). https://doi.org/10.1007/978-3-030-05755-8_15
4. Sarma, R., Barbhuiya, F.A.: Internet of Things: attacks and defences. In: 2019 7th International Conference on Smart Computing and Communications (ICSCC), Sarawak, Malaysia, pp. 1–5 (2019). https://doi.org/10.1109/ICSCC.2019.8843649
5. Goldwasser, S., Micali, S., Rackoff, C.: The knowledge complexity of interactive proof systems. SIAM J. Comput. **18**(1), 186–208 (1989)
6. Blum, M., Feldman, P., Micali, S.: Non-interactive zero-knowledge and its applications. In: STOC, pp. 103–112 (1988)
7. Groth, J.: On the size of pairing-based non-interactive arguments. In: Fischlin, M., Coron, J.-S. (eds.) EUROCRYPT 2016. LNCS, vol. 9666, pp. 305–326. Springer, Heidelberg (2016). https://doi.org/10.1007/978-3-662-49896-5_11
8. Bitansky, N., Chiesa, A., Ishai, Y., Paneth, O., Ostrovsky, R.: Succinct non-interactive arguments via linear interactive proofs. In: Sahai, A. (ed.) TCC 2013. LNCS, vol. 7785, pp. 315–333. Springer, Heidelberg (2013). https://doi.org/10.1007/978-3-642-36594-2_18
9. Hopwood, D., Bowe, S.: Zcash protocol specification. https://github/zcash/zips/blob/master/protocol/protocol.pdf
10. Abbas, Z.A., Sulaiman, N.B., Yunus, N.A.M., Wan Hasan, W.Z., Ahmed, M.K.: An FPGA implementation and performance analysis between Radix-2 and Radix-4 of 4096 point FFT. In: 2018 IEEE 5th International Conference on Smart Instrumentation, Measurement and Application (ICSIMA), Songkla, Thailand, pp. 1–4 (2018). https://doi.org/10.1109/ICSIMA.2018.8688777
11. Hariri, A., Reyhani-Masoleh, A.: Bit-serial and bit-parallel montgomery multiplication and squaring over GF(2^m). IEEE Trans. Comput. **58**(10), 1332–1345 (2009). https://doi.org/10.1109/TC.2009.70
12. Zhang, S., Chen, Y., Zhao, G., Guo, K.: A new elliptic curve cryptosystem algorithm based on the system of chebyshev polynomial. In: 2014 IEEE 7th Joint International Information Technology and Artificial Intelligence Conference, Chongqing, pp. 350–353 (2014). https://doi.org/10.1109/ITAIC.2014.7065068
13. Virza, M.: Libsnark. https://github.com/scipr-lab/libsnark
14. Kotobi, K., Sartipi, M.: Efficient and secure communications in smart cities using edge, caching, and blockchain. In: 2018 IEEE International Smart Cities Conference (ISC2), Kansas City, MO, USA, pp. 1–6 (2018). https://doi.org/10.1109/ISC2.2018.8656946

Energy-Efficient Optimization Design for UAV-Assisted Wireless Powered MEC Systems

Jin Wang[1,2], Caiyan Jin[1], Yiming Wu[1], Qiang Tang[1], and Neal N. Xiong[3]($\boxtimes$)

[1] Department of Computer Science and Communication Engineering, Changsha University of Science and Technology, Changsha 410114, China
{jinwang,tangqiang}@csust.edu.cn, {caiyan_jin, wuyiming_cs}@stu.csust.edu.cn
[2] School of Information Science and Engineering, Fujian University of Technology, Fujian 350118, China
[3] Department of Mathematics and Computer Science, Northeastern State University, Tahlequah, OK 74464, USA
xiong31@nsuok.edu

Abstract. The quality of services requirements of wireless networks are always limited by devices' on-board computation capabilities and batteries. Considering the advantages of Unmanned aerial vehicles, such as high mobility and probability of line-of-sight links, this paper proposes a UAV-assisted wireless powered mobile edge computing system in 5G networks, where the UAV is dispatched as a cloud-let and a flying energy source to provide the energy transferring and task offloading services to IoTDs. To promote the green communication and prolong the duration of system service, we intend to minimize the weighted system energy consumption by jointly optimizing the working mode decisions, IoTD associations, charging resource allocations, which is a non-convex problem with the mixed integer variables. Therefore, an efficient alternative algorithm is developed to obtain the high-quality sub-optimal solutions based on the block-coordinate descent optimization technique. Additional, the experimental results show that the proposed algorithm is superior to the benchmarks.

Keywords: Mobile edge computing · Wireless power transferring · Network optimization · Quality of service improvement · Resource sharing and allocation · Unmanned aerial vehicle

1 Introduction

With the development of wireless sensor networks (WSN) [4], the number of the Internet of Things devices (IoTDs) have been explosive increased to provide the ubiquitous services accessibility in the fifth generation (5G) networks, which could support at least 100 billion devices [6]. But for now, it is difficult to satisfy the massive data processing demands and quality of service (QoS) requirements over the limited on-board capacity

© Springer Nature Switzerland AG 2021
M. Qiu (Ed.): SmartCom 2020, LNCS 12608, pp. 146–155, 2021.
https://doi.org/10.1007/978-3-030-74717-6_16

of devices and terrestrial base stations [7]. Faced with the data processing, mobile edge computing (MEC) [5] has been widely regarded as an emerging computing paradigm in 5G to improve the performance of resource-constrained IoTDs. Faced with the energy supplement, wireless power transferring (WPT) [13] has been considered as a promising solution to provide power for the energy-constrained devices.

Considering their high mobility, line-of-sight (LoS) links, flexible and cost-effective deployment, unmanned aerial vehicles (UAVs) have drawn great concern in 5G [2], as flexible communication platforms (e. g., moving relays [1, 3, 20] and flying base stations (BSs) [9, 10, 17]). For instance, Zhang *et al.* in [22] considered a UAV as the relay between a single-cell cellular BS and ground users under spectrum sharing to improve the system communication performance. In addition, Wan *et al.* in [14] investigated the path planning and resource management of UAV-BSs in IoT. Although a few recent research began to focus on the UAV-enabled WPT-MEC [8, 18], it continues to be challenges for aerial platforms to cooperate with devices in the terrestrial networks and to satisfy a variety of users' QoS requirements efficiently by the limited system resources.

Against this background, we conceive a UAV-assisted wireless powered mobile edge computing (MEC) system in 5G networks, where the UAV is deployed as a moving power source and a flying BS. In particular, we formulate the joint powering resource allocation and task offloading process as an optimization problem with the aim of minimizing the weighted system power consumption, which is intractable due to mixed integer variables, as well as the non-convexity of objective function and constraints. Therefore, we develop an efficient algorithm based on the block-coordinate descent (BCD) optimization technique to find a high-quality sub-optimal approximate solution. The experimental results show that our proposed algorithm is superior to the baseline schemes.

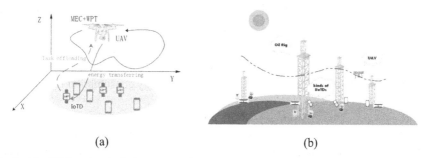

(a)   (b)

Fig. 1. The UAV-assisted wireless powered MEC system in 5G networks: (a) the proposed system model; (b) one feasible example of application scenarios.

2 System Model and Problem Formulation

In this paper, we propose a UAV-assisted wireless powered MEC system in 5G networks, as shown in Fig. 1(a), where the UAV is deployed to provide energy supplement and task offloading services for IoTDs. In addition, to illustrate the practicability and realistic feasibility of our research, we present the desert oil fields as a certain application scenario, as shown in Fig. 1(b).

Assumed that the UAV flies around at a fixed altitude H and hovers at M given positions, denoted by $q(m) = (X(m), Y(m), H)$, $m \in M = \{1, 2, ..., M\}$, while N IoTDs are distributed at the xy-plane, denoted by $w_k = (x_k, y_k, 0)$, $k \in K = \{1, 2, ..., K\}$. Each IoTD has an individual task $U_k = (I_k, C_k)$, where I_k is the amount of transmission data and C_k is the CPU cycles needed for computing. In addition, each IoTD can select one and only one hovering positon to get access to UAV. Therefore, we define $a_k(m) = \{0, 1\}$ as the connection indicator variables, where $a_k(m) = 1$ denotes that the IoTD connected with the UAV at m-th hovering position, otherwise, $a_k(m) = 0$.

The channel power gain between k-th IoTD and the UAV at m-th hovering position is $h_k(m) = \frac{h_0}{\|q(m) - w_k\|^2}$, where h_0 denotes the channel power gain at the reference distance as 1 m. By defining B as the channel bandwidth, p_k^{iotd} as the transmission power of k-th IoTD, and σ^2 as the noise power, the achievable transmission rate for k-th IoTD at m-th hovering position can be given as:

$$r_k(m) = B \log_2 \left(1 + \frac{p_k^{iotd} h_k(m)}{\sigma^2} \right) \tag{1}$$

2.1 Energy Consumption Based on Working Pattern

We assume p_k^l denotes the CPU cycles for task U_k in local processing, while p_k^o denotes the CPU cycles allocated to U_k in offloading computing. Therefore, the energy consumption for U_k is:

$$E_k^\zeta = \rho T_k^\zeta \left(f_k^\zeta \right)^v = \rho \frac{C_k}{f_k^\zeta} \left(f_k^\zeta \right)^v = \rho C_k \left(f_k^\zeta \right)^{v-1} \tag{2}$$

where $\zeta = \{l, o\}$, $\rho \geq 0$ denotes the effective switched capacitance. And $v \geq 1$ is a positive constant, which is seted as $v = 3$ in this paper refering to the realistic application.

In offloading execution mode, the data migration energy consumption for U_k is:

$$E_k^m(m) = p_k^{iotd} T_k^m(m) = p_k^{iotd} \frac{I_k}{r_k(m)} \tag{3}$$

Define $\eta_k = \{0, 1\}$ as the offloading indicator variable, where $\eta_k = 1$ denotes that k-th IoTD performs local execution mode, otherwise performs offloading execution mode. By adopting the linear energy transferring model, considering the local energy consumption of U_k (e.g., local processing and data offloading) all from the harvested energy, one can be obtained:

$$T_k^h(m) = \frac{E_k^h(m)}{\varphi^e h_k(m) p_k^e(m)} = \frac{(1 - \eta_k) E_k^l + \eta_k E_k^h(m)}{\varphi^e h_k(m) p_k^e(m)} \tag{4}$$

Where φ^e is the energy harvesting efficiency and $p_k^e(m)$ is the charging power allocated to k-th IoTD at m-th hovering position. Denoting P^H as the UAV hovering power and ϕ

as the weight, the weighted system energy consumption can be expressed as:

$$E = E^T + E^H = \sum_{k \in K} \sum_{m \in M} (1 - \eta_k)E_k^l + \eta_k \left(E_k^h(m) + E_k^o \right)$$
$$+ \phi P^H \sum_{k \in K} \sum_{m \in M} a_k(m) \left(T_k^h(m) + \eta_k T_k^m(m) \right) \tag{5}$$

2.2 Problem Formulation

Let $\boldsymbol{P}^e = \{p_k^e(m), \forall k \in K, \forall m \in M\}$, $A = \{a_k(m), \forall k \in K, \forall m \in M\}$ and $\eta = \{\eta_k, \forall k \in K\}$. With the aim of minimizing the weighted system energy consumption, the system process can be formulated as the following optimization problem:

$$\min_{\boldsymbol{P}^e, A, \eta} E \tag{6a}$$

$$s.t. \sum_{k \in K} a_k(m)p_k^e(m) \leq p_{max}, \quad \forall m \in M, \tag{6b}$$

$$p_k^e(m) \geq 0, \forall k \in K, \forall m \in M, \tag{6c}$$

$$\sum_{k \in K} a_k(m) = 1, \forall k \in K, \tag{6d}$$

$$a_k(m) = \{0, 1\}, \forall k \in K, \quad \forall m \in M, \tag{6e}$$

$$\eta_k = \{0, 1\}, \forall k \in K, \tag{6f}$$

3 Proposed Solution

3.1 Charging Resources Allocation Optimization

With fixed η and A, the sub-problem with regard to charging resources allocation is:

$$\min_{\boldsymbol{P}^e} \sum_{k \in K} \sum_{m \in M} a_k(m) \frac{D_k(m)}{p_k^e(m)}, \tag{7}$$
$$s.t. \quad (6b) \text{ and } (6c).$$

where $D_k(m) = \phi P^H \frac{(1-\eta_k)\rho C_k(f_k^l)^2 + \eta_k p_k^{iotd} T_k^m(m)}{\varphi^e h_k(m)}$ for simplification.

By taking the first derivative for the objective function of the sub-problem (7), it is quite clear that the sub-problem is convex. Therefore, we can obtain a high-quality solution by adopting the Lagrange dual method. Further, by denoting the Lagrange multipliers associated with constraints (6b) as $\mu = \{\mu_m \geq 0\}_{m \in M}$, the partial Lagrangian is given as:

$$\mathcal{L}(\boldsymbol{P}^e, \mu) = \sum_{k \in K} \sum_{m \in M} a_k(m) \frac{D_k(m)}{p_k^e(m)}$$

$$+ \sum_{m \in M} \mu_m \left(\sum_{k \in K} a_k(m) p_k^e(m) - p_{max} \right). \tag{8}$$

Then, the Lagrange dual problem of sub-problem (7) is given as:

$$\max_{\mu} \quad g(\mu) = \min_{P^e} \mathcal{L}(P^e, \mu), \tag{9}$$
$$s.t. \quad \mu_m \geq 0, \quad \forall m \in M.$$

Since the sub-problem satisfies the Slater's condition, problem (7) can be solved by tackling the dual problem (9) equivalently. For any given μ, $g(\mu)$ can be decoupled into $K \times M$ sub-problems:

$$\min_{P^e} \quad a_k(m) \frac{D_k(m)}{p_k^e(m)} + \mu_m a_k(m) p_k^e(m), \tag{10}$$
$$s.t. \quad (6c).$$

According to the monotonicity, the optimal solution of problem (10) are given as:

$$p_k^e(m)^* = \begin{cases} \sqrt{\frac{D_k(m)}{\mu_m}}, & \mu_m > 0, \\ p_{max}, & \mu_m = 0. \end{cases} \tag{11}$$

Then, by plugging in (11) the problem (9) can be decoupled into the M sub-problems and solved similarly:

$$\mu_m^* = \left(\frac{\sum_{k \in K} a_k(m) \sqrt{D_k(m)}}{p_{max}} \right)^2 = \frac{\sum_{k \in K} a_k(m) D_k(m)}{(p_{max})^2}. \tag{12}$$

Therefore, the optimal solution of sub-problem (7) can be obtained by:

$$p_k^e(m)^* = \arg \max_{P^e, \mu} g \left(p_k^e(m)^*, \mu_m^* \right)$$
$$= \begin{cases} p_{max} \sqrt{\frac{D_k(m)}{\sum_{k \in K} a_k(m) D_k(m)}}, & \mu_m > 0, \\ p_{max}, & \mu_m = 0. \end{cases} \tag{13}$$

3.2 IoTDs Associations Optimization

With any fixed η and P^e, the sub-problem of IoTDs associations optimization is:

$$\min_{A} \quad \sum_{k \in K} \sum_{m \in M} a_k(m) S_k(m), \tag{14a}$$

$$s.t. \quad 0 \leq a_k(m) \leq 1, \quad \forall k \in K, \forall m \in M, \tag{14b}$$
$$(6b) \text{ and } (6d).$$

where $S_k(m) = \phi P^H T_k^h(m) + (\phi P^H + p_k^{iotd})\eta_k T_k^m(m)$ for simplification.

Similar to the sub-problem (7), the convex sub-problem (14a and 14b) could be tackled by the Lagrange dual method. By denoting $\boldsymbol{\beta} = \{\beta_m \geq 0\}_{m \in M}$ as the Lagrange multipliers associated with constraints (6b), the partial Lagrangian is given as:

$$
\mathcal{L}(\boldsymbol{A}, \boldsymbol{\beta}) = \sum_{k \in K} \sum_{m \in M} a_k(m) S_k(m)
$$

$$
+ \sum_{m \in M} \mu_m \left(\sum_{k \in K} a_k(m) p_k^e(m) - p_{max} \right). \tag{15}
$$

It is quite apparently that function (15) is a linear combination of $a_k(m)$. Therefore, the optimal IoTDs associations and updating procedure of $\{\beta_m \geq 0\}_{m \in M}$ are given as:

$$
a_k(m)^* = \begin{cases} 1, & if \ \ m = \arg \max_{m \in M} S_k(m) + \beta_m p_k^e(m); \\ 0, & otherwise. \end{cases} \tag{16}
$$

$$
\beta_m^{(l+1)} = \left[\beta_m^{(l)} - \omega \left(\sum_{k \in K} a_k(m) p_k^e(m) - p_{max} \right) \right]^+. \tag{17}
$$

where $[x]^+ = \max\{x, 0\}$, l denotes the number of updating iterations of the Lagrange multipliers, and ω denotes the updating step-size sequence.

3.3 Working Mode Decisions Optimization

With obtained $\boldsymbol{A}$ and $\boldsymbol{P}^e$, the sub-problems of the working mode decisions is:

$$
\min_{\eta} \sum_{k \in K} \sum_{m \in M} \eta_k \left\{ \phi P^H a_k(m) \left(1 + p_k^{iotd} B_k(m) \right) T_k^m(m) + E_k^o \right. \tag{18a}
$$

$$
\left. - \phi P^H a_k(m) B_k(m) E_k^l \right\},
$$

$$
s.t. \ \ 0 \leq \eta_k \leq 1, \quad \forall k \in K, \tag{18b}
$$

Algorithm 1 The integral optimization algorithm

1: Initialize $(\boldsymbol{P}^e)^0, (\boldsymbol{A})^0, (\boldsymbol{\eta})^0$ and define r = 1.
2: **repeat**
3:    Use (13) to obtain $(\boldsymbol{P}^e)^r$.
4:    **repeat**
5:        Obtain $(\boldsymbol{A})^r$ according to (16).
6:        Update $\{\beta_m \geq 0\}_{m \in M}$ based on (17).
7:    **until** The objective function (14) converge.
8:    Use CVX tool box to obtain $(\boldsymbol{\eta})^r$.
9:    Set r = r + 1.
10: **until** The objective function (8) converge.
11: **return** $(\boldsymbol{P}^e)^*, (\boldsymbol{A})^*, (\boldsymbol{\eta})^*$.

---

where $B_k(m) = 1 + \frac{1}{\varphi^e h_k(m) p_k^e(m)}$ for simplification.

Apparently, the sub-problem is a linear problem which can be effectively tackled by the well-constructed optimization toolbox such as CVX. Fortunately, for any given A and P^e, the working mode decisions depends on the achievable minimum weighted system energy consumption. In other words, $\eta_k = 1$ if and only if the coefficient of sub-problem (18a and 18b) is minimum, otherwise, $\eta_k = 0$.

4   Experimental Results and Analysis

In this section, we compare our proposed algorithm with several benchmark schemes and the experimental results demonstrate the performance of our proposed algorithm, presented as *Algorithm 1*, in comparison with these benchmarks. We suppose that there are $K = 50$ IoTDs distributed within the 1 km^2 area, while the UAV hovers $M = 3$ times at $H = 40$ m. Moreover, I_k and C_k are set as 100 KBits and 1 MHz, respectively. The maximum computation and charging capacity of the UAV are set as 2 MHz and 0.1 W respectively. We set $h_0 = -50$ dB, $B = 1$ MHz, and $\sigma^2 = -60$ dBm. The transmission power and computation of IoTDs are set as 2.83 mw and 1.2–1.6 MHz, respectively. We set the energy conversation effiency $\varphi^e = 80\%$ and the computation effective $\rho = 10^{-27}$. The UAV hovering power P^H is set as 59.2 W and the weight is set as 3×10^{-11}.

We set the schemes of random selection and random offloading as our benchmarks, where the random selection scheme shows that the IoTDs get access to the UAV at any hovering position randomly, while the random offloading scheme shows that the IoTDs choose their working mode randomly. Figure 2 shows the energy-effectiveness of our proposed algorithm by compared with above two benchmarks.

Figure 2(a) and Fig. 2(b) illustrate the weighted system energy consumption versus the number of IoTDs in the target area N and the maximum charging capacity of the UAV.

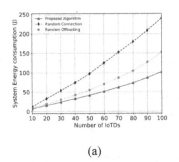
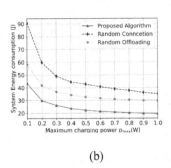

| (a) | (b) |

Fig. 2. The system energy consumption versus: (a) the number of IoTDs; (b) the maximum charging capacity of the UAV and p_{max}, respectively.

In Fig. 2, we can see that with the increase of both N and p_{max}, the weighted system energy consumption increases as expected. In Fig. 2(a), one can be seen that the pace of change accelerated as the number of IoTDs rose while the proposed algorithm is superior

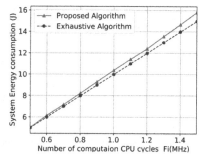

Fig. 3. The gap between the proposed algorithm and optimal solution with the increasing of the number of required task computation CPU cycles

with the lowest slope. In Fig. 2(b), we can observe that the pace of change decelerated as the maximum charging power of UAV rose, because of the diminishing energy harvesting resource competition for IoTDs. One can also be seen that when the maximum charging capacity of the UAV increase to a larger value, the decreasing variation of the weighted system energy consumption becomes small, due to the less time consumption on energy harvesting and smaller charging resource competition, similarly. The figures show that our proposed algorithm is superior to the benchmarks on the energy conservation.

Figure 4 illustrates the weighted system energy consumption versus the number of required task computation CPU cycles C_k, in which we set the computation capacity of IoTDs f_k^l fixed as 1.6 MHz. In Fig. 4, we compare the proposed algorithm with the exhaustive scheme, which is considered as the optimal solution in general. We can see that the two algorithm are close in performance, which demonstrates that our proposed algorithm can get a high-quality approximate optimal solution. In fact, the principle of the exhaustive search scheme is searching all the feasible solution under the constraints, which is inefficient obviously. Above all, our algorithm is effective and efficient with much less complexity.

5 Conclusion

In this paper, we propose UAV-assisted wireless powered MEC system in 5G networks, in which the UAV is equipped to provide energy supplement and assisting IoTDs in task processing. Specifically, we formulate an optimization problem with the goal of minimizing the weighted system energy consumption to describing the system process. Due to the non-convexity of the formulated problem, we decouple it into several sub-problems and design an effective iterative algorithm to obtain the approximate optimal solution. Experimental results validate that our algorithm can enhance the system performance in energy consumption compared with the benchmarks.

References

1. Ahmed, S., Chowdhury, M.Z., Jang, Y.M.: Energy-efficient UAV relaying communications to serve groundnodes. IEEE Commun. Lett. **24**(4), 849–852 (2020)

2. Cao, X., Yang, P., Alzenad, M., Xi, X., Wu, D., Yanikomeroglu, H.: Airborne communication networks: a survey. IEEE J. Sel. Areas Commun. **36**(9), 1907–1926 (2018)
3. Fan, R., Cui, J., Jin, S., Yang, K., An, J.: Optimal node placement and resource allocation for UAV relaying network. IEEE Commun. Lett. **22**(4), 808–811 (2018)
4. Huang, K., Zhang, Q., Zhou, C., Xiong, N., Qin, Y.: An efficient intrusion detection approach for visual sensor networks based on traffic pattern learning. IEEE Trans. Syst. Man Cybern. Syst. **47**(10), 2704–2713 (2017)
5. Ju, X., et al.: An energy conserving and transmission radius adaptive scheme to optimize performance of energy harvesting sensor networks. Sensors **18**(9), 2885 (2018)
6. Li, B., Fei, Z., Zhang, Y.: Uav communications for 5g and beyond: Recent advances and future trends. IEEE Internet Things J. **6**(2), 2241–2263 (2019)
7. Liao, Z., et al.: Distributed probabilistic offloading in edge computing for 6g-enabled massive internet of things. IEEE Internet of Things J. **8**(7), 5298–5308 (2021)
8. Liu, Y., Xiong, K., Ni, Q., Fan, P., Letaief, K.B.: Uav-assisted wireless powered cooperative mobile edge computing: joint offloading, CPU control, and trajectory optimization. IEEE Internet Things J. **7**(4), 2777–2790 (2020)
9. Mei, H.B., Wang, K.Z., Zhou, D.D., Yang, K.: Joint trajectory-task-cache optimization in UAV-enabled mobile edge networks for cyber-physical system. IEEE Access **7**, 156476–156488 (2019)
10. Mei, W., Wu, Q., Zhang, R.: Cellular-connected UAV: uplink association, power control and interference coordination. IEEE Trans. Wireless Commun. **18**(11), 5380–5393 (2019)
11. Sun, L., Wan, L., Liu, K., Wang, X.: Cooperative-evolution-based WPT resource allocation for large-scale cognitive industrial IoT. IEEE Trans. Industr. Inf. **16**(8), 5401–5411 (2020)
12. Tang, Q., Wang, K., Song, Y., Li, F., Park, J.H.: Waiting time minimized charging and discharging strategy based on mobile edge computing supported by software-defined network. IEEE Internet Things J. **7**(7), 6088–6101 (2020)
13. Tang, Q., Wang, K., Yang, K., Luo, Y.: Congestion-balanced and welfare-maximized charging strategies for electric vehicles. IEEE Trans. Parallel Distrib. Syst. **31**(12), 2882–2895 (2020)
14. Wan, S., Lu, J., Fan, P., Letaief, K.B.: Toward big data processing in IoT: path planning and resource management of UAV base stations in mobile-edge computing system. IEEE Internet Things J. **7**(7), 5995–6009 (2020)
15. Wang, J., et al.: LogEvent2vec: LogEvent-to-vector based anomaly detection for large-scale logs in Internet of Things. Sensors **20**(9), 2451 (2020)
16. Wang, J., et al.: A probability preferred priori offloading mechanism in mobile edge computing. IEEE Access **8**, 39758–39767 (2020)
17. Wu, Q., Xu, J., Zhang, R.: Capacity characterization of UAV-enabled two-user broadcast channel. IEEE J. Sel. Areas Commun. **36**(9), 1955–1971 (2018)
18. Yang, Y., Xiong, N., Chong, N.Y., Défago, X.: Joint resources and workflow scheduling in UAV-enabled wirelessly-powered MEC for IoT systems. IEEE Trans. Veh. Technol. **68**, 10187–10200 (2019)
19. Yang, Y., Xiong, N., Chong, N.Y., Defago, X.: A decentralized and adaptive flocking algorithm for autonomous mobile robots. In: International Conference on Grid and Pervasive Computing Workshops (2008)
20. Yu, Z., Gong, Y., Gong, S., Guo, Y.: Joint task offloading and resource allocation in UAV-enabled mobile edge computing. IEEE Internet Things J. **7**(4), 3147–3159 (2020)
21. Zhang, Q., Huang, T., Zhu, Y., Qiu, M.: A case study of sensor data collection and analysis in smart city: provenance in smart food supply chain. Int. J. Distrib. Sens. Netw. **9**(11), 382132 (2013)
22. Zhang, S.H., Zhang, H.L., Di, B.Y., Song, L.Y.: Cellular UAV-to-X communications: design and optimization for multi-UAV networks. IEEE Trans. Wireless Commun. **18**(2), 1346–1359 (2019)

23. Zhang, Q., Zhou, C., Xiong, N., Qin, Y., Li, X., Huang, S.: Multimodel-based incident prediction and risk assessment in dynamic cybersecurity protection for industrial control systems. IEEE Trans. Syst. Man Cybern. Syst. **46**(10), 1429–1444 (2017)

24. Zhao, C., Cai, Y., Liu, A., Zhao, M., Hanzo, L.: Mobile edge computing meets mmwave communications: joint beamforming and resource allocation for system delay minimization. IEEE Trans. Wirel. Commun. **19**(4), 2382–2396 (2020)

25. Zhou, F., Wu, Y., Hu, R.Q., Qian, Y.: Computation rate maximization in UAV-enabled wireless-powered mobile-edge computing systems. IEEE J. Sel. Areas Commun. **36**(9), 1927–1941 (2018)

Privacy-Preserving Accelerated Clustering for Data Encrypted by Different Keys

Jun Zhang$^{(\boxtimes)}$, Yuanyuan Wang, Bing Li, and Shiqing Hu

Shenzhen University, Shenzhen, Guandong Province, People's Republic of China
{zhangjuniris,yywang,libingice}@szu.edu.cn

Abstract. k-Means is a widely used clustering algorithm that divides data points into k groups. Previous studies focus on implementing secure k-Means clustering either for single user or for multiple users. The common way is to operate on outsourced and encrypted datasets in the cloud. Two difficulties exist in this procedure - how to handle with different keys and how to compute distance, comparison and division on the ciphertexts. On the other hand, k-Means algorithm includes many unnecessary distance calculations. In this paper, we aim to construct privacy-preserving protocol which accelerates k-Means algorithm for data encrypted by different keys in the cloud. Experiments show that our scheme is effective.

Keywords: Privacy protection · Cloud computing · Accelerated clustering · Homomorphic encryption · Multiple keys · Secure protocol

1 Introduction

Lloyd's algorithm [6] is the standard way to conduct k-Means clustering, which is an iterative process. Initially, the algorithm selects k data records as the cluster centers. One round of clustering consists of three steps - (i) assignment (ii) update (iii) termination. In the assignment step, the algorithm calculates the Euclidean distance between each data record and k cluster centers. Then the algorithm assigns the data record to the cluster which has the minimum Euclidean distance. In the update step, centers of the clusters are updated through computing the new means of all the data records in the same clusters. In the termination step, the algorithm verifies whether a pre-defined termination condition holds. Despite of the popularity of Lloyd's algorithm, most distance calculations are redundant. Utilizing triangle inequality, lower and upper bounds, acceleration schemes are proposed [1–4]. Elkan [3] keeps 1 upper bound and k lower bounds. Elkan's algorithm runs very fast for high-dimensional data. Hamerly [4] uses only 1 lower bound for each data point. Hamerly's algorithm is better than Elkan's algorithm on low and medium datasets, for example, less than 50 dimensions. Drake and Hamerly [4] choose b lower bounds ($1 < b < k$). This algorithm is faster than

© Springer Nature Switzerland AG 2021
M. Qiu (Ed.): SmartCom 2020, LNCS 12608, pp. 156–162, 2021.
https://doi.org/10.1007/978-3-030-74717-6_17

other methods on datasets of $20 - 120$ dimensions. Ding et al. [1] maintain three filters (local, global and group) and accelerate both the assignment and center update steps.

There exist different implementations for privacy-preserving k-means clustering in the cloud. In most cases, data records are encrypted to guarantee data security and privacy. Current studies can be generally classified into two categories - single user [5,9,11] and multiple users [7,8,10]. With one single key, Liu et al. [5] rely on fully homomorphic encryption without bootstrapping and have to keep an updatable distance matrix. Yuan et al. [11] integrate MapReduce framework into their design to support large-scale datasets. One disadvantage of their system is that data owner are responsible for updating cluster centers in each iteration. The scheme in [9] reduces the workload of the data owner based on a trusted server, which is a strong assumption. Although the solutions in [7,10] can cope with multiple users, they are single-key schemes in essence. All the users use the same pair of public and secret keys. Nobody is permitted to own the secret key, as revealing it to one user equals to disclosing datasets of other users. Rong et al. [8] allow the data users to encrypt data with their own keys. To handle with different keys, they need to transform ciphertexts under different keys into a common key. The underlying encryption method they use is multiplicatively homomorphic. Their secure addition protocol is too complicated. The theoretical computational cost of secure addition is high, which also leads to expensive secure distance computation. As far as we know, all the existing secure protocols for k-means clustering focus on Lloyd's algorithm. Even though the researchers already tried to pick up efficient homomorphic encryption to reduce computational complexity or communication overhead, they overlook the importance of designing schemes on accelerated algorithms.

Considering the acceleration algorithms in [1–4] include common operations such as distance computation and comparison. In this paper, we propose secure distance computation and comparison protocols for accelerated clustering on encrypted data. Our contributions can be summarized as follows.

- Our scheme support multiple users and facilitate accelerated clustering on data encrypted by different keys.
- We construct secure protocols for distance computation and comparison.
- We conduct experimental evaluation and the results show the efficiency of our scheme.

The rest of this paper is organized as follows. Section 2 presents some preliminary information. Section 3 introduces our system model and threat model. The details of our scheme are clarified in Sect. 4. Then we perform experimental evaluation in Sect. 5. Finally, we conclude this paper in Sect. 6.

2 Preliminaries

2.1 k-Means Clustering

Given that we have a dataset $\{x_i\}_{i=1}^n$ where $x_i \in R^m$, k-Means clustering aims to partition n data records into k disjoint sets $C = \{C_1, C_2, \cdots, C_k\}$, each of which has a center. The objective is to minimize within-cluster sum of squares.

$$\arg \min_C \sum_{i=1}^{k} \sum_{x \in C_i} ||x - \mu_{c_i}||^2 \tag{1}$$

where μ_{c_i} is the mean of the records in C_i.

Definition 1. *Let $C_i = \{x_1, \cdots, x_t\}$ be a cluster where $x_1, \cdots, x_t$ are data records with m attributes. The center of cluster C_i is defined as a vector μ_{c_i}:*

$$\mu_{c_i}[j] = \frac{x_1[j] + \cdots + x_t[j]}{|C_i|}, \qquad 1 \le j \le m \tag{2}$$

where $x_i[j]$ is the jth attribute of data record x_i and $|C_i|$ denotes the number of data records in cluster C_i.

2.2   Avoiding Divisions

The cluster center μ_{c_i} is a vector of which the attributes may be rational numbers. However, the existing encryption schemes only support integer values. Therefore, the cluster center μ_{c_i} is represented as $(sum_i, |C_i|)$, where sum_i and $|C_i|$ denote the sum vector and the number of data records in cluster C_i, respectively.

Definition 2. *Given a data record x_i and cluster C_i with cluster center μ_{c_i}. The Euclidean distance between x_i and μ_{c_i} is computed as*

$$||x_i - \mu_{c_i}|| = \sqrt{\sum_{j=1}^{m}(x_i[j] - \mu_{c_i}[j])^2} \tag{3}$$

To avoid division operations, the above Euclidean distance is scaled and squared:

$$
\begin{aligned}
Distance_{ND} &= (||x_i - \mu_{c_i}|| \cdot |C_i|)^2 \\
&= \left(\sqrt{\sum_{j=1}^{m}(x_i[j] - \frac{sum_i[j]}{|C_i|})^2} \cdot |C_i|\right)^2 \\
&= \sum_{j=1}^{m}(|C_i| \cdot x_i[j] - sum_i[j])^2
\end{aligned} \tag{4}
$$

3   Model Description

3.1   System Model

Our system model consists of data users and two cloud servers. The data users encrypt their databases with their own keys and outsource the encrypted databases to the cloud. The two cloud servers are non-colluding and represented by S_1 and S_2. A trusted party distributes public and private keys in our system.

3.2 Threat Model

We assume that all parties in our system are semi-honest. All the participants will follow the protocol step by step honestly, but they are curious about the true value of encrypted data. They try to figure out the plaintexts of encrypted data by observing the input, intermediate results or final output. The semi-honest assumption makes sense in real applications.

4 Our Scheme

4.1 Secure Distance Computation

We utilize the cryptosystem $\mathcal{E}_{BCP}$ proposed in [12]. Assume that we have one encrypted record $\mathcal{E}_{BCP}(x_1) = (x_1 - b_1, E_a(b_1))$ from user Alice encrypted by key s_a, and another encrypted record $\mathcal{E}_{BCP}(x_2) = (x_2 - b_2, E_b(b_2))$ from user Bob encrypted by key s_b, the Secure Squared Euclidean distance (SSED) is computed as Eq. (5), where $\mathcal{E}_{BCP}(x_1 - x_2) = [(x_1 - x_2) - (b_1 - b_2), E_a(b_1 - b_2)]$.

$$
\begin{aligned}
SSED(x_1, x_2) =& \mathcal{E}_{BCP}(x_1 - x_2) \times \mathcal{E}_{BCP}(x_1 - x_2) \\
=& E_{ab}[(x_1 - x_2)^2 - (b_1 - b_2)^2]
\end{aligned}
\tag{5}
$$

4.2 Secure Comparison

Given two encrypted squared Euclidean distances such as $SSED(x_1, x_2)$ and $SSED(x_1, x_3)$, the goal of our secure comparison protocol is to obtain an encrypted indicator $E(u)$ to show whether $SSED(x_1, x_2) \geq SSED(x_1, x_3)$ or $SSED(x_1, x_2) < SSED(x_1, x_3)$. If $u = 0$, it shows $SSED(x_1, x_2) \geq SSED(x_1, x_3)$; and if $u = 1$, $SSED(x_1, x_2) < SSED(x_1, x_3)$.

Firstly, we remove the random item and complete partial decryption from SSED through cooperation between S_1 and S_2. For example, $SSED^*(x_1, x_2)$ is derived from $SSED(x_1, x_2)$ and $SSED^*(x_1, x_2) = E_a(x_1 - x_2)^2$. S_1 calculates $E(d_{12})$ and $E(d_{13})$.

$$
\begin{aligned}
E(d_{12}) &= E(2SSED^*(x_1, x_2) + 1) \\
E(d_{13}) &= E(2SSED^*(x_1, x_3))
\end{aligned}
\tag{6}
$$

Next, S_1 flips a coin c randomly. If $c = 1$, S_1 computes $E(l) = E(d_{12} - d_{13})$. If $c = 0$, S_1 computes $E(l) = E(d_{13} - d_{12})$. We use $\mathcal{L}(x)$ to denote the bit-length of x. S_1 then chooses a random number r subject to $\mathcal{L}(r) < \mathcal{L}(N)/4$ and calculates $E(l') = E(l)^r$. S_1 partially decrypt $E(l')$ and sends the result to S_2.

S_2 decrypts and gets l'. If $\mathcal{L}(l') > \frac{N}{2}$, S_2 sets $u^* = 1$. Otherwise, S_2 sets $u^* = 0$. Then S_2 encrypts u^* and sends $E(u^*)$ back to S_1.

Once $E(u^*)$ is received, S_1 returns $E(u) = E(u^*)$ as the final output if $c = 1$. Otherwise, S_1 returns $E(u) = E(1 - u^*)$.

REMARK The secure comparison protocol can be easily extended to compute secure minimum among k squared distances (SMIN$_k$). Assume that S_1 holds a set of encrypted squared Euclidean distances $SSED(x_i, c_1), SSED(x_i, c_2), \cdots, SSED(x_i, c_k)$, we calculate the minimum by invoking the secure comparison protocol with two inputs each time sequentially.

4.3   Privacy Preserving Accelerated Clustering

Given homomorphic addition, secure distance computation and secure comparison, privacy-preserving accelerated clustering can be constructed. To be specific, secure distance computation is obviously a must in clustering algorithm as distance is chosen as the similarity metric. For distance comparison between lower or upper bound, our secure comparison protocol works. As for updating lower and upper bounds, homomorphic addition can complete this task. We do not list concrete algorithm details here. For calculating new cluster center, we also describe how to avoid divisions in Sect. 2.2. Readers can select appropriate acceleration algorithm according to the statistics of their dataset and design their own privacy preserving accelerated clustering based on our building blocks proposed in this paper.

4.4   Security Analysis

The underlying cryptosystem $\mathcal{E}_{BCP}$ is already proved secure in [12]. Our secure distance computation and secure comparison protocols run on encrypted data. As long as the underlying encryption is secure, the secure protocols we proposed are privacy-preserving as well. Otherwise, it contradicts with the fact of $\mathcal{E}_{BCP}$'s semantic security. The security of our scheme can be proved using the Real and Ideal Paradigm and Composition Theorem. The main idea is to use a simulator in the ideal world to simulate the view of a semi-honest adversary in the real world. If the view in the real world is computationally indistinguishable from the view in the ideal world, the protocol is secure. Due to space limit, we do not go into details here.

5   Experimental Evaluation

The configuration of our PC is Windows 10 Enterprise 64-bit Operating System with Intel(R) Core(TM) i5-7500 CPU (4 cores), 3.41 GHz and 16 GB memory. To provide platform independence, we use Java to implement our scheme. We use three security parameters - 1024-bit, 1536-bit and 2048-bit. We record the time to run secure distance computation and the results are shown in Table 1. Moreover, we also test the performance of secure comparison. On average, for 1024-bit parameter, it takes 0.045 s. For 1536-bit parameter, the execution time is 0.139 s. For 2048-bit parameter, the running time is 0.322 s.

Table 1. Time for secure distance computation under different parameters (**Second**)

Dimension	1024-bit	1536-bit	2048-bit
10	0.029 s	0.018 s	0.015 s
20	0.041 s	0.046 s	0.049 s
30	0.063 s	0.078 s	0.091 s
40	0.122 s	0.131 s	0.132 s
50	0.131 s	0.177 s	0.193 s
60	0.146 s	0.205 s	0.294 s
70	0.159 s	0.298 s	0.415 s
80	0.180 s	0.353 s	0.562 s
90	0.201 s	0.552 s	0.709 s
100	0.253 s	0.599 s	0.895 s

6   Conclusions

In this paper, we put forward privacy-preserving accelerated clustering. Compared to traditional k-means clustering, our method have great advantage by avoiding unnecessary distance computations. Our scheme handle with multiple users and operate on dataset encrypted under different keys. We construct building blocks for secure accelerated clustering, which include secure distance computation and secure comparison. The security of our scheme relies on the underlying cryptosystem, of which the security is already proved. The experimental evaluation validates the efficiency of our scheme.

References

1. Ding, Y., Zhao, Y., Shen, X., Musuvathi, M., Mytkowicz, T.: Yinyang k-means: a drop-in replacement of the classic k-means with consistent speedup. In: International Conference on Machine Learning, pp. 579–587 (2015)
2. Drake, J., Hamerly, G.: Accelerated k-means with adaptive distance bounds. In: 5th NIPS Workshop on Optimization for Machine Learning, vol. 8 (2012)
3. Elkan, C.: Using the triangle inequality to accelerate k-means. In: International Conference on Machine Learning (ICML), pp. 147–153 (2003)
4. Hamerly, G.: Making k-means even faster. In: SIAM International Conference on Data Mining, pp. 130–140 (2010)
5. Liu, D., Bertino, E., Yi, X.: Privacy of outsourced k-means clustering. In: 9th ACM Symposium on Information, Computer and Communications Security, pp. 123–134 (2014)
6. Lloyd, S.: Least-Squares Quantization in PCM. IEEE Trans. Inf. Theor. **28**(2), 129–137 (1982)
7. Rao, F.Y., Samanthula, B.K., Bertino, E., Yi, X., Liu, D.: Privacy-preserving and outsourced multi-user k-means clustering. In: IEEE Conference on Collaboration and Internet Computing (CIC), pp. 80–89 (2015)

8. Rong, H., Wang, H., Liu, J., Hao, J., Xian, M.: Privacy-preserving-means clustering under multiowner setting in distributed cloud environments. Secur. Commun. Networks **2017**, (2017)

9. Sakellariou, G., Gounaris, A.: Homomorphically encrypted k-means on cloud-hosted servers with low client-side load. Computing **101**(12), 1813–1836 (2019)

10. Wu, W., Liu, J., Wang, H., Hao, J., Xian, M.: Secure and efficient outsourced k-means clustering using fully homomorphic encryption with ciphertext packing technique. IEEE Trans. Knowl. Data Eng. (2020)

11. Yuan, J., Tian, Y.: Practical privacy-preserving mapreduce based k-means clustering over large-scale dataset. IEEE Trans. Cloud Comput. **7**(2), 568-579 (2017)

12. Zhang, J., He, M., Zeng, G., Yiu, S.M.: Privacy-preserving verifiable elastic net among multiple institutions in the cloud. J. Comput. Secur. **26**(6), 791–815 (2018)

Blockchain Technology in Automobile Insurance Claim Systems Research

Han Deng[1], Chong Wang[2(⊠)], Qiaohong Wu[3], Qin Nie[1], Weihong Huang[4],
and Shiwen Zhang[4]

[1] Big Data Development and Research Center, Guangzhou College of Technology
and Business, Guangzhou 528138, China
[2] College of Computer Science and Electronic Engineering, Hunan University,
Changsha 410000, China
f2010w0139@hnu.edu.cn
[3] Department of Accounting, Guangzhou College of Technology and Business, Guangzhou
528138, China
[4] School of Computer Science and Engineering, Hunan University of Science and Technology,
Xiangtan 411201, China
shiwenzhang@hnust.edu.cn

Abstract. Recently, academics and industries are increasingly paying attention to blockchain technology because it is used to improve traditional companies. Insurance company is one type of the traditional and inflexible companies. Its operation is not transparent, based on paper contracts, relying on human intervention and other characteristics resulting in low efficiency. The transparency, tamper-proof, decentralization and other advantages of blockchain-based Internet of Things and smart contract technology enable people to improve the traditional processes and structures of the insurance industry. This paper summarizes the application of block chain technology in the auto insurance claim systems. By comparing the implementations of various systems, the existing achievements and their advantages and disadvantages are described. In addition, this paper prospects the key directions of future research on the implementation of blockchain.

Keywords: Blockchain · Internet of Things · Smart contract · Auto insurance · Vehicle system

1 Introduction

The essence of the auto insurance industry is to collect information, analyze information, and make rates or claims based on the information. Although insurance companies have been trying to establish a perfect information management system with various digital technologies, the whole industry still has the following problems due to the way of information storage and exchange:

The more roles involved, the lower the efficiency: the roles that may be involved in the traffic accident includes policyholders of both vehicles, insurers of both vehicles, carmakers of both parties, government authorities and vehicle damage assessment agencies.

© Springer Nature Switzerland AG 2021
M. Qiu (Ed.): SmartCom 2020, LNCS 12608, pp. 163–172, 2021.
https://doi.org/10.1007/978-3-030-74717-6_18

And with the spread of smart vehicles, car crashes can also involve providing Internet of vehicles service providers. Information asymmetry between multiple actors leads to mutual distrust, which requires huge labor costs to reach a consensus and is prone to fraud in the process. Insurance clause design: Unreasonable content of clauses and the insurance rate lacking personalization can't fully meet customers' needs. Claims process problem: Relying on manual verification processes which are cumbersomely expensive and extremely depending on the subjective judgment of investigators accounting for 25% of insurance companies' costs [1].

Some insurance companies hope blockchain technology will transform the traditional insurance industry. As the core technology of distributed ledger, blockchain can be used to securely store and share pure digital records. By using the knowledge of distributed system, cryptography, game theory, network protocol and other disciplines, blockchain system has the following advantages: ① Traceability of information ② Encryption makes the created record untamable ③ Asymmetric encrypted digital signatures protect privacy ④ Decentralized, no longer dependent on the central server has a strong resistance to aggression ⑤ Automated transactions, managed payments and contract term modifications can be programmed into software programs to automate them. In the field of auto insurance, blockchain technology can simultaneously act on two aspects: one is vehicle information monitoring system based on the Internet of things; the other is intelligent contract to improve efficiency.

Blockchain 3.0, Smart networking, has gone beyond finance to provide a decentralized system for industries. In 2017, Sharma proposed DisBlockNet, a distributed blockchain security architecture based on blockchain [2]. The difference between cars and factory machines is that vehicle may be moving at a high speed, so network design is important in workshop communication. CAN [3] signal packaging algorithm improves bandwidth utilization. With the development of intelligent vehicles, Oham proposed a responsibility attribution framework for fully autonomous vehicles [4]. The sensors on the car provide reliable driving data efficiently and stored in the block chain, making it easy for intelligent contracts to trace the cause and make decisions when accidents and disputes occur. The data collected by the sensors could be stored and processed in cloud servers [5, 6].

The concept of smart contracts was first proposed in 1994. The contract is designed to be secure, efficient and automatic without relying on third parties. The blockchain technology is naturally compatible with the intelligent contract, which can seal the intelligent contract loader to complete the complex behavior of distributed nodes [7]. An intelligent contract can effectively deploy an infinite state of emergency through If-Then and automatically complete the process once the event occurs. In the first quarter of 2016 alone, $116 million was invested in smart contract venture-related deals [8]. In 2017, a German insurer built Etherisc based on a smart contract, focusing on using blockchain to improve efficiency [9].

Therefore, this paper aims to research the automatic auto insurance claim system based on block chain and Internet of Things. Auto insurers can use the data in the blockchain to reduce customer acquisition costs, accelerate understanding of customer history information and significantly improve fraud-prevention capabilities. At the same time, smart contracts can cut operating costs. In general, block chain auto insurance companies in the whole region compared with traditional companies can be reduced by about 10%.

The organizational structure of this paper is as follows: The first section outlines the corresponding background and related technology. The second section introduces the design concept and hierarchical structure of the current mainstream system. In the third section, the detailed implementation of various systems is compared. The traditional implementation methods and relatively new technologies in the field are compared. In the fourth section, the existing problems of intelligent systems are analyzed, and the future development trend of this kind of system is prospected. The fifth section summarizes the content of this paper.

2   Related Work Research

In order to achieve transparency, simplify procedures, reduce costs, improve efficiency and prevent fraud, the automation system should have the following key points: ① Ordinary vehicles will be transformed into intelligent vehicles that can communicate in real time and have strong anti-jamming sensors that can automatically collect vehicle and surrounding data. This way would provide real-time and reliable data stored in the blockchain. ② Turn the central server into a distributed ledger to connect all the roles. Manufacturers, system service providers, vehicles, insurance companies and relevant government regulators all have equal access to the data. This not only improves the reliability of data, but also reduces the communication costs of multiple actors to facilitate traceability of data and reach consensus. ③ Transform the traditional contract into an intelligent contract, which can automatically reach consensus and automatically complete the transfer of claim amount according to the contract. And it can improve the degree of innovation and individuation of insurance terms. ④ Transform the paper money into Cryptocurrency. On the one hand, it is more consistent with the blockchain framework; on the other hand, it can improve the ability of privacy protection.

Therefore, according to the different overall design ideas, the automation systems proposed in the 14 articles are divided into three categories:

1) The policyholder does not directly contact the block chain, and the data transmission needs to be through the insurance agent. The main reason is considering the low throughput of blockchain and the limitation of on-board equipment. The advantage lies in lower requirement to vehicle network, the disadvantage lies in policy holders which are in weak position and insurance agents have bigger right. The structure is shown in Fig. 1:

2) With the enhancement of the workshop communication network and the popularization of the roadside unit (RSU), this system emphasizes that the vehicle itself can serve as a node. The vehicle-mounted system can collect data independently, encrypt data and send data in real time, and keep complete ledger or some key information locally. The advantages lie in the improvement of data authenticity, At the same time the disadvantages lie in the extremely high requirements for on-board network and sensors. Malicious behavior may occur if the device is damaged due to network delay or collision, or data authenticity is damaged due to signal detection attacks. The structure is shown in Fig. 2

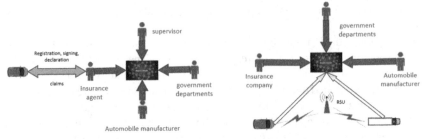

Fig. 1. Type one   **Fig. 2.** Type two

3) Double-chain design of supervised block chain/trading block chain (RBC/TBC) is adopted. ① Regulatory block chain: It is mainly composed of relevant regulatory departments and legal departments. The main function is to keep complete transaction data, provide intelligent contracts designed by insurance companies for trading blockchain, and solve issues such as transaction disputes. ② Transaction block chain: It is mainly used to store vehicle data, such as the location, speed, environment, and running state of vehicle internal parts, including maintenance records of vehicle maintainers, etc. The stored data will be provided to the monitoring blockchain for the determination of smart contracts. The advantage is to simplify the application structure and use different hardware facilities according to the emphasis of the chain. And the two chains can be parallel operation, the efficiency of operation has been improved. It has good scalability. The disadvantage is that the double chain coordination will appear certain problems. The specific framework design is shown in Fig. 3:

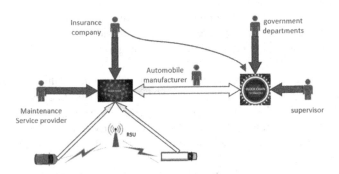

Fig. 3. Double chain design

3   Key Technology Contrast

In the previous section, mainstream systems were divided into 3 categories according to the differences of overall framework design. This section will select four key technical details of blockchain to compare the advantages and disadvantages of each framework in detail and compare them with relatively new technologies in the industry.

3.1 Public or Permissioned

CAIPY [1] use a common block chain to implement its own framework. Its trait is ① true decentralization, block records will not be privately held by the insurance company. ② The data is open, so anyone can theoretically have access to the vehicle block records. When it comes to privacy protection, a strong encryption system is needed. ③ Low throughput, high vehicle ownership rate, large data volume. Public blockchain cannot be timely packaged. The R3CEV report points out that public blockchain cannot be used as a solution for financial institutions.

B4F [14], B-FICA [4] adopted permissioned blockchain. Features: ① Being owned by a part of the organization is not really decentralized. ② The transaction speed is fast, and the consensus algorithm selection is more flexible compared with the common block chain. This part will be compared in detail in point 3. ③ Better privacy protection and high information security. ④ Helps to protect the insurance company's products from being damaged.

[22] two models of vehicle off-line detection framework are proposed. The authors hope that not only will cars and government traffic police be able to record data, but pedestrians will also be rewarded by offering real rewards [16]. It is also mentioned that users can upload records by taking photos (Table 1).

Table 1. Three types contrast

	Advantage	Disadvantage	Foreground
Inaccessible blockchain for policy holders [10–12]	Less demanding in network, because the data is stored in the black boxes	Insured person at a disadvantage Non-dynamic distributed ledger keeping	The model conception is in its infancy, Hard to commercialize
Vehicle as a separate node [9, 14–20]	Real-time data collection and packaging Information reliability	Shop floor communication Data storage problem Package node selection Throughput problem Consensus problem	Commercialized on a small scale
RBC/TBC double chain [4, 21]	Simplified structure Improved parallel computing High scalability	Double chain coordination problem	High probability

3.2   Node Selection and Consensus Algorithm

Traditional blockchain is faced with high overhead and low scalability. The increasing number of vehicles is a severe challenge to the blockchain framework. It is one of the most critical links for all systems to reach consensus quickly by selecting consensus algorithms [26].

Common consensus algorithms in the public block chain are ① PoW (proof of work). Waste of environment and power is extremely unlikely to be adopted by vehicle network [23]. ② PoS (proof of work)The advantage is energy saving, attackers need to pay a very high price; The disadvantage is that there is no issue of interest attack. ③ dPoW. Only blockchain using PoW and PoS can be adopted. Common in the permissioned blockchain are ① PBFT; ② dBFT ③ PoA ④ PoET.

[13] A lightweight scalable block chain (LSB) is proposed. The LSB example uses distributed trust algorithm instead of PoW to reduce the workload proof overhead of traditional block chain [15]. The fruit fly optimization algorithm (FOA) is used to prevent the packing of blocks from being completed by a single node for a long time. Experiments show that the model using this algorithm can generate more blocks in the same time. B-FERL [4] The concept of an additional block is adopted in the permissioned blockchain, where transactions are added to a block by the owner of its public key. Therefore, b-FERL framework does not require consensus algorithm to complete block packaging [24]. The proposed systems use the Stellar Consensus Protocol (SCP) and has the advantages of low latency, decentralized control, flexible trust and asymptotic safety [12]. PoA authoritative proof method is adopted.

The mainstream automatic claim system adopts the permissioned network, so there is more flexibility in the choice of consensus algorithm. Some frameworks even omit consensus algorithms through hash values and dynamic blocks.

3.3   Encryption Algorithm

The data stored on the blockchain is public, so digital encryption is extremely important. The main application scenarios of asymmetric encryption technology in block chain include information encryption, digital signature and login authentication.

SHA-512 hash function is adopted in [12] to store the encrypted fingerprint in the block chain, so that the judge can verify whether the fingerprint is correct, that is, whether the data has been tampered with. [16] relies on images or video to record the status of the vehicle. The MD5 algorithm can generate a 16-byte image data digest, thus limiting the amount of data stored in the block. [18] adopted SHA-256 algorithm to verify data integrity. PRIDE [19] designed a multiplication encryption method to encrypt the driver data. The problem is that uploading real-time driving data can reveal information about how the car is driving, and the car needs to upload data randomly when it stops [22]. In order to protect the identity privacy of witnesses around the use of ring signature, a kind of unconditional anonymous electronic signature. [25] adopted SHA-256. [1] adopted Keccak-256.

The encryption algorithm is used for digital signature and information encryption on the one hand, and for data compression and storage on the other hand. Currently, the most used algorithm is SHA-256, but because of the requirements of the framework, a more suitable algorithm is adopted.

3.4 The Data Analysis

On the one hand, the application of intelligent contract is to simplify the process and automatically complete information verification and claim settlement. On the other hand, the collected vehicle information is analyzed to facilitate the automatic generation of insurance policies and the formulation of more targeted terms.

[24] Cognitive engine is introduced to manage intelligent contract, payment process, point-to-point communication, learn and predict user behavior information, predict insurance rate and generate relevant insurance policies. At the same time it adopted INS-COIN [17]. Automatic policy generation based on the real-time query of past data based on the user's selected path is a flexibility that cannot be achieved manually alone. [22] Users can choose a specified route, which runs 45% of the time compared to a pure smart contract. It is also vital to find a highly efficient, user-friendly interface design method for showing information [27, 28].

4   Challenges for Blockchain

As an emerging technology, block chain itself has its own immaturity. When faced with the gradual development of the auto insurance industry, how to better empower blockchain through self-improvement is the focus of attention in the future. There are the following problems:

4.1 Throughput and Scalability

The efficiency of blockchain solutions cannot be compared with centralized solutions for centralized systems do not operate in transparent environment. Any measure is decided and completed autonomously by its central system. In permissioned networks, the solution is to seek consensus through multiple roles, which can be a burden. During 2012, the maximum number of bitcoin transactions supported by blockchain was 62,000 per day. In China, one person is injured in a car accident every minute. If the insurance coverage reaches 100% in the future, Today's blockchain throughput can't support such a huge demand for insurance business. The blockchain ledger, too, will grow enormous over time, which is certainly not desirable.

4.2 User Privacy Issues

The blockchain, which uses asymmetric cryptography, is safe at the current level of technology. But with developing of quantum computing, SHA encryption algorithm also needs to be improved. And the user information of all nodes is essentially transparent, once the private key is leaked or lost, it will be enormous loss. Although the decentralized feature of blockchain solves the trust problem between subjects, distributed ledger has the risk that the private information of each node will be spread maliciously.

4.3 Smart Contract

The former intelligent contract can only handle some simple IF-Then process issues. If complex multi-role consensus and large data references are involved, human roles are still required. Moreover, once the intelligent contract is written into the block chain, the whole network is visible and difficult to modify, and if there is a design error, it is easy to attract malicious attacks.

Blockchain has its own limitations, but this does not affect its combination with other technologies to innovate traditional industries. In the future with the popularization of intelligent driving, car forensics should not only detect parts problems, but also pay attention to network attacks, system failures and driver errors. As for blockchain 3.0, practitioners should start to solve these pain points, to meet the growing insurance industry, and provide more possibilities for the combination of the two.

5 Prospects

The modern insurance claims process is long not because of the limitations of payment technology per se, but because the content of claims requires manual verification, exchange of information and agreement with multiple parties. The emphasis should therefore be on the enhancement of smart contract capabilities, and payment technologies could be combined with bank credit cards. It is not clear whether decentralization or not, decentralized smart claims framework can still adopt a centralized business approach.

Blockchain technology itself is unable to detect the authenticity of input information, and once written to the blockchain is difficult to change. If the data entered at the beginning is falsified, it will have a big impact on the follow-up. Therefore, automatic inspection of data by sensors should be strengthened and artificial subjective inspection by a third party should be abandoned. Technologies involved include on-board event recorders (EDRS), roadside units (Rsus), and shop floor communications (V2V). The goal should be a smart urban Internet of things that can automatically collect data about cars from the time they are made, sold, run and destroyed. Car insurance is just one application of the data collected and can be extended to areas such as car rental and second-hand car transactions.

Traditional insurance companies tend to hide their insurance data. The completion of this system needs to break down the communication barriers between insurance companies and make all insurance terms open and transparent. And the need for government leadership and the improvement of relevant laws and regulations.

6 Conclusions

The application of blockchain technology in automobile insurance industry is a relatively new direction. Although there have been relevant insurance blockchain experiments, commercial automatic payment use cases are still being explored. The industry has not yet appeared directional system research. Therefore, this paper lists a variety of system designs and divides them into three categories according to the different overall design ideas. And according to the common technical difficulties from different directions are analyzed. This paper discusses the advantages and disadvantages of various systems of the framework and sorts out the unresolved problems and the future development trend.

Acknowledgment. This work is supported by the National Natural Science Foundation of China (No. 61702180) and the Doctoral Scientific Research Foundation of Hunan University of Science and Technology (No. E52083).

References

1. Bader, L., et al.: Smart contract-based car insurance policies. In: 2018 IEEE Globecom Workshops (GC Wkshps). IEEE (2018)
2. Sharma, P.K., et al.: DistBlockNet: a distributed blockchains-based secure SDN architecture for IoT networks. IEEE Commun. Mag. **55**(9), 78–85 (2017)
3. Xie, Y., Liang, W., Li, R.F., Wu, K.S., Hong, C.Q.: Signal packing algorithm for in-vehicle CAN in Internet of vehicles. Ruan Jian Xue Bao/J. Softw. **27**(9), 2365–2376 (2016). (in Chinese)
4. Oham, C.F.: A liability attribution and security framework for fully autonomous vehicles (2019)
5. Dai, W., Qiu, M., Qiu, L., Chen, L., Wu, A.: Who moved my data? Privacy protection in smartphones. IEEE Commun. Mag. **55**(1), 20–25 (2017). https://doi.org/10.1109/MCOM.2017.1600349CM
6. Gai, K., Qiu, M., Zhao, H.: Security-aware efficient mass distributed storage approach for cloud systems in big data. In: 2016 IEEE 2nd International Conference on Big Data Security on Cloud (BigDataSecurity) (2016)
7. Ouyang, L.-W., et al.: Intelligent contracts: architecture and progress. Acta Automata **45**(3), 445–457 (2019)
8. Cohn, A., West, T., Parker, C.: Smart after all: blockchain, smart contracts, parametric insurance, and smart energy grids. Georgetown Law Technol. Rev. **1**(2), 273–304 (2017)
9. Aleksieva, V., Hristo, V., Huliyan, A.: Application of smart contracts based on Ethereum blockchain for the purpose of insurance services. In: 2019 International Conference on Biomedical Innovations and Applications (BIA). IEEE (2019)
10. Demir, M., Turetken, O., Ferworn, A.: Blockchain based transparent vehicle insurance management. In: 2019 Sixth International Conference on Software Defined Systems (SDS), pp. 213–220. IEEE (2019)
11. Raikwar, M., Mazumdar, S., Ruj, S., et al.: A blockchain framework for insurance processes. In: 2018 9th IFIP International Conference on New Technologies, Mobility and Security (NTMS), pp. 1–4. IEEE (2018)
12. Morano, F., Ferretti, C., Leporati, A., et al.: A blockchain technology for protection and probative value preservation of vehicle driver data. In: 2019 IEEE 23rd International Symposium on Consumer Technologies (ISCT), pp. 167–172. IEEE (2019)
13. Dorri, A., Steger, M., Kanhere, S.S., Jurdak, R.: Blockchain: a distributed solution to automotive security and privacy. IEEE Commun. Mag. **55**(12), 119–125 (2017)
14. Cebe, M., Erdin, E., Akkaya, K., et al.: Block4Forensic: an integrated lightweight blockchain framework for forensics applications of connected vehicles. IEEE Commun. Mag. **56**(10), 50–57 (2018)
15. Sharma, P.K., Kumar, N., Park, J.H.: Blockchain-based distributed framework for automotive industry in a smart city. IEEE Trans. Industr. Inf. **15**(7), 4197–4205 (2018)

16. Lamberti, F., Gatteschi, V., Demartini, C., et al.: Blockchains can work for car insurance: using smart contracts and sensors to provide on-demand coverage. IEEE Consum. Electron. Mag. **7**(4), 72–81 (2018)

17. Vo, H.T., Mehedy, L., Mohania, M., et al.: Blockchain-based data management and analytics for micro-insurance applications. In: Proceedings of the 2017 ACM on Conference on Information and Knowledge Management, pp. 2539–2542 (2017)

18. Saldamli, G., Karunakaran, K., Vijaykumar, V.K., et al.: Securing car data and analytics using blockchain. In: 2020 Seventh International Conference on Software Defined Systems (SDS), pp. 153–159. IEEE (2020)

19. Wan, Z., Guan, Z., Cheng, X.: PRIDE: a private and decentralized usage-based insurance using blockchain. In: 2018 IEEE International Conference on Internet of Things (iThings) and IEEE Green Computing and Communications (GreenCom) and IEEE Cyber, Physical and Social Computing (CPSCom) and IEEE Smart Data (SmartData), pp. 1349–1354. IEEE (2018)

20. Kokab, S.T., Javaid, N.: Blockchain based insurance claim system for vehicles in vehicular network

21. Oham, C., Jurdak, R., Kanhere, S.S., et al.: B-FICA: blockchain based framework for auto-insurance claim and adjudication. In: 2018 IEEE International Conference on Internet of Things (iThings) and IEEE Green Computing and Communications (GreenCom) and IEEE Cyber, Physical and Social Computing (CPSCom) and IEEE Smart Data (SmartData), pp. 1171–1180. IEEE (2018)

22. Davydov, V., Bezzateev, S.: Accident detection in internet of vehicles using blockchain technology. In: 2020 International Conference on Information Networking (ICOIN), pp. 766–771. IEEE (2020)

23. Xu, J., Wu, Y., Luo, X., et al.: Improving the efficiency of blockchain applications with smart contract based cyber-insurance. In: Proceedings of the IEEE International Conference on Communications (2020)

24. Vahdati, M., HamlAbadi, K.G., Saghiri, A.M., et al.: A self-organized framework for insurance based on Internet of Things and blockchain. In: 2018 IEEE 6th International Conference on Future Internet of Things and Cloud (FiCloud), pp. 169–175. IEEE (2018)

25. Oham, C., Kanhere, S.S., Jurdak, R., et al.: A blockchain based liability attribution framework for autonomous vehicles. arXiv preprint arXiv:1802.05050 (2018)

26. Yong, Y., et al.: Development status and prospect of blockchain consensus algorithm. **44**(11), 2011–2022 (2018). (in Chinese)

27. Qiu, M.K., Zhang, K., Huang, M.: An empirical study of web interface design on small display devices. In: IEEE/WIC/ACM International Conference on Web Intelligence (WI 2004). IEEE (2004)

28. Qiu, M., Jia, Z., Xue, C., Shao, Z., Sha, E.H.M.: Voltage assignment with guaranteed probability satisfying timing constraint for real-time multiproceesor DSP. J. VLSI Sig. Process. Syst. Sig. Image Video Technol. **46**(1), 55–73 (2007)

Birds Classification Based on Deep Transfer Learning

Peng Wu[1,2], Gang Wu[1(✉)], Xiaoping Wu[3], Xiaomei Yi[2], and Naixue Xiong[4]

[1] School of Information, Beijing Forestry University, Beijing 100083, China
wugang@bjfu.edu.cn
[2] School of Information Engineering,
Zhejiang A & F University, Lin'an 311300, Zhejiang, China
[3] School of Information Engineering, Huzhou University, Huzhou 313000, Zhejiang, China
[4] Department of Mathematics and Computer Science, Northeastern State University,
Tahlequah, OK 74464, USA

Abstract. At present, there is no huge data set on bird classification, and the classic CUB-200-2011 data set has only 11 788 images, which are still unable to train a generalized classification and recognition model compared with ImageNet and other large data sets with millions of data. Therefore, using deep transfer learning, after tuning for bird recognition, is very valuable with large data sets training model parameters. In this paper, by comparing the training effect of common benchmark models in CUB-200-2011 dataset, ResNeXt model is selected as the transfer learning benchmark model for its well performance. Through optimizing the loss function and reducing the learning rate, the proposed model provides better performance for the data augmentation and adding the full connection layer. Compared with the benchmark model, the recognition rate of the proposed model can reach 84.43%.

Keywords: Transfer learning · Birds classification · Model Fine-Tuning · Data augmentation

1 Introduction

1.1 Research Background

China has a vast territory and a wide variety of biological species, among which 1,445 species of birds (2344 species and subspecies) belong to 497 genera, 109 families, 26 orders [1]. Using deep learning to conduct bird type identification is a recent research hotspot. The mainstream deep learning models include AlexNet [2], VGGNet [3], Inception v1-v4 [4–7], Xception [8], ResNet [9], ResNeXt [9], MobileNet [11, 12], etc. All these models have achieved good classification effect in ImageNet [13] and can effectively distinguish different objects. However, they were trained to classify all birds into one category, rather than subdividing them. As is known to all, there are great similarities between birds, some of the differences are even very subtle. Hence, people have great difficulties in the identification of birds.

© Springer Nature Switzerland AG 2021
M. Qiu (Ed.): SmartCom 2020, LNCS 12608, pp. 173–183, 2021.
https://doi.org/10.1007/978-3-030-74717-6_19

To address this problem, researchers had proposed many improved algorithms and models for fine-grained image recognition. For example, Xiu-Shen Wei et al. [14] designed a new model based on the Mask-CNN, which includes not only fully connected layer but also the fully convolutional layer, and achieved good effect in fine-grained bird classification. He-liang Zheng et al. [15] proposed a new component learning method based on MA-CNN, which enhances the recognition effect by mutually reinforcing of component generation and feature learning. Ning Zhang et al. [16] proposed an assuming boundary box annotation in the test model for fine-grained classification derived from R-CNN, which can overcome the limitation of assuming boundary box annotations in the test by using the deep convolution feature calculated in the bottom-up region proposal, and had achieved good results in the end-to-end evaluation. Yili Zhao [17], Xinye Li [18], Yuchen Win [19], Qiu [20–22] et al. also made improvements on fine-grained bird recognition based on existing models.

The training of deep learning model requires many samples as training data. The network structure reaches hundreds of layers, in which the model parameters can reach tens of millions of levels, so the training process will be very time-consuming [23–35]. Therefore, in practical applications, it is difficult to set up the sample data as large as ImageNet. In 2009, inspired by migration learning proposed by Yang Qiang et al. [36], deep migration learning [27] has been proposed to solve the problem of high cost of data collection and annotation and the difficulty of building large and well-annotated data sets.

In this paper, we propose a method of bird classification and recognition based on deep migration learning. In the proposed method, ResNeXt model is pretrained by ImageNet to extract features. Then it will put the extracted feature vectors into the classifier for training, so we can get the classification results.

1.2 Frame of the Thesis

1. Brief introduction of relevant research background.
2. Introduce the work foundation of this paper: convolution neural network, ResNeXt, model fine-tuning and transfer learning.
3. The experimental design model is proposed, and then the model is optimized to obtain the experimental results.
4. Analysis of experimental results: The precision, recall rate and F1 score are used to analyze the experimental results.
5. Get the conclusion.

2 Work Foundation

2.1 Convolutional Neural Network

The Convolutional Neural Network originates from the LeNet class model, in which convolution operation is used to extract the feature, the Max pooling operation is used for spatial subsampling. The model depth is strengthened, and the style of the CNN become more for the subsequent continuous convolution and pooling operations. The typical training process of Convolutional Neural Network is shown in Fig. 1:

The typical network structure of convolutional neural network is shown in Fig. 2:

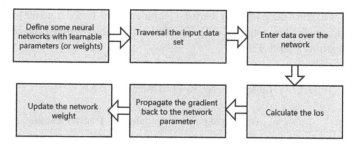

Fig. 1. Typical training process of convolutional neural network

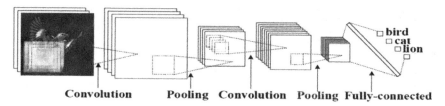

Fig. 2. Typical structure of a convolutional neural network

2.2 ResNeXt

ResNeXt is a combination of ResNet and Inception, unlike InceptionV4, ResNeXt introduces Group convolution into ResNet to get fewer parameters. And there is no need to manually design complex Inception structure details, but each branch uses the same topological structure. Its network block is shown in Fig. 3.

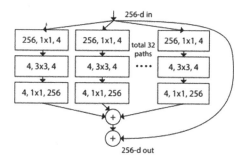

Fig. 3. ResNeXt network structure block (Number of groups = 32)

2.3 Deep Transfer Learning

By availing of the similarity between the data and the domain, deep transfer learning applies the knowledge learned previously to the new terra incognita. The core problem of transfer learning is to find the similarities between two different domains. And there are two main ways to learn deep transfer learning include model-based transfer learning and feature-based transfer learning [38, 39]. The following table outlines the requirement for transfer learning (Table 1).

Table 1. Necessity of transfer learning.

Contradiction	Traditional machine learning	Transfer learning
Big data and annotation lacking	Add manual annotation but are expensive and time-consuming	Migration annotation of data
Big data and weak computing	Only depending on the powerful computation ability, but audience is exceedingly small	Model migration
Pervasive models and personalized requirements	Generic models do not meet personalized requirements	Model can adjust adaptive
Specific application	The problem of cold boot cannot be solved	Data migration

2.4 Model Fine-Tuning

By fine-tuning the model, it can optimize the model while training to achieve the optimal training effect. Model fine-tuning typically involves three approaches.

Feature Extraction. We can use the pretraining model as feature extraction device. The specific practice is to remove the output layer and treat the rest of the network as a fixed feature extractor, and thus applied to the new data set.

Using the Structure of the Pre-training Model. We can use the structure of the pre-training model, but need to randomize all the weights firstly, and then according to your own data set for training.

Train Specific Layers Freeze Other Layers. Another way to use the pre-training model is to partially train it. And the specific practice is keeping the weights of some layers at the beginning of the model and retrain the back layers to gain new weights. In this process, we can try multiple times, and thus to find the best collocation between Frozen layers and Retrain Layers according to the results.

3 Algorithm

In this paper we use Adam [40] algorithm to fine-tune the model. Adam is a method for efficient stochastic optimization that only requires first-order gradients with little memory requirement. The method computes individual adaptive learning rates for different parameters from estimates of first and second moments of the gradients. The method is designed to combine the advantages of two recently popular methods: AdaGrad [41], which works well with sparse gradients, and RMSProp [42], which works well in on-line and non-stationary settings (Fig. 4).

Require: α: Stepsize
Require: $\beta_1, \beta_2 \in [0, 1)$: Exponential decay rates for the moment estimates
Require: $f(\theta)$: Stochastic objective function with parameters θ
Require: θ_0: Initial parameter vector
  $m_0 \leftarrow 0$ (Initialize 1st moment vector)
  $v_0 \leftarrow 0$ (Initialize 2nd moment vector)
  $t \leftarrow 0$ (Initialize timestep)
  while θ_t not converged **do**
    $t \leftarrow t + 1$
    $g_t \leftarrow \nabla_\theta f_t(\theta_{t-1})$ (Get gradients w.r.t. stochastic objective at timestep t)
    $m_t \leftarrow \beta_1 \cdot m_{t-1} + (1 - \beta_1) \cdot g_t$ (Update biased first moment estimate)
    $v_t \leftarrow \beta_2 \cdot v_{t-1} + (1 - \beta_2) \cdot g_t^2$ (Update biased second raw moment estimate)
    $\hat{m}_t \leftarrow m_t/(1 - \beta_1^t)$ (Compute bias-corrected first moment estimate)
    $\hat{v}_t \leftarrow v_t/(1 - \beta_2^t)$ (Compute bias-corrected second raw moment estimate)
    $\theta_t \leftarrow \theta_{t-1} - \alpha \cdot \hat{m}_t/(\sqrt{\hat{v}_t} + \epsilon)$ (Update parameters)
  end while
  return θ_t (Resulting parameters)

Fig. 4. The algorithm of adam (Source: arXiv:1412.6980v9, p. 2)

Adam is an algorithm for stochastic optimization. g_t^2 indicates the elementwise square $g^t \odot g^t$. Good default settings for the tested machine learning problems are $\alpha = 0.001$, $\beta_1 = 0.9$, $\beta_2 = 0.999$ and $\epsilon = 10^{-8}$. All operations on vectors are elementwise. With β_1^t and β_2^t we denote β_1 and β_2 to the power t.

4 Experimental Design

4.1 Data Set

In our experiments, CUB-200-2011 [43] Data set is used. As is known that it is a representative and challenging bird image data set in fine-grained image categorization. An excellent data set includes 200 species of birds like Phoebastria nigripes, Phoebastria immutabilis, Geothlypis trichas, etc. The data is divided as follows: the training data set contains 5994 pictures, testing data set contains 5794 pictures, and 11788 pictures in total.

4.2 Experimental Environment

The experimental environment is Ubuntu 16.04 operating system. The CPU is a two-channel Intel(R) Xeon(R) CPU E5-2680, 2.4ghz, and 56 cores, internal storage capacity is 128G. Based on PyTorch deep learning framework, the video card used is NVIDIA GeForce GTX 1080 TI, video memory is 11 GB, and use CUDA 10.1 and cuDNN to accelerate.

4.3 Experimental Procedure

Step 1: Select the benchmark model. To verify the effect of transfer learning, we chose 7 models including GoogleNet, ResNeXt and so on as alternative benchmark models for transfer learning. In the case without image-enhance, we used cross entropy as loss function for training. And the training results were measured by 4 indicators, including Accuracy, Precision, Recall and F1 Score. Among them:

Accuracy means accuracy rate, that is, the proportion of the correct predictions in the total number.

Precision: Precision rate, prediction is a positive number, and the proportion of the correct sample size in the number of samples predicted to be positive, also called precision ratio.

Recall: recall rate, prediction is a positive number, and the proportion of the correct sample size in the number of actually to be positive, also called recall ratio.

F1 Score is the harmonic average of the model's accuracy and recall rates.

The training results are shown in Fig. 5.

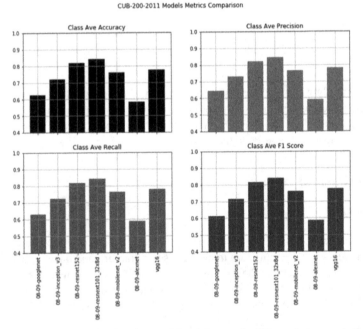

Fig. 5. Alternative benchmark migration model training results (Figure A represents the average classification accuracy, Figure B represents the average training precision, Figure C represents the average recall, and Figure D represents F1 Score)

By comparing the training results of the above alternative benchmark model, we could find that ResNeXt model is superior to other models in the statistics of these key indicators, so we chose this model as deep transfer learning benchmark model for the subsequent operation.

Step 2: To determine the loss function. Loss function is used to evaluate the difference between the predicted value and the real value in the proposed model. The better the loss function is, the better performance the model has in general. Aimed to achieve multi-classification network model, this paper adopts cross entropy loss as loss function. The computational formula of cross entropy is listed as follows:

$$H(p, q) = - \sum_{x} (p(x)logq(x)) \tag{1}$$

p and q are probability distributions in a sample set. p is the real distribution and q is unreal distribution.

Step 3: Train the ResNeXt model. As mentioned above, ResNeXt model is trained based on ImageNet data set, but ImageNet has 1000 categories, most of which are unrelated to birds. Therefore, in the process of model training, we set all the pretraining parameter to True, namely, retraining model parameters.

Set the training process to 40 Epochs, Batch-Size to 16. The training results showed that after 8 Epochs training, in the case of which the loss in the training data set tends to 12, accuracy tends to 1, but verification data set loss is around 0.7, accuracy around 0.8, overfitting phenomenon appeared in the model (Fig. 6).

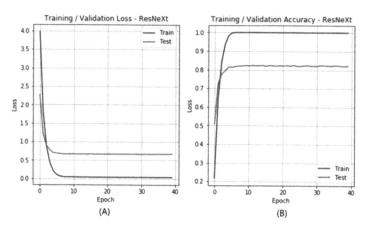

Fig. 6. ResNeXt Model training results (Figure A represents the loss of training data set and validation data set, and Figure B represents the accuracy of training data set and validation data set)

Step 4: Model tuning. Firstly, the benchmark model structure is loaded. Then all the weights are randomized. Using Image horizontal flip enhances the data by choosing the cross entropy as loss function. And next, repeat step 3, the training results are shown in Fig. 7. We could see the result tends to be stable after 10 Epochs, but there is no overfitting phenomenon appeared. Instead, as the number of training sessions increased, the loss of training and validation data sets was decreased, and the accuracy of validation data set was increased slowly at the same time. It is showed that the adjustment effect of the model is better than that of the benchmark model.

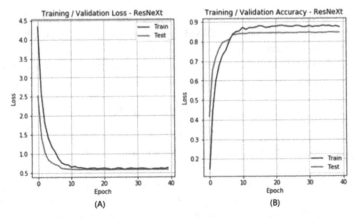

Fig. 7. Training results of ResNeXt model after fine-tuning (Figure A represents the loss of training data set and validation data set, and Figure B represents the accuracy of training data set and validation data set)

5 Analysis of Experimental Results

Testing data set contains 200 categories and 5794 pictures. Table 2 and Table 3 show the results of the benchmark model validation and the model validation after tuning. It is not hard to be seen by comparison, the model after tuning is superior to benchmark model in the statistics of key indicators include accuracy, recall and F1 Score. Among them, F1 Score has a very obvious advantage, the classification with the score of 1 was increased from 3 to 6, the classification with the score of 0.9 was increased from 80 to 88, had increased by 50% and 10% respectively. Table 4 shows that the mean value of the key indexes in the model after tuning is superior to the benchmark model. The improved ratio of F1 Score is best among them and has reach 1.45% points.

Table 2. Verification results of the benchmark model (unit: Type)

Precision range and indicator	1	0.9–1	0.8–0.9	0.7–0.8	0.6–0.7	Below 0.6
Precision	19	72	37	38	19	15
Recall	24	67	51	21	15	22
F1 score	3	80	52	29	18	18

Table 3. Model validation results after tuning (unit: type)

Precision range and indicator	1	0.9–1	0.8–0.9	0.7–0.8	0.6–0.7	Below 0.6
Precision	21	74	46	27	17	15
Recall	25	76	46	22	13	18
F1 score	6	88	49	24	18	15

Table 4. Comparison table of mean value of key indicators of the model

Model	Precision	Recall	F1 score
ResNeXt	0.8305	0.8259	0.8244
The model in this paper	0.8443	0.8408	0.8389

6 Conclusion

In this paper, ResNeXt is considered as benchmark model. The concept of Cardinality is introduced in ResNeXt, so its classification effect is better than ResNet. Meanwhile, ResNeXt simplifies the difficulty of network design and improves its generalization performance. In this work, we adopt such methods, like increasing the fully connection layer and optimizing the loss function learning rate. Using training data enhancement can reach the recognition rate of 84.43 percent obtained by this model in CUB-200-2011 data set. Therefore, our proposed model has certain practical application value.

Acknowledgements. This work is supported the Fundamental Research Funds for the Central Universities (TD2014-02).

References

1. Guangmei, Z.: A Checklist on the Classification and Distribution of the Birds of China. Beijing Science Press, Beijing (2017)
2. Mamalet, F., Garcia, C.: Simplifying ConvNets for Fast learning. In: Villa, A.E.P., Duch, W., Érdi, P., Masulli, F., Palm, G. (eds.) ICANN 2012. LNCS, vol. 7553, pp. 58–65. Springer, Heidelberg (2012). https://doi.org/10.1007/978-3-642-33266-1_8
3. Simonyan, K., Zisserman, A.: Very deep convolutional networks for large-scale image recognition. arXiv preprint arXiv:1409.1556 (2014)
4. Szegedy, C., Liu, W., Jia, Y., et al.: Going deeper with convolutions. In: Proceedings of the IEEE Conference on Computer Vision and Pattern Recognition, pp. 1–9 (2015)
5. Ioffe, S., Szegedy, C.: Batch normalization: Accelerating deep network training by reducing internal covariate shift (2015). arXiv preprint arXiv:1502.03167
6. Szegedy, C., Vanhoucke, V., Ioffe, S. et al.: Rethinking the inception architecture for computer vision. In: Proceedings of the IEEE Conference on Computer Vision and Pattern Recognition, pp. 2818–2826 (2016)

7. Szegedy, C., Ioffe, S., Vanhoucke, V., et al.: Inception-v4, inception-resnet and the impact of residual connections on learning (2016). arXiv preprint arXiv:1602.07261
8. Chollet, F.: Xception: deep learning with depthwise separable convolutions. In: Proceedings of the IEEE Conference on Computer Vision and Pattern Recognition, pp. 1251–1258 (2017)
9. He, K,. Zhang, X., Ren, S., et al.: Deep residual learning for image recognition. In: Proceedings of the IEEE Conference on Computer Vision and Pattern Recognition, pp. 770–778 (2016)
10. Xie, S., Girshick, R., Dollár, P., et al.: Aggregated residual transformations for deep neural networks. In: Proceedings of the IEEE Conference on Computer Vision and Pattern Recognition, pp. 1492–1500 (2017)
11. Howard, A.G., Zhu, M., Chen, B., et al.: Mobilenets: Efficient convolutional neural networks for mobile vision applications (2017). arXiv preprint arXiv:1704.04861
12. Sandler, M., Howard, A., Zhu, M., et al.: Mobilenetv2: inverted residuals and linear bottlenecks. In: Proceedings of the IEEE Conference on Computer Vision and Pattern Recognition, pp. 4510–4520 (2018)
13. Deng, J., Dong, W., Socher, R., et al.: Imagenet: a large-scale hierarchical image database. In: 2009 IEEE conference on computer vision and pattern recognition, pp. 248–255. IEEE (2009)
14. Wei, X.-S., Xie, C.-W., Wu, J., et al.: Mask-CNN: Localizing parts and selecting descriptors for fine-grained bird species categorization. Pattern Recognit. **76**, 704–714 (2018)
15. Zheng, H., Fu, J., Mei, T., et al.: Learning multi-attention convolutional neural network for fine-grained image recognition. In: Proceedings of the IEEE International Conference on Computer Vision, pp. 5209–5217 (2017)
16. Zhang, N., Donahue, J., Girshick, R., Darrell, T.: Part-Based R-CNNs for Fine-Grained Category Detection. In: Fleet, D., Pajdla, T., Schiele, B., Tuytelaars, T. (eds.) ECCV 2014. LNCS, vol. 8689, pp. 834–849. Springer, Cham (2014). https://doi.org/10.1007/978-3-319-10590-1_54
17. Yili, Z.: Fine-grained Recognition of Yunnan Wild Bird Images Based on Deep Learning. Yunnan University, Kunming Yunnan (2018)
18. Xin-ye, L., Guang-bi, W.: Fine-grained bird recognition based on convolution neural network semantic detection. Sci. Technol. Eng. **18**(10), 240–244 (2018)
19. Yuchen, W.: Fine-grained Bird Recognition Based on Deep Learning Beijing Forestry University, Beijing (2018)
20. Gai, K., Qiu, M., Zhao, H.: Security-aware efficient mass distributed storage approach for cloud systems in big data. In: 2016 IEEE 2nd international conference on big data security on cloud (BigDataSecurity). In: IEEE International Conference on High Performance and Smart Computing (HPSC), and IEEE International Conference on Intelligent Data and Security (IDS). IEEE, pp. 140–145 (2016)
21. Qiu, M., Jia, Z., Xue, C., et al.: Voltage assignment with guaranteed probability satisfying timing constraint for real-time multiproceesor DSP. J. VLSI Signal Proc. Syst. Signal Image Video Technol. **46**, 55–73 (2007). https://doi.org/10.1007/s11265-006-0002-0
22. Zhang, Q., Huang, T., Zhu, Y., et al.: A case study of sensor data collection and analysis in smart city: provenance in smart food supply chain. Int. J. Distrib. Sens. Netw. **9**, (2013)
23. Gao, Y., Xiang, X., Xiong, N., et al.: Human action monitoring for healthcare based on deep learning. IEEE Access **6**, 52277–52285 (2018)
24. Gao, Y., Xiong, N., Yu, W., et al.: Learning identity-aware face features across poses based on deep siamese networks. IEEE Access **7**, 105789–105799 (2019)
25. Huang, K., Zhang, Q., Zhou, C., et al.: An efficient intrusion detection approach for visual sensor networks based on traffic pattern learning. IEEE Tran. Syst. Man Cybern. Syst. **47**, 2704–2713 (2017)
26. Li, L.-F., Wang, X., Hu, W.-J., et al.: Deep Learning in Skin Disease Image Recognition: A Review. IEEE Access (2020)

27. Liu, M., Zhou, M., Zhang, T., et al.: Semi-supervised learning quantization algorithm with deep features for motor imagery EEG recognition in smart healthcare application. Appl. Soft Comput. **89**, (2020)

28. Mao, Z., Su, Y., Xu, G., et al.: Spatio-temporal deep learning method for ADHD fMRI classification. Inf. Sci. **499**, 1–11 (2019)

29. Shahzad, A., Lee, M., Lee, Y.-K., et al.: Real time MODBUS transmissions and cryptography security designs and enhancements of protocol sensitive information. Symmetry **7**, 1176–1210 (2015)

30. Wu, C., Ju, B., Wu, Y., et al.: UAV autonomous target search based on deep reinforcement learning in complex disaster scene. IEEE Access **7**, 117227–117245 (2019)

31. Wu, C., Luo, C., Xiong, N., et al.: A greedy deep learning method for medical disease analysis. IEEE Access **6**, 20021–20030 (2018)

32. Wu, W., Xiong, N., Wu, C.: Improved clustering algorithm based on energy consumption in wireless sensor networks. Iet Netw. **6**, 47–53 (2017)

33. Yang, Y., Xiong, N., Chong, N.Y., et al.: A decentralized and adaptive flocking algorithm for autonomous mobile robots. In: 2008 The 3rd International Conference on Grid and Pervasive Computing-Workshops, pp. 262–268. IEEE (2008)

34. Yi, B., Shen, X., Liu, H., et al.: Deep matrix factorization with implicit feedback embedding for recommendation system. IEEE Trans. Ind. Inf. **15**, 4591–4601 (2019)

35. Zhang, Q., Zhou, C., Xiong, N., et al.: Multimodel-based incident prediction and risk assessment in dynamic cybersecurity protection for industrial control systems. IEEE Trans. Syst. Man Cybern. Syst. **46**, 1429–1444 (2015)

36. Pan, S.J., Yang, Q.: A survey on transfer learning. IEEE Trans. Knowl. Data Eng. **22**, 1345–1359 (2009)

37. Tan, C., Sun, F., Kong, T., Zhang, W., Yang, C., Liu, C.: A survey on deep transfer learning. In: Kůrková, V., Manolopoulos, Y., Hammer, B., Iliadis, L., Maglogiannis, I. (eds.) ICANN 2018. LNCS, vol. 11141, pp. 270–279. Springer, Cham (2018). https://doi.org/10.1007/978-3-030-01424-7_27

38. Zhuang, F.Z., Luo, P., He, Q., Shi, Z.Z.: Survey on transfer learning research. Ruan Jian Xue Bao/J. Softw. **26**(1), 26–39 (2015)

39. Wang, J.: Transfer Learning Tutorial. (2019) In p https://tutorial.transferlearning.xyz/index

40. Kingma, D.P., Ba, J.: Adam: a method for stochastic optimization (2014). arXiv preprint arXiv:1412.6980

41. Duchi, J., Hazan, E., Singer, Y.: Adaptive subgradient methods for online learning and stochastic optimization. J. Mach. Learn. Res. **12**, 2121–2159 (2011)

42. Hinton, G., Srivastava, N., Swersky, K.: Neural networks for machine learning. Coursera Video Lect. **264**, (2012)

43. Wah, C., Branson, S., Welinder, P., et al.: The caltech-ucsd birds-200–2011 dataset (2011)

Research on Security Consensus Algorithm Based on Blockchain

Minfu Tan[1], Jing Zheng[2(⊠)], Jinshou Zhang[1], Yan Cheng[1], and Weihong Huang[2]

[1] Guangzhou College of Technology and Business, Guangzhou 528138, China
[2] Hunan University of Science and Technology, Xiangtan 411100, China

Abstract. As a distributed shared ledger and database technology, blockchain has the characteristics of decentralization, tamper-proof, openness and transparency. With the continuous development of blockchain technology, the technology has received high attention from government departments, finance, technology and other industries. The key technology of blockchain is the design of consensus algorithm, which affects the overall efficiency and performance of the blockchain system. How to achieve consistency and efficiency among nodes in a simple and efficient blockchain distributed system. This paper divides consensus algorithms into CFT and BFT according to fault tolerance types, and analyzes the basic principles and consensus process of each algorithm in detail. In five aspects of block generation speed, degree of decentralization, security, consistency and resource consumption, the mainstream consensus algorithms in the blockchain are compared and analyzed, the advantages and disadvantages of each consensus algorithm are summarized. Finally, the future blockchain consensus algorithms are summarized and prospected.

Keywords: Blockchain · Shared ledger · Consensus algorithm · Consistency · Distributed system

1 Introduction

In 2008, a mysterious figure named Satoshi Nakamoto published a paper Bitcoin: A Peer-to-Peer Electronic Cash System [1] in the community. In this paper, the concept of blockchain was proposed for the first time. With the release of white paper by Google's Libra [2] project, the value of Bitcoin ushered in a substantial increase, which makes more and more people pay attention to blockchain technology [3].

Blockchain is a distributed shared ledger and database [4] and a more open and decentralized system [5]. It uses a series of technologies such as distributed data storage, mathematics, cryptography, and P2P networks to conduct decentralized trust transactions and cooperation on mutually untrusted network nodes. In view of the shortcomings of insecure data storage and low efficiency in many traditional centralized systems, the emergence of the new computing paradigm of blockchain can solve these problems. With the rapid development of the digital economy, blockchain technology has begun to receive the attention of the government [6], and it has begun to integrate with politics, finance, etc.

© Springer Nature Switzerland AG 2021
M. Qiu (Ed.): SmartCom 2020, LNCS 12608, pp. 184–193, 2021.
https://doi.org/10.1007/978-3-030-74717-6_20

Consensus algorithm is one of the cores of blockchain technology. It determines the performance and efficiency of the entire blockchain system. In a distributed system, the consensus algorithm is to keep the recorded data on each node consistent. With the emergence of the PoW (Proof of Work) proposed in Bitcoin, the consensus algorithm suitable for blockchain gradually has new ideas and development. In the early stages of blockchain development, mainstream blockchain networks were based on PoW consensus algorithms, including Bitcoin, Ethereum, Litecoin, Zcash, etc. Due to the waste of mining resources in PoW, many scholars have proposed some new consensus algorithms.

In addition to the PoW algorithm, the mainstream consensus algorithms today include proof of stake [7], delegated proof of stake [8], practical Byzantine fault tolerance [9], etc. Although they have been recognized by the blockchain industry, these algorithms still have different problems.

This paper elaborates the working principles of six consensus algorithms, summarizes the advantages and disadvantages of each algorithm, and compares and evaluates them from the aspects of resource consumption, block generation speed, degree of decentralization, safety and consistency, hoping to provide some new ideas for the development of block chain consensus algorithm.

The paper is structured as follows. Section 2, Some consensus algorithms of CFT and BFT are introduced. Section 3, The algorithm is compared from five aspects, and some common problems of BFT and CFT algorithms are listed. Section 4 is Summary and outlook.

2 Introduction of Several Consensus Algorithms

Consensus algorithms enable nodes in a highly fragmented and distrustful network environment to agree on a transaction without bifurcation. According to the fault tolerance type of consensus algorithm, this paper divides the algorithm into CFT (Crash Fault Tolerance) and BFT (Byzantine Fault Tolerance).

2.1 CFT Consensus Algorithms

The CFT type consensus algorithm solves the non-Byzantine general problem. It is an algorithm that makes a distributed system that contains only faulty nodes (nodes downtime, etc.) and does not contain malicious nodes (maliciously tamper with nodes and send data) to reach a consensus. Currently, the mainstream CFT consensus algorithm mainly includes Paxos and Raft.

Paxos Consensus Algorithm. Lamport proposed the Paxos in the paper "Paxos Made Simple" [10] published in 1998. There are three roles in the Paxos: Proposer, Acceptor, and Learner. Proposer is the initiator of the proposal. The proposal consists of a number and a resolution value. The number is globally unique and monotonically increasing. Acceptor is to accept the proposal, and it is responsible for voting on the proposal. Learner is the synchronizer, and it passively accepts the received proposals. The Paxos is divided into two phases, the preparation stage and the acceptance stage.

Prepare stage: the proposer selects a proposal number n and sends acceptance requests to most nodes (more than half of the nodes). Acceptor will judge the proposal number n currently received and the largest proposal number it has responded to. If the number n currently received is large, then the acceptor will respond to the proposal with the highest number it has accepted (if any), and this acceptor promises not to accept any proposal with a number less than n. Accept stage: When the proposer receives more than half of the response replies, it will continue to make an acceptance (n, value) request to the acceptor. The value is of the proposal with the largest number received in the response. If not, the value is determined by the proposer itself. After acceptor receives the acceptance request sent by proposer, as long as it does not respond to a proposal greater than N, it records the proposal and responds to the request to accept. As long as proposer receives more than half of the requests, it means the proposal has been accepted, otherwise it will re-initiate the proposal.

Paxos has high performance, low resource consumption and high system fault tolerance, but it requires a lot of network communication to form a resolution, not suitable for high-concurrency environments. Its algorithm principle is also complicated and difficult to implement.

Raft Consensus Algorithm. Because the Paxos is difficult to understand and implement, Diego Ongaro and John Ousterhoutfrom of Stanford University proposed the Raft [11] algorithm for the purpose of easy understanding, which uses log replication to maintain the system consistency. Raft divides the entire system into three roles: leader, follower, and candidate. Each term will have a leader, which is responsible for managing the entire cluster, receiving client requests and synchronizing request logs from follower. It also continuously sends heartbeat information to follower to prove its current operating status is normal. Follower is responsible for accepting and persisting the log synchronized by the leader, and submitting the log after the leader informs the submission. Follower also take responsibility for participating in the election of leader voting. Candidate is the leader candidate, participating in the Leader competition.

The entire Raft algorithm is divided into leader election and log replication. At the Leader election stage, the initial state of each node is follower. In order to avoid election conflicts, they will all be assigned a different sleep value. After the sleep value ends and the heartbeat information from the leader is not received, they will become a Candidate and initiate a voting request to other nodes. After getting more than half of the nodes' votes, they will become Leader. If the election for leader fails, the identity of follower will be regained. Log replication stage: When the client sends a log request to the leader, it updates the log and sends the client's log request to all followers for synchronization.

Raft is an improvement of the Paxos. It takes into account the simplicity and completeness of the entire algorithm and easier to implement in practical applications and with high fault tolerance.

2.2   BFT Consensus Algorithms

The BFT consensus algorithm can solve the consistency problem of the Byzantine nodes in the system. Its process can be divided into master selection, block building, verification, chaining, and the master selection and verification are its core [12].

PoW Consensus Algorithm. Proof of Work is used by Satoshi Nakamoto in the Bitcoin system which was originally a computer technology used to solve spam [13]. Its core idea is to use computer computing power to solve a SHA256 mathematical problem to obtain accounting rights. Through the computing power of the computer, different random Nonce values are continuously tried to make the hash value of the entire block header data smaller than a certain difficulty target value. When the Nonce value of this random number is found, the block will be broadcast to the entire network. After other nodes have passed the verification, they will join the main chain of the blockchain. At this time, all nodes on the blockchain have reached a consensus on the transaction information. The miner who generates the block will also receive the block reward and the transaction fee in the block transaction. System stability and security attack detection [14] are very important to the operation of the system. In order to ensure the stability and security of the Bitcoin system, the blockchain system can flexibly adjust the difficulty value to ensure that the average block time of the system is 10 min (Fig. 1).

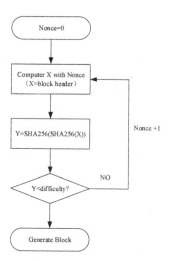

Fig. 1. The mining process of the Bitcoin system

PoW competes for the accounting rights of nodes by means of computing power, which not only realizes the decentralization of the Bitcoin system, but also effectively prevents the tampering of system node transaction data and ensures the security of the system. The fork problem is solved by selecting the longest chain as the legal chain,and The reward mechanism in the Bitcoin system also promotes the enthusiasm of miners for mining and also maintains the security of the system.

Of course, PoW also has many shortcomings. It uses computing power resources to compete for the node's bookkeeping rights, resulting in wasting a lot of computing resources and power resources [15]. Its block generation speed and transaction volume recorded in each block limit the transaction throughput of the entire system, and it is not suitable for application scenarios with a huge number of transactions and high real-time requirements. In order to obtain node accounting rights, more and more mining pools appear. These mining pools make the computing power too centralized, which runs counter to Satoshi Nakamoto's concept of decentralization.

PoS Consensus Algorithm. In order to solve the defect that PoW wastes a lot of computing resources, Quantum Mechainic [16] proposed the proof of stake consensus algorithm. It still needs to mine to obtain the accounting rights, but the difficulty of mining will change with the amount of held by the user node, that is, the greater the miner's equity, the less difficult the mining. The stake in PoS represents the coin age of the digital currency owned by the miner. The coin age is the product of the digital currency obtained by the miner in the transaction and the time of owning these digital currency. When the miner succeeds in mining and obtains the accounting right, the coin age will be cleared. PeerCoin [17] is the first digital currency system to use the PoS consensus algorithm. At present, more and more digital currency systems have also begun to use the proof-of-stake mechanism, such as BlackChain, EOS, etc.

Compared with PoW, PoS saves more resources. It dose not require excessive computing resources to compete for the right to bookkeeping. The PoS uses the coin age to represent the size of the equity, but the coin age will be cleared after obtaining the node's bookkeeping rights. This allows miners with a small number of digital currencies to also have the opportunity to compete with those with a large amount of currency for the bookkeeping rights which reflects the fairness of the system and avoids the problem of concentration of computing power at the same time. PoS can also provide users with interest at a certain annual interest rate based on their currency age to increase their enthusiasm and avoid deflation.

Although PoS reduces the time for the entire system to reach a consensus through the proof of stake, it is more prone to fork problems, which may cause a transaction to need more time before it can be confirmed. Its security is relatively low, and malicious nodes are prone to launch attacks.

DpoS Consensus Algorithm. DpoS is also known as the Delegated proof of Stake. In order to solve the problems of PoS, that some nodes with rights and interests passively participate in the accounting rights competition, and the user nodes can also obtain security issues such as coin age [18] when they are offline, Daniel Larimer proposed the DpoS in 2014. Its core idea is to allow nodes with rights and interests (shareholders) to elect delegates to participate in bookkeeping through voting. The votes of shareholders are related to their own stake, The top N nodes with the number of votes will act as delegates, delegates will take turns to obtain the right to keep accounts and verify the newly generated blocks. Each time a block is generated, the delegates will receive transaction fees as a reward. All behaviors of delegate nodes will be supervised by shareholders. Once a delegate node is found to fail to fulfill their contracts, obligations, and responsibilities, the delegate node will be kicked out as an ordinary shareholder, and the deposit [19] submitted by it will be confiscated. The list of delegates will also be replaced according to a cycle time to allow more nodes to participate.

The DpoS mechanism uses voting to select representatives to participate in accounting, which not only saves computing resources, but also reduces the pressure and consensus time of the entire network by reducing the number of election nodes, and improves the transaction throughput of the entire system. Before a node becomes a delegate, it must pay a huge amount of deposit. The serious performance of the delegates' responsibilities has been related to their vital interests, and the system can maintain fairness and stability.

DpoS also has shortcomings. Due to the implementation of the election process, there are many intermediate steps, making the implementation more complicated and prone to security loopholes. Since the node accounting power is always in the hands of the representatives, it reduces its decentralization.

PBFT Consensus Algorithm. Practical Byzantine Fault Tolerance is a variant of the Byzantine fault-tolerant consensus mechanism. It was proposed by Castro and Liskov in the paper Practical Byzantine Fault Tolerance and Proactive Recovery in 1999. The PBFT consensus algorithm is different from proof-based algorithms such as PoW and PoS, and it is a consensus algorithm based on voting. It solves the shortcomings of the original Byzantine algorithm that is not high in efficiency, reduces the complexity of the algorithm from exponential to polynomial, making PBFT simpler in practical applications.

The following is the specific consensus process of PBFT: a client sends a Request message to the master node.After receiving the request and passing the signature verification, the master node will assign a number n to the request operation and broadcast the Pre-Preprape message to all replica nodes. In the Pre-prepare phase, the master node will broadcast the Pre-preprare message to all replica nodes.In the Preprare phase, when a backup node receives the Pre-prerare message broadcast by the master node, it will verify it. After the verification is passed, it will enter the preparation phase. The node will send a Prepare message. If the verification fails, no response will be made. The replica node will also record the messages of Pre-prerare and Prerare to the local log. In the Commit phase: When the master node and the replica node receive the Preprare message, they need to verify it, check whether the signature of the message backup node is correct, and whether the node has received the message request n operations in the current view. If the verification fails, it is discarded. When the node receives $2f + 1$ Preprare messages that have passed the verification, it sends a Commit message to all other nodes. If a node currently receives $2f + 1$ validated Commit messages, it means that most nodes in the network have reached a consensus, start to perform the client's request operation and return a Replay message to the client. The specific PBFT process is shown in Fig. 2:

PBFT is an optimized version of the BFT consensus algorithm, which reduces the complexity of the algorithm and makes it easier to implement in specific application scenarios. Compared with algorithms such as PoW and PoS, it consumes less energy and does not have security issues such as forks.

PBFT also have shortcomings. A large number of messages need to be transmitted between nodes, resulting in a large network load. The more nodes in the network, the longer it takes for the system to reach a consensus. This makes it unsuitable for some high-concurrency and fast transaction application scenarios, and can only be applied to some consortium chains or private chains with small nodes.

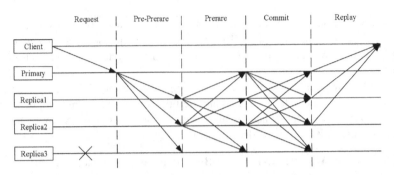

Fig. 2. PBFT consensus process

3   Comparison of Consensus Algorithm Performance

This paper will compare the performance of each algorithm in terms of resource consumption, block generation speed, dgree of decentralization, security, and consistency. Finally, some problems of CFT and BFT algorithms are summarized (Table 1).

Through the analysis and comparison of several consensus algorithms in this paper, we can clearly know that each of them has different problems:

For CFT consensus algorithms, Paxos and Raft can only be applied to network systems with a limited number of nodes and relative reliability. The big obstacle of CFT consensus algorithms is that they cannot accommodate the existence of Byzantine nodes. In order to reach a consensus, the nodes need to conduct a lot of communication, and the entire blockchain network has a large load, which cannot meet the high concurrency requirements. In addition, their design ideas are more complicated and difficult to implement, so their actual application scenarios are relatively limited.

For BFT consensus algorithms, they generally have the following problems:

① High resource consumption: Both PoW and PoS mechanisms use computing power to compete for the right to bookkeeping, resulting in a lot of waste of calculation and power resources.

② Decentralization: Although PoW and PoS and their deformation mechanisms have a high degree of decentralization, the problems of PoW with huge computing power and currency holding centralization of PoS problem will lead to the mastery of accounting rights in the hands of a few nodes, it goes against the idea of decentralization.

③ Long consensus time: PoW, PoS, and PBFT take too much time to reach a consensus. Although some DpoS algorithms can achieve second-level verification, they still cannot meet many application scenarios with huge transaction throughput.

④ Security issues: Regardless of the consensus algorithm, there are different types of security issues, such as 51% computing power attacks, double spending attacks, currency age accumulation and other security threats. In order to meet specific application scenarios, some aspects of the algorithm may be improved and innovated, but many new security issues also follow.

Table 1. Consensus algorithm performance comparison

	Resource consumption	Block generation speed (time/block)	Degree of decentralization	Security	Consistency
Paxos [10]	Low	–	High	Tolerance to Faulty Nodes Less than 1/2 and Intolerance to Malicious Nodes	Final Consistency
Raft [11]	Low	–	High	Tolerance to Faulty Nodes Less than 1/2 and Intolerance to Malicious Nodes	Final Consistency
PoW [1]	Extremely High	10 min (Bitcoin)	High but in Violation of Decentralization with the Emergence of Large Scale Mining Pools	51% Computing Power Attacks and Double Spending Attacks	Prone to Bifurcation and without Final Consistency
PoS [7]	High	64 s (Blackcion)	High with Centralization of Currency Holdings Based on Equity for Booking Rights	Overcoming of Pow's 51% Computing Power Attacks but Existence of Vulnerabilities Such as Unprofitable Attacks	Prone to Bifurcation and without Final Consistency
DpoS [8]	Low	3s(Bitshares)	High, Centralized for Principal and Decentralized for Shareholder	Many Intermediate Links in the Algorithm, Prone to Security Loopholes and the Accumulation of Currency Age	Without Final Consistency

(continued)

⑤ Limited application scenarios: There is no consensus algorithm that can be applied to various application scenarios, and consensus algorithms with specific performance are designed in combination with specific needs and application scenarios.

Table 1. (*continued*)

	Resource consumption	Block generation speed (time/block)	Degree of decentralization	Security	Consistency
PBFT [9]	Low	Slow and Unsuitable for Large Scale Node Consensus	High	Low Security and Intolerance to Malicious Nodes More Than 1/3	Final Consistency Without Bifurcation

⑥ Low cost of malicious behaviors: Many consensus mechanisms do not consider the cost of penalizing malicious nodes, resulting in low costs of malicious behaviors.

4 Summary and Outlook

As the key technology of blockchain technology, the blockchain consensus algorithm has attracted the attention and research of a large number of experts and scholars. In different application scenarios of blockchain, the performance requirements of consensus algorithms are also different. How to design a consensus algorithm suitable for most blockchain application scenarios has become an important direction for the development of blockchain in the future. This paper makes a detailed analysis of six algorithms such as Paxos, Raft, PoW, PoS, DpoS, and PBFT, and expounds their implementation principles and consensus process. By comparing each algorithm in terms of resource consumption, block generation speed, degree of decentralization, security, etc., the advantages and disadvantages of each algorithm are summarized. Future work will first consider resources. Among them, mechanisms such as PoW and PoS require a lot of computing power. We can learn from the Ghost protocol to use these computing powers to solve prime number and matrix problems, instead of pure mining calculation or consider using more environmental-friendly and energy-saving resources to compete for the right to bookkeeping. In terms of security, penalties for malicious nodes can be increased to reduce malicious behavior, credit values can be set for each node's historical behavior, and the difficulty of mining can be adjusted according to different credit values. Problems such as the accumulation of coin age in the PoS mechanism can be solved by changing the accumulation method of coin age. In terms of system consistency, it is possible to consider combining traditional consensus algorithms and algorithms based on Byzantine fault tolerance in the consensus process, so that the new BFT algorithms have final consistency. Only by constantly updating and iterating the consensus algorithm and balancing the security, resources, and fairness of the consensus algorithm, can the blockchain develop better.

Acknowledgment. This work was supported by the Research on identification baesd on EEG high-dimensional data model (E52083) of PhD program of Hunan University of Science and Technology.

References

1. Nakamoto, S., Bitcoin, A.: A peer-to-peer electronic cash system. Bitcoin (2008). https://bit coin.org/bitcoin.pdf
2. Amsden, Z., Arora, R., et al.: The Libra Blockchain https://www.chainnode.com/doc/3631
3. Zeng, S., Ni, X.: A bibliometric analysis of blockchain research, pp. 102–107 (2018)
4. Liang, W., Zhang, D., Lei, X., et al.: Circuit copyright blockchain: blockchain-based homomorphic encryption for IP circuit protection. IEEE Trans. Emerg. Top. Comput. (99), 1 (2020)
5. Liang, W., Fan, Y., Li, K.-C., Zhang, D., Gaudiot, J.-L.: Secure data storage and recovery in industrial blockchain network environments. IEEE Trans. Industr. Inform. **16**, 6543–6552 (2020)
6. Ren, M., Tang, H.B., Si, X.M., et al.: Survey of applications based on blockchain in government department. Comput. Sci. **045**(002), 1–7 (2018)
7. Proof of Stake [EB/OL], 10 November 2017. https://en.bitcoin.it/wiki/ProofofStake. Accessed 17 Oct 2018
8. Bisola A. Delegated Proof-of-Stake (DPoS) Explained [EB/OL], 01 November 2018. https://www.mycryptopedia.com/delegated-proof-stake-dpos-explained/. Accessed 10 Mar 2020
9. Castro, M., Liskov, B.: Practical Byzantine fault tolerance and proactive recovery. ACM Trans. Comput. Syst. (TOCS) **20**(4), 398–461 (2002)
10. Lamport, L.: Paxos made simple. ACM SIGACT News **32**(4) (2016)
11. Ongaro, D., Ousterhout, J.: In search of an understandable consensus algorithm. In: 2014 {USENIX} Annual Technical Conference ({USENIX}{ATC} 2014), pp. 305–319 (2014)
12. Yong, Y., Xiao-Chun, N.I., Shuai, Z., et al.: Blockchain consensus algorithms: the state of the art and future trends. Acta Automatica Sinica (2018)
13. Back, A.: Hashcash-a denial of service counter-measure (2002)
14. Liang, W., Li, K.-C., Long, J., Kui, X., Zomaya, A.Y.: An industrial network intrusion detection algorithm based on multifeature data clustering optimization model. IEEE Trans. Industr. Inform. **16**(3), 2063–2071 (2020)
15. Das, D., Dutta, A.: Bitcoin's energy consumption: is it the Achilles heel to miner's revenue? Econ. Lett. **186**, 1085301–1085306 (2020)
16. Wikipedia. Proof-of-stake [EB/OL]. https://en.bitcoin.it/wiki/Proof_of_Stake
17. King, S., Nadal, S.: PPcoin: peer-to-peer crypto-currency with proof-of-stake, 19 August 2012. 1. self-published paper
18. Lampson, B.: How to build a highly available system using consensus. In: Babaoğlu, Ö., Marzullo, K. (eds.) Distributed Algorithms. LNCS, vol. 1151, pp. 1–17. Springer, Heidelberg (1996). https://doi.org/10.1007/3-540-61769-8_1
19. Schuh, F., Larimer, D.: BitShares 2.0: Financial Smart Contract Platform (2017). Accessed 15 Jan 2015

Design of Intelligent Detection and Alarm System for Accident Anti-collision Based on ARM Technology

Minfu Tan[1], Yujie Hong[2(✉)], Guiyan Quan[1], Yinhua Lin[1], Erhui Xi[3], Shiwen Zhang[4], and Weihong Huang[4]

[1] Big Data Development and Research Center, Guangzhou College of Technology and Business, Guangzhou 528138, China
[2] College of Computer Science and Electronic Engineering, Hunan University, Changsha 410000, China
[3] Department of Computer Science and Engineering, Guangzhou College of Technology and Business, Guangzhou 528138, China
[4] School of Computer Science and Engineering, Hunan University of Science and Technology, Xiangtan 411201, China

Abstract. With the increase of social transportation, traffic accidents are not uncommon, and various disputes often arise in accident responsibility. Law enforcement agencies can only understand the situation at the scene of the accident through the two parties who caused the accident, and the two sides of the accident often have their own opinions. This article designs an anti-collision intelligent detection and alarm system. When cars collide, the collision video recorded during driving can be used as direct evidence, saving a lot of time for both sides of the accident and law enforcement agencies, and improving the efficiency of law enforcement. This article first introduces the research significance of the ARM-based accident collision detection and alarm system, and then deeply analyzes the key technologies used in the system, including the main control A10, embedded system Melis, bus I^2C and acceleration sensor BMA250. Then analyze the entire hardware module of the system. After that, it analyzes in detail the initialization and control of the acceleration sensor BMA250 in the software, and the alarm processing process when an accident occurs. Finally, we summarized the design and research of the entire system, analyzed the system's shortcomings, and looked forward to the work that needs to be improved in the future.

Keywords: ARM-based · Acceleration sensor · Accident collision detection · Accident collision alarm

1 Introduction

Every year about 1.3 million of people dies in road accident, besides 20–50 million are injured or disabled [1]. Demand for safety and security increase in each and every sector [2]. In recent years, China's car ownership and driving population are in a period

© Springer Nature Switzerland AG 2021
M. Qiu (Ed.): SmartCom 2020, LNCS 12608, pp. 194–203, 2021.
https://doi.org/10.1007/978-3-030-74717-6_21

of rapid growth, which indirectly increase the vehicle density, leading to many road accidents resulting in injuries and sometimes lead to death [3, 4]. In 2011, about 211,000 people died or injured in traffic accidents [5]. And road traffic accidents are much more serious than in developed countries. Accident liability disputes are common, and the responsibilities are not clear.

With the rapid advance of the Internet technology, the Internet of Things (IoT) has subtly influenced our daily life [6, 7]. Security in Vehicular Sensor Networks (VSNs) is a critical subject [8]. This can be placed on any vehicle [9]. This article proposes an accident detection system based on the ARM system, which will always record the environment around the car during the operation of the system. It is a user friendly and versatile system [10]. When an accident occurs, a video of the accident process will be saved. It can be used as direct evidence to resolve liability disputes in traffic accidents. The system is controlled by the ARM-based Allwinner A10, and the Melis system is the operating environment. It controls and detects the acceleration sensor BMA250, drives it through the LCD and camera provided by Melis in the event of an accident, records the surrounding environment, and restores it, which is true and accurate. This article proposes an accident detection and restores accident scenarios [11]. It can provide solutions to detect accidents in real time and record the video scene at that time. It can provide the most authentic evidence for dispute settlement in traffic accidents, mainly for the judgment of accident responsibility. Tremendous progress has been made with regard to vehicle safety [12]. The research results can help relevant law enforcement agencies to improve law enforcement efficiency, and at the same time help both parties in the accident to quickly resolve accident disputes, saving both parties' time. The experimental results show that our method has achieved effective results [13]. At the same time, these data can be uploaded to the cloud for storage [14].

2 Research Status at Home and Abroad

In the 1970s, the European Community developed an electromechanical analog vehicle recorder. In the 1990s, computers began to become popular, and the United States and the European Community increased research costs and developed electronic vehicle recorders with better performance and more powerful functions. Leading automobile manufacturers such as General Motors and Ford Motor Company of the United States have also begun to develop on-board recorders, and they have been installed and used in several high-end cars developed by their companies. Generally speaking, the research and development technology of foreign vehicle recorders is relatively mature and has been widely used. In the event of a traffic accident, through the data acquisition sensor on the recorder, the driving data of the car will be automatically recorded in the magnetic card of the vehicle recorder. The collected data can be analyzed by PC to analyze the collision data, simulate the process before the car accident and the scene, reproduce and restore the accident and violation process, and provide a true and reliable legal basis for police, insurance, perpetrators and other departments and individuals.

China's driving recorder has expanded in recent years, and the industry's production capacity has continued to expand. The production capacity of the industry in 2013 was 11.56 million units, and the output in the same period was 10.65 million units, and the industry operating rate was 92.1%. The production capacity in 2017 was 29.94 million units, and the output in the same period was 28.17 million units, and the operating rate during the same period was 94.1%. In 2018, China's driving recorder industry produced about 33.554 million units, imported about 202,000 units, and exported about 7.547 million units. The domestic demand for driving recorder industry was about 26.209 million units.

3 Related Technology Introduction

3.1 Melis Window Introduction

Melis operating system has a simple, practical and efficient window mechanism. There are many types of Melis windows, mainly management windows, dialog boxes, control windows and layer windows. Figure 1 is a schematic diagram of window dependency:

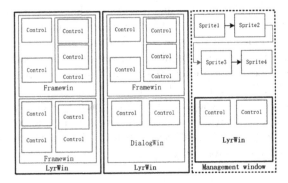

Fig. 1. Window dependency

Management window: The management window is a kind of virtual window with no actual display, but a mechanism window. It is mainly responsible for message processing and distribution, supporting window cross-layer management, coordinating the message processing of lower-level windows, and unifying the entire window tree into a whole.

Dialog box: entity window, with its own unique rectangular area, title bar and client bar. It is the most important and commonly used window type in window applications.

Control window: The entity window that has been encapsulated and tested. Each control window has its own unique properties and operations. Its main function is to facilitate application calls, reduce code reuse, and speed up development efficiency.

Layer window: The basic unit of screen display management, corresponding to a layer, is an entity window on the screen. It has its own display area and is also the carrier of dialog boxes and control windows.

3.2 Melis Message Mechanism

The processing of Melis internal operations and the transmission of information are all done through messages. The message types mainly include messages sent to the window manager by external events and messages triggered internally by the window manager. Figure 2 is a schematic diagram of the message routing structure:

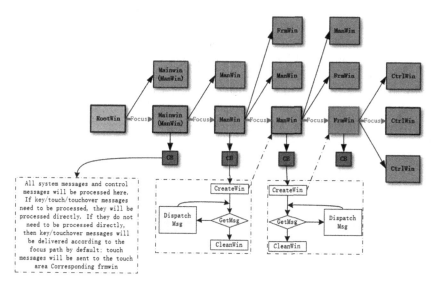

Fig. 2. Message routing structure

Through the message mechanism, the transmission of many internal signals can be completed quickly. Through message routing, each message can be accurately delivered to the sub-window, and various operations can be accurately completed.

4 Anti-collision System Software Design

After the system is powered on, first run the Bootloader of the Melis system after batch file processing. After the system enters, it starts to load LCD and camera drivers, and initialize the timer and G-Sensor. Recording starts after the initialization of the various functions of the system is completed. The G-Sensor pin connected to A10 is read every 0.5 s in the timer. When the read port signal is high, it indicates that a collision has occurred. At this time, the recording will be stopped, and then the first 10 s of recording will be read and used as the beginning of the collision accident video. The video will end after 1 min, and the accident video folder will also be saved.

4.1 System Main Program

It can be seen from the system flow chart of Fig. 3: After the system completes the Bootloader, the system jumps to the main program home_main(). At this point, the system hardware initialization has been completed, and all tasks will be handed over to the Melis operating system for execution. In home_main(), start to initialize all relevant settings of the system, such as the loading of the driver, the initialization of the interface and the initialization of the message manager. The main loaded drivers are camera and LCD drivers. The module driver is provided by Allwinner, which supports almost all related equipment products.

Load the camera and LCD drivers used by the system in home_misc_init().The camera model used by the system is NT99141. NT99141 is a high-performance image sensor with an on-chip 10-bit analog-to-digital conversion ADC with image signal processing functions required by an embedded image signal processor. Load the camera driver: esDEV_Plugin("\\drv\\csi.drv", 0, 0, 1);

A10 provides 3 GPIOs as LCD signal control, 1 PWM as backlight control and complete interface definition of LCD and A10 connection. The LCD display model used by the system is ILI9342. Load LCD driver: esDEV_Plugin("\\drv\\display.drv", 0, 0, 1);

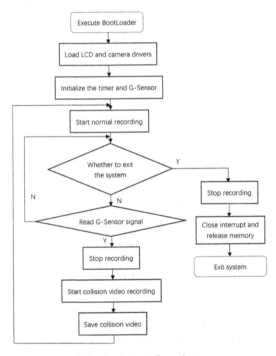

Fig. 3. System flow chart

4.2 G-Sensor Control Design

The control of G-Sensor mainly includes the initialization of G-Sensor, the configuration of related parameters and the detection of G-Sensor collision signal.

G-Sensor Hardware Connection. G-Sensor mainly controls the configuration when working through the I^2C bus and the data communication processing. Read the G-Sensor

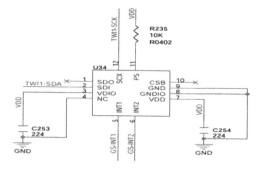

Fig. 4. G-Sensor schematic

status, and judge whether a collision event occurs by the level of the interrupt pin signal. Figure 4 is the schematic diagram of G-Sensor:

TWI1-SDA is used to connect to the A10 port register, corresponding to the I²C data control line. TWT1-SCK is connected to the I²C clock control line. GS-INT1 is the G-Sensor interrupt control pin port, connected to the A10 expansion IO port. In the Melis system, the value of this port can be read directly.

G-Sensor Initialization. After loading the driver, power on and initialize the G-Sensor, configure and start the BMA250 register through I²C. For the internal parameter configuration of G-Sensor, it is mainly to set bandwidth, sensitivity, interrupt and interrupt related configuration. The bandwidth setting register address is 0 x 10, the sensitivity setting register address is 0 x 0f, and the interrupt setting register address is 0 x 16.

(1) Power on the module: According to the BMA250 register table, the 8-bit control module at address 0 x 11 is powered on. Write 0 x 11 to 0 x 11 via I²C, BMA250 is powered on and starts to work. Table 1 is the register configuration function table at address 0 x 11:

Table 1. Power register (0 x 11)

BIT07	BIT06	BIT05	BIT04	BIT03	BIT02	BIT01	BIT00
Suspend	lowpower_on	Reserved	sleep_dur <3>	sleep_dur <2>	sleep_dur <1>	sleep_dur <0>	Reserved

(2) Write 0 x 0a to 0 x 10 via I²C and set the bandwidth to 31.25 Hz. Here we can set 31.25 Hz, the function of broadband setting is to send the range of the effective component frequency of the information contained in the G-Sensor signal (Table 2).

(3) There are many options for sensitivity. The sensitivity range is ±2 g when writing 0 x 03, the sensitivity range is ±4 g when writing 0 x 05, the sensitivity range is

Table 2. Bandwidth register (0 x 10)

BIT07	BIT06	BIT05	BIT04	BIT03	BIT02	BIT01	BIT00
Reserved	Reserved	Reserved	bw <4>	bw <3>	bw <2>	bw <1>	bw <0>

Table 3. Sensitivity register (0 x 1f)

BIT07	BIT06	BIT05	BIT04	BIT03	BIT02	BIT01	BIT00
Reserved	Reserved	Reserved	Reserved	range <3>	range <2>	range <1>	range <0>

±8 g when writing 0 x 08, and the sensitivity range is ±16 g when writing 0 x 0c. The sensitivity setting depends on different scenarios and applications (Table 3).

According to the above function of initializing the relevant register table, we can use the I²C bus to configure the value of the register to complete the initialization of the G-Sensor.

gSensorInitTblGrp[] defines the sensitivity selection and the array of bytes occupied. This array is mainly for the convenience of initialization and sensitivity setting. There are only 3 sensitivity options of ±4 g, ±8 g, and ±16 g. In the actual test, these three sensitivities have fully met the system requirements, and the collision signal can be detected in real time. RegInitTblxg[] is the register address that saves the parameters that BMA250 needs to set and the predetermined parameter value. Its variables are the same by default except for different sensitivity settings. Iic_write_8 is the 8-bit data written to I²C. According to the incoming I²C bus address, as well as the written register address and size, a specific data format is formed, and then written to the I²C bus through the system function.

When the I²C bus transmits data, it is specified that when SCL is high, the data on SDA remains unchanged; when SCL is low, the data on SDA can be changed. When SCL is high, the start signal and stop signal are the only ways to make the bus enter the transmission data or wait for transmission. Under other conditions, there will be no data changes on the bus, and the bus is in the "idle" state. When data needs to be transmitted, change the SCL level to enter the bus "busy" state.

(1) SCL generates 9 clock pulses, including the first 8 bits of data and 1 bit of response clock;
(2) As the data sender, the host controls the SDA to transfer 8-bit data to the slave;
(3) When in the data transmission state, the slave as the data receiver receives the 8-bit data transmitted on SDA;
(4) After the master releases the SDA, the master is converted to a slave and is controlled by the subsequent master;
(5) When the slave is converted from the receiver to the sender and becomes the master, it controls SDA to output ACK;

(6) When in the data input state, the host detects and receives the ACK on SDA as the receiver;

(7) The master and slave determine their role transitions and subsequent actions based on the ACK status.

Figure 5 is a complete communication process of the I²C bus. Among them, SCL represents the clock line; SDA represents the data line; ACK represents low level, with response signal; NACK represents high level, without response.

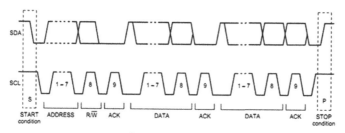

Fig. 5. I²C communication process

In the Melis system, no signal detection is required. As long as the data that needs to be transmitted is combined in a specific format, and then passed into the eLIBs_fioctrl() function, data communication on the I²C bus can be realized.

G-Sensor Settings. Sensitivity is one of the most critical parameters of collision detection. It is essential to set the sensitivity in the system. It is mainly through I²C to write different gSensorInitTblGrp[] values to set the sensitivity.

4.3 Anti-collision Design

When the collision signal is detected, the system ends the current ordinary recording, stops the ordinary recording file, and sets the collision file storage path and the collision file name. The timer processing function __mgui_timer_proc (msg MSG) completes the detection of collisions and the processing after the collision. Figure 6 is the processing flow chart of the anti-collision system:

4.4 Anti-collision System Test Analysis

System testing mainly includes drop test, crash test and flip test. The drop test is mainly the y-axis acceleration test in the vertical direction; The collision test mainly tests whether the system will generate a G-Sensor signal in the event of a collision. It mainly detects that the x-axis and y-axis acceleration in the horizontal direction is too high, exceeding the predetermined accident detection; In the flip test, all accelerations on the 3 axis will be detected. Table 4 shows the accident detection results of 30 simulations conducted by 3 test methods.

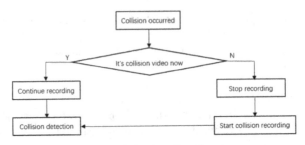

Fig. 6. Collision handling flowchart

Table 4. Test results

Test type	G-Sensor signal detected (times)	Not detected (times)
Drop test	28	2
Crash test	29	1
Flip test	27	3

The test result shows that the detection accuracy is collision > drop > flip under simulated conditions. This shows that collision detection is related to the direction of acceleration on the G-Sensor, and it is better than 3-axis in the 2-axis case. The sensitivity used in the test is ± 8 g. If you want to make the accuracy higher, you can set a higher sensitivity to complete.

5 Summary and Outlook

This paper proposes an accident detection and anti-collision alarm scheme, which can accurately record the surrounding situation of the accident when a collision accident occurs, and provide the most true, direct and efficient evidence for accident identification. This facilitates the work of law enforcement personnel, saves law enforcement time, improves law enforcement efficiency, and at the same time provides a convenient method for both parties in the accident to dispute liability.

This article first introduces the characteristics of the entire system and analyzes the background of the system; Secondly, the key technologies used in each module are analyzed, including the main control, sensors, bus, etc.; Then analyze the key programs and codes of the system software in detail; Finally, detailed steps and process analysis are made for the compilation and testing of the system. In the next step, we will modify the configuration of the acceleration sensor so that it can be detected even when the car has a small collision. In addition, how to use the acceleration sensor to increase the inclination measurement function of the car is also our key research direction.

Acknowledgment. This work is supported by the National Natural Science Foundation of China (No. 61702180) and the Doctoral Scientific Research Foundation of Hunan University of Science and Technology (No. E52083).

References

1. Taher, T., et al.: Accident prevention smart zone sensing system. In: 2017 IEEE Region 10 Humanitarian Technology Conference (R10-HTC). IEEE (2017)
2. Chakole, S.S., Ukani, N.A.: Low power smart vehicle tracking, monitoring, collision avoidance and antitheft system. In: 2020 Second International Conference on Inventive Research in Computing Applications (ICIRCA). IEEE (2020)
3. Kaur, P., et al.: Smart vehicle system using arduino. ADBU J. Electric. Electron. Eng. (AJEEE) 3(1), 20–25 (2019)
4. Kaur, P., Das, A., Borah, M.P.: Vehicles safety system using arduino. ADBU J. Electric. Electron. Eng. (AJEEE) 3(2), 37–43 (2019)
5. Bin, L.: Suggestions on over-the-horizon cloud warning system for collision avoidance of two intersections. In: IOP Conference Series: Earth and Environmental Science, vol. 587, no. 1. IOP Publishing (2020)
6. Dai, W., Qiu, M., Qiu, L., Chen, L., Wu, A.: Who moved my data? privacy protection in smartphones. IEEE Commun. Mag. 55(1), 20–25 (2017). https://doi.org/10.1109/MCOM. 2017.1600349CM
7. Qiu, M.K., Zhang, K., Huang, M.: An empirical study of web interface design on small display devices. In: IEEE/WIC/ACM International Conference on Web Intelligence (WI'04). IEEE (2004)
8. Al-Turjman, F., Lemayian, J.P.: Intelligence, security, and vehicular sensor networks in internet of things (IoT)-enabled smart-cities: an overview. Comput. Electric. Eng. 87, 106776 (2020)
9. Anjaneyulu, M., Mani Kumar, B.: Ultra Smart Integrated vehicle controlled system. Int. Res. J. Eng. Technol. (IRJET) 4(11), 1361–1366 (2017)
10. Sanjana, T., et al.: Automated anti-collision system for automobiles. In: 2017 International Conference on Electrical, Computer and Communication Engineering (ECCE). IEEE (2017)
11. Hacker, J.A., Swan, V.: Automated detection of traffic incidents via intelligent infrastructure sensors. U.S. Patent Application No. 16/735, 323 (2020)
12. Abraham, S., Luciya Joji, T., Yuvaraj, D.: Enhancing vehicle safety with drowsiness detection and collision avoidance. Int. J. Pure Appl. Math. 120(6), 2295–2310 (2018)
13. Qiu, M., Jia, Z., Xue, C., Shao, Z., Sha, E.H.M.: Voltage assignment with guaranteed probability satisfying timing constraint for real-time multiproceesor DSP. J. VLSI Signal Process. Syst. Signal Image Video Technol. 46(1), 55–73 (2007)
14. Gai, K., Qiu, M., Zhao, H.: Security-aware efficient mass distributed storage approach for cloud systems in big data. In: 2016 IEEE 2nd International Conference on Big Data Security on Cloud (BigDataSecurity), IEEE International Conference on High Performance and Smart Computing (HPSC), and IEEE International Conference on Intelligent Data and Security (IDS). IEEE (2016)

Resisting Adversarial Examples
via Wavelet Extension and Denoising

Qinkai Zheng[1], Han Qiu[2], Tianwei Zhang[3], Gerard Memmi[2], Meikang Qiu[4], and Jialiang Lu[1(✉)]

[1] Shanghai Jiao Tong University, 200240 Shanghai, China
{paristech_hill,jialiang.lu}@sjtu.edu.cn
[2] Telecom Paris, 91120 Palaiseau, France
{han.qiu,gerard.memmi}@telecom-paris.fr
[3] Nanyang Technological University, 639798 Singapore, Singapore
tianwei.zhang@ntu.edu.sg
[4] Texas A&M University Commerce, Commerce, TX 75428, USA
meikang.qiu@tamuc.edu

Abstract. It is well known that Deep Neural Networks are vulnerable to adversarial examples. An adversary can inject carefully-crafted perturbations on clean input to manipulate the model output. In this paper, we propose a novel method, WED (Wavelet Extension and Denoising), to better resist adversarial examples. Specifically, WED adopts a wavelet transform to extend the input dimension with the image structures and basic elements. This can add significant difficulty for the adversary to calculate effective perturbations. WED further utilizes wavelet denoising to reduce the impact of adversarial perturbations on the model performance. Evaluations show that WED can resist 7 common adversarial attacks under both black-box and white-box scenarios. It outperforms two state-of-the-art wavelet-based approaches for both model accuracy and defense effectiveness.

Keywords: Adversarial examples · Deep Learning · Model robustness · Wavelet transform · Image denoising

1 Introduction

With the revolutionary development of Deep Learning technology, Deep Neural Networks (DNN) has been widely adopted in many computer vision tasks and applications, e.g., image classification, objective detection, image reconstruction, etc. However, DNN models are well known to be vulnerable against Adversarial Examples (AEs) [5]. An adversary can add carefully-crafted imperceptible perturbations to the original images, which can totally alter the model results. Various methods have been proposed to generate AEs efficiently and effectively such as FGSM [5] and CW [2]. These adversarial attacks have been applied to physical scenarios [7] and real-world computer vision applications.

© Springer Nature Switzerland AG 2021
M. Qiu (Ed.): SmartCom 2020, LNCS 12608, pp. 204–214, 2021.
https://doi.org/10.1007/978-3-030-74717-6_22

It is extremely challenging to defend AEs. First, an adversary has different approaches to generate AEs. It is hard to train a attack-agnostic DNN model without pre-knowledge of the specific attack techniques. Second, correcting the model's results on AEs can usually alter its behaviors on normal inputs, which can lead to non-negligible performance loss. Third, due to the strong transferability of the adversarial examples, the adversary can compromise the models without having access to the model structures or parameters. Such black-box techniques significantly increase the difficulty of model protection via hiding or restricting model information.

There are a variety of works attempting to address those challenges, however, they all suffer certain drawbacks or practical issues. Up to now, there are no satisfactory solutions to defeat all adversarial attacks under both black-box and white-box scenarios. Specifically, (1) adversarial training [5,8] is a class of methods that consider generating AEs during training. Such methods require the knowledge of adversarial attacks to create AEs, which depends on the type of the attacks. (2) Defensive distillation [11] enhances the model robustness during training through the model distillation mechanisms. This approach has been shown ineffective when the adversary slightly modifies the AE generation algorithm. (3) Some approaches proposed to pre-process the input images [12] to remove the adversarial perturbations at the cost of accuracy degradation on clean images. The adversary can still easily defeat such methods by including the transformations into optimization procedure [1].

In this paper, we propose a novel approach, WED, to defend AEs via Wavelet Extension and Denoising. The key insight of our approach is that *by extending the input image into a higher dimensional tensor with non-differentiable wavelet transforms, it is extremely difficult for the adversary to generate perturbations that alter the model's output.* Specifically, WED consists of two innovations. The first innovation is Wavelet Extension. WED adopts wavelet transform to extend the input image into two scales: the original one and low-frequency one. These two scales are then combined as one tensor as the input, which makes it difficult to generate effective adversarial perturbations. The second innovation is to adopt wavelet denoising only at the inference step. The original image, as one scale, keeps the original visual content and elements to assist the model classification. WED also utilizes wavelet denoising during the inference phase to reduce the impact of the adversarial perturbations.

Our comprehensive experiments show that WED outperforms state-of-the-art defenses (e.g., Pixel Deflection (PD) [12] or Wavelet Approximation (WA) [13]) from different aspects. For robustness, WED shows higher robustness against almost all known adversarial attacks under black-box scenarios (e.g. 97% accuracy for AEs generated by CW) compared with PD. For performance, WED has higher classification accuracy on clean image samples (97% accuracy on the test compared with 92% from PD, more details see Sect. 4.2).

The roadmap of this paper is as follows. In Sect. 2, we briefly introduce the background of adversarial attacks and defenses. In Sect. 3, we presents the design of our proposed WED. In Sect. 4, we comprehensively evaluate the robustness and performance of WED and compare it with state-of-the-art solutions under different scenarios. We conclude in Sect. 5.

2   Research Background

2.1   Adversarial Attacks

The goal of adversarial attacks is to make the DNN model give wrong predictions by adding imperceptible perturbations to the original input. Formally, for a clean image x, we denote its corresponding AE as $\tilde{x} = x + \delta$ where δ is the adversarial perturbation. Let F be the classifier mapping function of the DNN model, the process of generating AEs can be formulated as the following problem:

$$min\|\delta\|$$
$$s.t.\ F(\tilde{x}) = l', F(x) = l, l' \neq l \tag{1}$$

where l is the correct class of clean image x, l' is the target class misled by $\tilde{x}$. The adversary aims to find the optimal perturbation to mislead the classifier.

Various techniques have been proposed to generate AEs, and there are two main categories. (1) *Naive Gradient-based* approaches: the adversary generates AEs by calculating the model gradients with pre-set constraints of how much modification can be made on input samples. For instance, Fast Gradient Sign Method (FGSM) [5] calculates the perturbations based on the sign of the gradient of the loss function with respect to the input sample. This kind of methods, also including I-FGSM [7] and MI-FGSM [4], aim at iteratively calculating the perturbations with a small step or with momentum. (2) *Optimized Gradient-based* approaches: the adversary adopts optimization algorithms [2] to find optimal perturbations by considering the gradients of the predictions with respect to input images. This kind of solutions are very powerful under white-box scenario since the attackers can adjust the AE generation according to defense methods. Recently, there are many optimized gradient-based approaches to generate AEs including JSMA [10], DeepFool [9], LBFGS [15], CW [2].

2.2   Defense Strategies

There exists an arms race between adversarial attacks and defenses. Different solutions were proposed to defend different attacks, which are mainly following two directions. The first direction is to apply preprocessing like transformations on input images to reduce the impact of carefully-crafted adversarial perturbations. As shown in Eq. 2, the defense applies a transformation function τ on the input image. It tries to maximize the probability of classifying the adversarial examples $\tilde{x}$ to the correct class of the original image x. The transformation function can be non-differentiable and non-invertible, which makes it difficult for adversaries to get the gradients through back-propagation. Signal processing techniques are commonly adopted in this case, e.g., wavelet transformation [13] and denoising [12]. However, there exists a trade-off between effectiveness and performance: weak transformation fails to remove the impact of small adversarial perturbations; while intensive transformation can result in an obvious accuracy loss on the clean images. Also, under the white-box scenario, such methods are vulnerable to the optimized gradient-based adversarial attacks. If the adversary includes the preprocessing in the optimization procedure, the AEs generated can still be effective to mislead the DNN model.

$$\max_{\tau} P(F(\tau(\widetilde{x})) = F(\tau(x))) \tag{2}$$

The second direction is to modify the target model to increase its robustness. As shown in Eq. 3, F' denotes the classifier mapping function of the modified model. Typical examples include adversarial training [5,8] and network distillation [11]. These approaches are effective under some conditions, however, there will be a high cost due to model retraining.

$$\max_{F'} P(F'(\widetilde{x}) = F'(x)) \tag{3}$$

3 Methodology

3.1 Overview

The key idea of our approach WED is to extend the input image x into two parts $(\tau_1(x), \tau_2(x))$ with two transformation functions (τ_1 and τ_2), as shown in Eq. 4.

$$\max_{F',\tau_1,\tau_2} P(F'(\tau_1(\widetilde{x}), \tau_2(\widetilde{x})) = F'(\tau_1(x), \tau_2(x))) \tag{4}$$

Equation 4 combines two defenses strategies in Eq. 2 and Eq. 3 at the same time by introducing transformations (τ_1 and τ_2) and modifying the mapping function of model F'. We then try to maximize the probability of classifying the AE $\widetilde{x}$ to the correct class as the original image x.

These two transformations must be carefully designed to satisfy two requirements: (1) it should prevent the adversary from affecting the processed images by perturbing the input; (2) it should maintain high prediction accuracy on clean images. To achieve those goals, we adopt the wavelet transformations (extension and denoising) for τ_1 and τ_2 respectively. Figure 1 shows the overview of our approach. During the training phase (Fig. 1(a)), each image is extended to a higher dimensional tensor of two scales, τ_1: an identity mapping of the original image; τ_2: the low-frequency scale extracted by wavelet decomposition. During the inference phase (Fig. 1(b)), the input image is pre-processed in a similar way, except that τ_1 is now a wavelet denoising function applied to the original image.

Both of these two transformations play critical roles in increasing robustness. In order to fool the target model, the two parts must be affected in a sophisticated and collaborative way at the same time. However, since the adversary can only make changes to the original image x, it is difficult to affect the model output by small perturbations. Below, we elaborate and validate each technique.

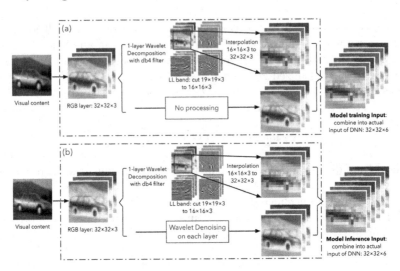

Fig. 1. Methodology overview. (a) Training phase; (b) Inference phase.

3.2 Wavelet Extension

As shown in Fig. 1, WED adopts Wavelet Extension (WavExt) to extract low-frequency information and extend the input. This process can build an image extension that represents the basic visual structures, which can be hardly influenced by adversarial perturbations and can assist the model prediction. This is done by extending the RGB image from $N \times N \times 3$ into $N \times N \times 6$ by adding an image obtained by a reconstruction algorithm based on wavelet transform.

Generally, the wavelet transform represents any arbitrary signal as a superposition of wavelets. Discrete Wavelet Transformation (DWT) decomposes a signal into different levels. DWT decomposes one signal into a low-frequency band (L band) and a high-frequency band (H band) with equal size. For image processing, the DWT is normally processed in a two-dimensional manner as 2D-DWT in two directions: vertical and horizontal. There will be four sub-bands generated as shown in Fig. 2: LL band, LH band, HL band, and HH band. With the proper choice of the wavelet filter, the image can be decomposed into different frequency bands representing various elements such as basic structures, details, etc. As we can observe in Fig. 2, the LL band is an abstract of the basic image structures while the rest three bands represent the image details.

Algorithm 1 shows the detailed steps to process the images. After one level of two-dimensional DWT (Line 1), there are four sub-bands generated each of size $floor(\frac{N-1}{2}) + n$, with $n = 4$ as the filter length. We crop the LL band of each RGB layer to $\frac{N}{2} \times \frac{N}{2}$ (Line 2), and then perform the Bicubic interpolation to resize the extracted low-frequency component back to $N \times N$ (Line 4).

We visually show the effectiveness of the wavelet extension by the saliency map [14]. The saliency map is obtained by first calculating gradients of outputs of the penultimate layer with respect to the input images, and then by normalizing

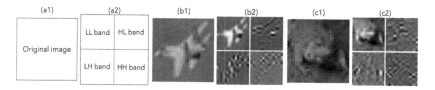

Fig. 2. Examples of 2D-DWT decomposition. 2D-DWT operations: (a1) and (a2); two images as examples: (b1) and (b2), (c1) and (c2).

ALGORITHM 1: WED: Wavelet Extension and Wavelet Denoising.

Input: Image x, size $N \times N \times 3$
Output: processed tensor $x_{Extended}$, size $N \times N \times 6$

1: LL, LH, HL, HH = wavelet-decomposition(x)
2: The size of LL is cropped to $N/2 \times N/2 \times 3$
3: The value of LL is re-scaled to $[0, 255]$
4: x_{Ext} = Bicubic-interpolation(LL)
5: **if** Inference phase **then**
6:    $x_{Extended}$ = concat(wavelet-denoising(x), x_{Ext})
7: **else**
8:    $x_{Extended}$ = concat(x, x_{Ext})
9: **end if**

---

the absolute values of those gradients. A brighter pixel in the saliency map means that the corresponding pixel in the original image has more influence on the model's output. Figure 3 shows that although the saliency maps of a clean image (e) and its AE (f) are very different, the saliency maps of their extended components are nearly identical (see (g) and (h)). This indicates that the wavelet extension can effectively remove the effects of adversarial perturbations.

During training phase, we concatenate the extended component with the original input as the actual input tensor of size $N \times N \times 6$ (Line 8). We adjust the model structure to accept this input size only by changing the dimension of weights in the first layer. The reason that we include the original image is that they still keep details of the images, which can assist the classification and maintain accuracy. However, the existence of those original images provides adversaries with opportunities to manipulate the model behaviors. We introduce another mechanism to eliminate this threat during inference phase.

3.3 Wavelet Denoising

During inference phase, an extra non-differentiable transformation is applied to the original input (Fig. 1(b)). Due to the different pre-processing steps between training phase and inference phase, the AEs generated using the gradients of the trained weights will not have the optimized result as desired by the adversary.

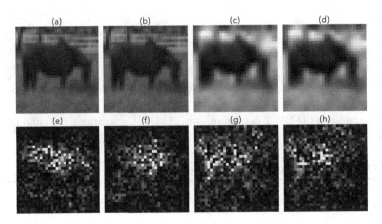

Fig. 3. Similar saliency maps between wavelet extension results ((g) and (h)). (a) and (e): a clean image and its saliency map; (b) and (f): corresponding AE image and its saliency map; (c) and (g): wavelet extension of image (a) and its saliency map; (d) and (h): wavelet extension of AE image (b) and its saliency map.

We implement the wavelet denoising method by combining two approaches, VisuShrink and BayesShrink [3]. Normally wavelet denoising relies on the basic assumption that the noise tends to be represented by small values in the frequency domain. These small values can be removed by setting coefficients below a given threshold to zero (hard threshold) or by shrinking different coefficients to zero by a soft threshold. First, we use the VisuShrink to set a threshold to remove additive noise. For an image X with N pixels, this threshold is given by $\sigma\sqrt{2\log N}$, where the σ is normally smaller than the standard deviation of noise. Second, we adopted the method from [12] and use the BayesShrink based on a soft threshold. We model the threshold for each wavelet coefficient as a Generalized Gaussian Distribution (GGD). The optimal threshold can be further approximated by $\frac{\sigma^2}{\sigma_x}$ where σ_x and β are parameters of the GGD for each wavelet sub-band (Eq. 5). Normally, an approximation of T_h, as shown on the right side of Eq. 5, is used to adapts to the amount of noise in the given image. We adopted the parameter settings from [12].

$$T_h^*(\sigma_x, \beta) = \underset{T_h}{\mathrm{argmin}}\, E(\widehat{X} - X)^2 \approx \frac{\sigma^2}{\sigma_x} \tag{5}$$

4   Evaluations

4.1   Experimental Settings and Implementations

Dataset and Models. We consider an image classification task on CIFAR-10 with 50,000 images for training and 10,000 images for testing. Each image ($32 \times 32 \times 3$) belongs to one of ten classes. All pixel values are normalized to the range $[0, 1]$. The target model is ResNet-29 [6]. It consists of 29 layers containing three bottleneck residual blocks with channel sizes 64, 128, 256, respectively. We use Keras package with Tensorflow backend to implement the model. The training is done by using Adam optimization with its hyper-parameters $\beta_1 = 0.9, \beta_2 = 0.999$. The model reaches the Top-1 accuracy of 92.27% over the testing set after about 150 epochs. Experiments are done on a platform with Intel(R) Core(TM) i7-8700K CPU @ 2.40 GHz and NVIDIA GeForce GTX 1080 Ti GPU.

Attack and Defense Implementations. We test our defense by 7 common AE generation methods: FGSM [15], I-FGSM [7], MI-FGSM [4], L-BFGS [5], JSMA [10], DeepFool [9], CW [2]. All attacks are implemented with the help of the CleverHans library (v3.0.1). We consider two attack scenarios. (1) Black-box scenario: the adversary does not have access to the model parameters, defense mechanism, etc. (2) White-box scenario: the adversary has detailed knowledge of the target model including the trained parameters and defense mechanism. Since the size of input after wavelet extension becomes $N \times N \times 6$, we train another ResNet-29 model by only changing the dimension of weights of the first layer to accept the extended input. The Top-1 testing accuracy reaches 91.96%, which means that WED does not influence the performance of the target model.

Under different attacks, we compare WED with two wavelet transform-based defense solutions: Wavelet Approximation (WA) [13] and Pixel Deflection (PD) [12]. WA uses level-1 wavelet approximation on the input image to get low-resolution images. PD randomly replaces pixels by other pixels randomly selected within a small window and use Bayeshrink wavelet denoising to reduce adversarial noise. For each experiment, we consider the targeted attack, where a new label different from the correct one is selected as the adversary's target. We randomly select 100 samples which are correctly predicted by the original model from the test set, generate the corresponding AEs and measure the top-1 accuracy.

4.2 Black-Box Scenario

In the black-box scenario, the adversary does not know the model parameter. He trains another shadow model with the same network architecture (ResNet-29) to generate AEs. Table 1 shows the top-1 prediction accuracy on the generated AEs for the models without any protection (baseline), with WA, PD and WED, respectively. To clearly show the magnitude of AEs distortion compared with the original images, we calculate the average normalized L_{inf} and L_2 distance.

We observe that our defense is effective towards all kinds of attacks, and outperforms both WA and PD in most cases. Particularly, for attacks with larger L_{inf} distortion (e.g., I-FGSM, and MI-FGSM), PD is less effective. In contrast, our defense still shows strong resistance. Moreover, WED has better accuracy on clean images than WA or PD. Such high prediction accuracy on both clean images and AEs is due to the concatenation of two scales of information, details from the wavelet denoised part and structures from the wavelet extended part.

Table 1. The Top-1 accuracy on **baseline**, WA, PD, WED in the black-box scenario.

Attack	L_{inf}	L_2	**Baseline**	WA	PD	WED
Clean	0.0	0.0	1.0	0.89	0.92	**0.97**
FGSM	0.005	0.28	0.39	0.84	0.84	**0.94**
I-FGSM	0.005	0.21	0.21	0.85	0.86	**0.93**
MI-FGSM	0.005	0.25	0.29	0.84	0.86	**0.92**
JSMA	0.832	4.12	0.0	0.60	0.49	**0.68**
DeepFool	0.015	0.12	0.0	0.87	0.95	**0.95**
LBFGS	0.018	0.15	0.0	0.86	0.92	**0.97**
CW	0.011	0.09	0.0	0.86	0.95	**0.97**

4.3   White-Box Scenario

In this scenario, the adversary knows the exact values of the model parameters. Then he can directly generate AEs based on the target model. In WED, the model input are the tensor ($N \times N \times 6$) extended by wavelet transformations, while the adversary can only provide AEs that have three channels ($N \times N \times 3$). Due to the non-differentiability of the transformations, it is hard for the adversary to adjust the original input with small perturbations to change the output tensors to the malicious ones. Thus in this scenario, we assume the adversary will directly use half of the calculated adversarial tensors as the input.

Table 2. Top-1 accuracy on **WavExt** and WED in the white-box scenario.

Attack	L_{inf}	L_2	**WavExt** (without denoising)	WED
FGSM	0.005	0.28	0.48	**0.62**
I-FGSM	0.005	0.21	0.24	**0.48**
MI-FGSM	0.005	0.24	0.28	**0.50**
JSMA	0.898	4.95	0.19	**0.68**
DeepFool	0.015	0.12	0.85	**0.95**
LBFGS	0.017	0.15	0.77	**0.97**
CW	0.012	0.09	0.87	**0.97**

Existing approaches like WA or PD cannot fit into this scenario as the model input and pre-processed input have different dimensions. So we only compare the effectiveness of WED and the one with only Wavelet Extension (Sect. 3). The results are shown in Table 2. The comparison results indicate that the robustness is increased with the wavelet denoising at the inference phase. We observe that our defense still shows strong resistance against some attacks even the adversary knows the parameters, especially for DeepFool, LBFGS, and CW.

5   Conclusion

In this paper, we propose a novel approach to defend DNN models against adversarial examples. First, we apply a wavelet transform to extend the input images with their structures and basic elements. Second, we utilize wavelet denoising to further reduce the impact of the adversarial perturbations. These two non-differentiable operations can increase the difficulty of generating adversarial perturbations while maintaining the model performance. Our approach provides better robustness and effectiveness over other wavelet-based solutions in defeating different popular adversarial attacks under different scenarios.

References

1. Athalye, A., Carlini, N., Wagner, D.: Obfuscated gradients give a false sense of security: circumventing defenses to adversarial examples. In: Proceedings of the 35th International Conference on Machine Learning, ICML (2018)
2. Carlini, N., Wagner, D.: Towards evaluating the robustness of neural networks. In: 2017 IEEE Symposium on Security and Privacy (SP), pp. 39–57. IEEE (2017)
3. Chang, S.G., Yu, B., Vetterli, M.: Adaptive wavelet thresholding for image denoising and compression. IEEE Trans. Image Process. **9**(9), 1532–1546 (2000)
4. Dong, Y., et al.: Boosting adversarial attacks with momentum. In: Proceedings of the IEEE Conference on Computer Vision and Pattern Recognition, pp. 9185–9193 (2018)
5. Goodfellow, I.J., Shlens, J., Szegedy, C.: Explaining and harnessing adversarial examples. arXiv preprint arXiv:1412.6572 (2014)
6. He, K., Zhang, X., Ren, S., Sun, J.: Identity mappings in deep residual networks. In: Leibe, B., Matas, J., Sebe, N., Welling, M. (eds.) ECCV 2016, Part IV. LNCS, vol. 9908, pp. 630–645. Springer, Cham (2016). https://doi.org/10.1007/978-3-319-46493-0_38
7. Kurakin, A., Goodfellow, I., Bengio, S.: Adversarial examples in the physical world. arXiv preprint arXiv:1607.02533 (2016)
8. Madry, A., Makelov, A., Schmidt, L., Tsipras, D., Vladu, A.: Towards deep learning models resistant to adversarial attacks. arXiv preprint arXiv:1706.06083 (2017)
9. Moosavi-Dezfooli, S.M., Fawzi, A., Frossard, P.: Deepfool: a simple and accurate method to fool deep neural networks. In: Proceedings of the IEEE Conference on Computer Vision and Pattern Recognition, pp. 2574–2582 (2016)
10. Papernot, N., McDaniel, P., Jha, S., Fredrikson, M., Celik, Z.B., Swami, A.: The limitations of deep learning in adversarial settings. In: 2016 IEEE European Symposium on Security and Privacy (EuroS&P), pp. 372–387. IEEE (2016)
11. Papernot, N., McDaniel, P., Wu, X., Jha, S., Swami, A.: Distillation as a defense to adversarial perturbations against deep neural networks. In: 2016 IEEE Symposium on Security and Privacy (SP), pp. 582–597. IEEE (2016)
12. Prakash, A., Moran, N., Garber, S., DiLillo, A., Storer, J.: Deflecting adversarial attacks with pixel deflection. In: Proceedings of the IEEE Conference on Computer Vision and Pattern Recognition, pp. 8571–8580 (2018)
13. Shaham, U., et al.: Defending against adversarial images using basis functions transformations. arXiv preprint arXiv:1803.10840 (2018)

14. Simonyan, K., Vedaldi, A., Zisserman, A.: Deep inside convolutional networks: Visualising image classification models and saliency maps. arXiv preprint arXiv:1312.6034 (2013)
15. Szegedy, C., et al.: Intriguing properties of neural networks. arXiv preprint arXiv:1312.6199 (2013)

Privacy-Preserving Computing Framework for Encrypted Data Under Multiple Keys

Jun Zhang[1]([⊠]), Zoe L. Jiang[2], Ping Li[3], and Siu Ming Yiu[4]

[1] Shenzhen University, Shenzhen, China
jzhang3@cs.hku.hk
[2] Harbin Institute of Technology (Shenzhen), Shenzhen, China
[3] South China Normal University, Guangzhou, China
liping26@mail2.sysu.edu.cn
[4] The University of Hong Kong, Pok Fu Lam, Hong Kong
smyiu@cs.hku.hk

Abstract. With the popularity of cloud computing, data owners encrypt their data with different keys before uploading data to the cloud due to privacy concerns. Homomorphic encryption makes it possible to calculate on encrypted data but is usually restricted to single key. Based on an additively homomorphic encryption supporting one multiplication, we propose a general computing framework to execute common arithmetic operations on encrypted data under multiple keys such as addition, multiplication, exponentiation. Our scheme rely on two non-colluding servers and is proven to be secure against semi-honest attackers. The experimental evaluations demonstrate the practicality of our computing framework in terms of lower computational complexity and communication overhead.

Keywords: Privacy-preserving computing framework · Cloud computing · Homomorphic encryption · Multiple keys

1 Introduction

Cloud computing has already been a popular paradigm for various applications. The cloud service providers (i.e., Amazon, Google) offer abundant storage and computing resources, which are delivered on demand over the internet as a service. Compared to the traditional paradigm (local database), cloud computing shows great advantages such as flexibility, scalability, stability and cost-effectiveness. More and more institutions buy cloud service to support their technical implementations. They outsource their datasets to the cloud and perform data mining remotely with the cloud's computing power. Despite of the advantages of cloud computing, the public are concerned about data security and privacy [5,6].

M. Qiu (Ed.): SmartCom 2020, LNCS 12608, pp. 215–225, 2021.
https://doi.org/10.1007/978-3-030-74717-6_23

Encryption is a common privacy-preserving technique to protect data security and privacy. Homomorphic encryption allows computation on encrypted data and generates an encrypted output which is the same as being performed on decrypted data. A lot of privacy-preserving data mining schemes are based on homomorphic encryption [3,8,15–17]. Zhang et al. propose a secure dot product protocol and demonstrate an application of the SVM classifier [16]. They also construct a privacy preserving regression protocol [15]. Secure schemes are put forward in [8] to run association rule mining. Zou et al. show how to conduct k-Means clustering with cloud computing [17]. Privacy-preserving schemes for deep neural networks on encrypted data are designed in [3]. We find that the researchers choose different homomorphic encryption methods for different data mining algorithms and each privacy-preserving scheme is tailored for a specific data mining algorithm. However, the data owners cannot predict in advance which data mining algorithm they would like to run on their outsourced data in the future. On the other hand, it is impractical for the data owners to run different encryption methods before outsourcing and keep several ciphertexts for the same data in the cloud. Therefore, a general computing framework is required which allows different data mining algorithms to execute on the ciphertexts encrypted under the same encryption method.

Most homomorphic encryption methods can only guarantee the correctness of computations on the ciphertexts under single key. However, cooperation between different institutions are very common in this big data era. For example, the hospitals might collaborate with each other to figure out the causes of diseases. A big and varied dataset contribute to improving the performance of data mining. Although the hospitals are willing to build data mining together, they still need to make sure the data privacy of their patients. Encrypting data with the same key equals to sharing data in plaintext among the hospitals, which compromises data privacy. We cannot force the hospitals to encrypt their dataset with the same key so that single-key homomorphic encryption can work correctly. As a result, different institutions should encrypt their datasets with different keys.

In this paper, we propose a **P**rivacy-preserving **C**omputing **F**ramework for encrypted data under **M**ultiple **K**eys (PCFMK). We depend on a cryptosystem which supports homomorphic additions and one multiplication under multiple keys [14]. Our contributions can be summarized as follows:

- Different data owners are allowed to encrypt their data with different keys and outsource their encrypted datasets to the cloud for storage and data mining.
- We construct secure protocols in our PCFMK for common operations such as addition, multiplication and exponentiation.
- We perform experimental evaluation and the results show that our PCFMK performs better than the existing solutions, in terms of computational complexity and communication overhead between cloud servers.

The rest of this paper is organized as follows. Section 2 lists the existing studies related to this paper. Section 3 presents some preliminary information. Section 4 introduces our system model and threat model. The details of our scheme are clarified in Sect. 5. Then we perform security analysis in Sect. 6 and experimental evaluation in Sect. 7. Finally, we conclude this paper in Sect. 8.

2   Related Work

Generally speaking, there are three ways to cope with multiple encryption keys. The first one is multi-key fully homomorphic encryption, which allows any computable polynomial function to be executed on the encrypted data under multiple keys. As far as we know, there exist two multi-key fully homomorphic encryption solutions [4,10], but the efficiency of them remains an open question. The second approach is ciphertext transformation. Ciphertext transformation changes the keys of encrypted data to a common one. This procedure depends on decryption and re-encryption [11], proxy re-encryption [13], or key-switching matrix [16]. The third method is ciphertext extension, which achieves multi-key homomorphism at the cost of expanding the size of ciphertexts [14,15]. It is observed that different privacy-preserving schemes use different underlying encryption methods and are designed for different data mining algorithms. Moreover, these schemes handle with multiple keys in different ways.

Liu et al. [9] propose a toolkit for privacy-preserving outsourced calculation under multiple keys. Their cryptographic primitive is an additively homomorphic encryption with double trapdoor decryption [1]. Their privacy-preserving outsourced calculation toolkit includes basic operations of integer numbers. The splitting of strong secret key causes potential security leak, of which the details are explained in [7]. Rong et al. [12] put forward two sets of building blocks to support outsourced computation over encrypted data under multiple keys. The first set of building blocks needs lookup table to complete decryption which restricts the size of message space. Two operations (addition and exponentiation) in the second set of building blocks lead to many interactions between cloud servers.

3   Preliminaries

3.1   Additively Homomorphic Encryption (AHE)

Homomorphic encryption allows computation on encrypted data and generates an encrypted output which is the same as being performed on decrypted data. Supposed that we are given two ciphertexts $E(m_1)$ and $E(m_2)$, we have $E(m_1+m_2) = E(m_1)E(m_2)$ for an additively homomorphic encryption (denoted as E). In this paper, **we focus on an additively homomorphic encryption method called BCP cryptosystem** (also known as Modified Paillier Cryptosystem) [1].

3.2 AHE Under Multiple Keys

Zhang et al. showed how to make the BCP cryptosystem support multiple keys by extending the ciphertext [14,15]. **To achieve additive homomorphism under n keys, the ciphertext should be modified to include n + 1 parts.** For example, assume that we have two ciphertexts under different keys, $E_a(m_1) = (C_{m_1}^{(1)}, C_{m_1}^{(2)})$ under key a and $E_b(m_2) = (C_{m_2}^{(1)}, C_{m_2}^{(2)})$ under key b, where $C^{(1)}$ and $C^{(2)}$ denote the first and second parts of the ciphertext. We can compute $E_{ab}(m_1 + m_2) = (C_{m_1}^{(1)}, C_{m_2}^{(1)}, C_{m_1}^{(2)} C_{m_2}^{(2)})$.

3.3 AHE Supports One Multiplication

Catalano and Fiore [2] put forward a scheme to enable existing additively homomorphic encryption to compute one multiplication (denoted as $\mathcal{E}$). Given two "multiplication friendly" ciphertexts $\mathcal{E}(m_1) = (m_1 - b_1, E(b_1))$ and $\mathcal{E}(m_2) = (m_2 - b_2, E(b_2))$, the multiplication $\mathcal{E}(m_1 m_2)$ is calculated as follows. **Two non-colluding servers are used to store $\mathcal{E}(m)$ and b, respectively.**

$$\mathcal{E}(m_1 m_2) = E[(m_1 - b_1)(m_2 - b_2)]E(b_1)^{m_2 - b_2} E(b_2)^{m_1 - b_1}$$
$$= E(m_1 m_2 - b_1 b_2) \tag{1}$$

3.4 AHE Supports One Multiplication Under Multiple Keys

The BCP cryptosystem supporting multiple keys (see Sect. 3.2) can be combined with the scheme that allows additively homomorphic encryption to compute one multiplication (see Sect. 3.3). The final additively homomorphic encryption (AHE) supporting one multiplication under multiple keys are denoted as $\mathcal{E}_{BCP}$.

4 Model Description

4.1 System Model

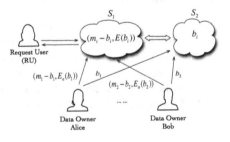

Fig. 1. System model

Our system model consists of data owner, request user and cloud server in Fig. 1. A trusted party handles with the distribution of public and private keys in our system, which is not shown in the figure. We assume that there are n data owners in our scheme. To make it clear, we only list two data owners in the figure - Alice and Bob. Alice owns message m_1 and Bob owns message m_2. We need two non-colluding servers - S_1 and S_2. Alice sends $(m_1 - b_1, E_a(b_1))$ to cloud server S_1 where E_a represents an encryption under key a. Alice sends b_1 to cloud server S_2. Likewise, Bob sends $(m_2 - b_2, E_b(b_2))$ under key b to cloud server S_1. Bob sends b_2 to cloud server S_2. Therefore, cloud server S_1 stores $(m_i - b_i, E(b_i))$ and cloud server S_2 stores b_i. A request user (RU) requests the cloud to run some calculations on encrypted data under multiple keys. Once the cloud completes those calculations, RU receives an encrypted result returned by the cloud. One data owner can also act as a request user. Our goal is to construct a general computing framework to support different requests of calculations on encrypted data under multiple keys.

4.2   Threat Model

We assume that all parties in our system model are semi-honest, which include data owners, request users, cloud servers. Semi-honest can also be described as honest-but-curious. All the participants will follow the protocol step by step honestly, but they are curious about the true value of encrypted data. They try to figure out the plaintexts of encrypted data by observing the input, intermediate results or final output. The semi-honest assumption makes sense in real applications. A request user should get authorization from data owners before launching a computing request to the cloud. Data owners would not allow a complete stranger compute or analyze on their encrypted data. Data owners generate signatures with their private keys based on the identity of the request user. The cloud can verify the validity of these signatures from a request user with the public keys of data owners. Moreover, we also assume that S_1 and S_2 are non-colluding, meaning they would never collude with each other. This assumption is reasonable as we can set up S_1 and S_2 using different cloud service providers.

5   Privacy Preserving Computing Framework

We present the underlying protocols of our privacy-preserving computing framework for encrypted data under multiple keys: Secure Addition Protocol (**SA**), Secure Multiplication Protocol (**SM**) and Secure Exponentiation Protocol (**SE**). The cryptosystem we use is $\mathcal{E}_{BCP}$, which is introduced in Sect. 3.4. For a message m_i, we have $\mathcal{E}_{BCP}(m_i) = (m_i - b_i, E(b_i))$ where E denotes the BCP cryptosystem described in Sect. 3.1. The cloud server S_1 stores $\mathcal{E}_{BCP} = (m_i - b_i, E(b_i))$. The cloud server S_2 stores b_i in plaintext. The secret key of each data owner is shared between S_1 and S_2.

5.1   Secure Addition Protocol (SA)

Given two ciphertexts $\mathcal{E}_{BCP}(m_1) = (m_1 - b_1, E_a(b_1))$ under key a and $\mathcal{E}_{BCP}(m_2) = (m_2 - b_2, E_b(b_2))$ under key b. The secure addition protocol calculates $\mathcal{E}_{BCP}(m_1 + m_2)$ as Eq. (2), where b_1 and b_2 are random numbers.

$$\begin{aligned}
\mathcal{E}_{BCP}(m_1 + m_2) &= \mathcal{E}_{BCP}(m_1) + \mathcal{E}_{BCP}(m_2) \\
&= [(m_1 + m_2) - (b_1 + b_2), E_{ab}(b_1 + b_2)]
\end{aligned} \tag{2}$$

5.2   Secure Multiplication Protocol (SM)

Given two ciphertexts $\mathcal{E}_{BCP}(m_1) = (m_1 - b_1, E_a(b_1))$ under key a and $\mathcal{E}_{BCP}(m_2) = (m_2 - b_2, E_b(b_2))$ under key b. The secure multiplication protocol calculates $\mathcal{E}_{BCP}(m_1 m_2)$ as Eq. (3).

$$\begin{aligned}
\mathcal{E}_{BCP}(m_1 m_2) &= E_a[(m_1 - b_1)(m_2 - b_2)] E_a(b_1)^{m_2 - b_2} E_b(b_2)^{m_1 - b_1} \\
&= E_{ab}(m_1 m_2 - b_1 b_2)
\end{aligned} \tag{3}$$

According to Sect. 3.3, the scheme in [2] can only make an additively homomorphic encryption support one multiplication. To remove this restriction, we should modify the structure of $\mathcal{E}_{BCP}(m_1 m_2)$. Our goal is to recover an encrypted multiplication $\mathcal{E}_{BCP}(m_1 m_2) = E_{ab}(m_1 m_2 - b_1 b_2)$ back to the form of $\mathcal{E}_{BCP}(m) = (m - b, E(b))$. We rely on interactions between two non-colluding servers to complete decryption and encryption. To be specific, the secret keys are shared between two cloud servers and we have $a = a_1 + a_2$ and $b = b_1 + b_2$. The cloud server S_1 stores a_1 and b_1. The cloud server S_2 stores a_2 and b_2.

(1) The cloud server S_1 sends $E_{ab}(m_1 m_2 - b_1 b_2)$ to S_2.
(2) S_2 partially decrypts $E_{ab}(m_1 m_2 - b_1 b_2)$ with a_2 and b_2. S_2 computes $b_1 b_2$, encrypts $b_1 b_2$ with key a and gets $E_a(b_1 b_2)$.
(3) S_2 sends the partially decrypted result $E_{ab}^{pd}(m_1 m_2 - b_1 b_2)$ and $E_a(b_1 b_2)$ to S_1, where E^{pd} denotes partially decryption.
(4) S_1 decrypts $E_{ab}^{pd}(m_1 m_2 - b_1 b_2)$ with a_1 and b_1, and obtains $m_1 m_2 - b_1 b_2$. $\mathcal{E}_{BCP}(m_1 m_2)$ are represented as $(m_1 m_2 - b_1 b_2, E_a(b_1 b_2))$.

Secure Constant Multiplication (SCM): Given a positive constant t and a ciphertext $\mathcal{E}_{BCP}(m) = (m - b, E(b))$, $\mathcal{E}_{BCP}(tm) = [t(m - b), E(b)^t]$. Given a negative constant $-t$, $\mathcal{E}_{BCP}(-tm) = [-t(m - b), E(b)^{N-t}]$.

5.3   Secure Exponentiation Protocol (SE)

Given two ciphertexts $\mathcal{E}_{BCP}(m_1) = (m_1 - b_1, E_a(b_1))$ under key a and $\mathcal{E}_{BCP}(m_2) = (m_2 - b_2, E_b(b_2))$ under key b. The secure exponentiation protocol calculates $\mathcal{E}_{BCP}(m_1^{m_2})$ as follows.

(1) The cloud server S_1 computes $E_a(m_1) = E_a(m_1 - b_1)E_a(b_1)$ and $E_b(m_2) = E_b(m_2 - b_2)E_a(b_2)$. S_1 generates two positive random numbers - $t_1 \in \mathbb{Z}_N^+$ and $t_2 \in \mathbb{Z}_N^+$. S_1 computes $E_a(t_1 m_1) = E_a(m_1)^{t_1}$ and $E_b(t_2 m_2) = E_b(m_2)^{t_2}$. S_1 partially decrypts $E_a(t_1 m_1)$, $E_b(t_2 m_2)$ with a_1 and b_1. S_1 sends the partially decrypted result $E_a^{pd}(t_1 m_1)$, $E_b^{pd}(t_2 m_2)$ to S_2.

(2) S_2 decrypts $E_a^{pd}(t_1 m_1)$, $E_b^{pd}(t_2 m_2)$ with a_2 and b_2. S_2 generates a positive random numbers - $t_3 \in \mathbb{Z}_N^+$. S_2 calculates $(t_1 m_1)^{t_2 m_2}$, $-t_2 m_2 t_3$ and send them to S_1.

(3) S_1 computes $[(t_1 m_1)^{t_2 m_2}]^{t_2^{-1}} = (t_1 m_1)^{m_2}$ and $(t_1^{-t_2 m_2 t_3})^{t_2^{-1}}$. S_1 sends $(t_1^{-m_2 t_3})$ and $\mathcal{E}_{BCP}[(t_1 m_1)^{m_2}]$ to S_2.

(4) S_2 first calculates $(t_1^{-m_2 t_3})^{t_3^{-1}}$ and then executes $\mathbf{SCM}(\mathcal{E}_{BCP}[(t_1 m_1)^{m_2}]; t_1^{-m_2})$, where $\mathbf{SCM}$ denotes secure constant multiplication.

6 Security Analysis

We consider the semi-honest model. Data owners, request user, and cloud server S_1 or S_2 follow the protocol step by step honestly, but they are curious to infer the value of encrypted data by observing the inputs, outputs and intermediate results. We perform security analysis of our scheme with the Real and Ideal Paradigm and Composition Theorem. We use a simulator in the ideal world to simulate the view of a semi-honest adversary in the real world. If the view in the real world is computationally indistinguishable from the view in the ideal world, the protocol is secure. According to the Composition Theorem, the entire scheme is secure if each step is proved to be secure. For sake of simplicity, we consider a scenario where data owner Alice, data owner Bob, cloud server S_1, and cloud server S_2 are involved. The possible adversaries are denoted as $\mathcal{A} = (\mathcal{A}_{Alice}, \mathcal{A}_{Bob}, \mathcal{A}_{S_1}, \mathcal{A}_{S_2})$.

Theorem 1. *The secure addition (**SA**) protocol computes addition on encrypted data and is secure against semi-honest adversaries $\mathcal{A}$.*

Proof. We set up simulators for the involved parties - Sim_{Alice}, Sim_{Bob}, Sim_{S_1} and Sim_{S_2}.

We assume that Sim_{Alice} computes $\mathcal{E}_{BCP}(1)$. The view of $\mathcal{A}_{Alice}$ is $\mathcal{E}_{BCP}(m_1)$. $\mathcal{A}_{Alice}$ cannot distinguish the real execution from the ideal simulation due to the semantic security of the cryptosystem $\mathcal{E}_{BCP}$.

Likewise, Sim_{Bob} computes $\mathcal{E}_{BCP}(2)$. The view of $\mathcal{A}_{Bob}$ is $\mathcal{E}_{BCP}(m_2)$. Similarly, $\mathcal{A}_{Bob}$ cannot distinguish the real execution from the ideal simulation due to the semantic security of the cryptosystem $\mathcal{E}_{BCP}$.

Sim_{S_1} simulates $\mathcal{A}_{S_1}$ with $\mathcal{E}_{BCP}(x)$ and $\mathcal{E}_{BCP}(y)$ where x and y are two random numbers. $\mathcal{A}_{S_1}$ cannot distinguish $\mathcal{E}_{BCP}(m_1)$, $\mathcal{E}_{BCP}(m_2)$ from $\mathcal{E}_{BCP}(x)$, $\mathcal{E}_{BCP}(y)$. Based on the semantic security of $\mathcal{E}_{BCP}$, the views of $\mathcal{A}_{S_1}$ are indistinguishable in the real and ideal world.

Sim_{S_2} simulates $\mathcal{A}_{S_2}$ in method 2 with $E_{ab}(b_1' + b_2')$ where b_1' and b_2' are two random numbers. $\mathcal{A}_{S_2}$ cannot distinguish $E_{ab}(b_1 + b_2)$ from $E_{ab}(b_1' + b_2')$. Due to the semantic security of the underlying BCP cryptosystem, the reviews of $\mathcal{A}_{S_2}$ are also indistinguishable in the real and ideal world.

The security proofs of SM and SE are similar to the above. Due to space limit, we do not go into details here.

7 Experimental Evaluation

7.1 Performance of Our Scheme

The configuration of our PC is Windows 10 Enterprise 64-bit Operating System with Intel(R) Core(TM) i5-7500 CPU (4 cores), 3.41 GHz and 16 GB memory.

To provide platform independence, we use Java to implement our scheme. We rely on the BigInteger class to process big numbers, which supports all the required basic operations. We use five security parameters - 1024-bit, 1280-bit, 1536-bit, 1792-bit and 2048-bit (following the same setting as in [9]). It is obvious that we achieve higher-level security with longer bit-length, but it takes more time to operate on the ciphertexts when the bit-length increases.

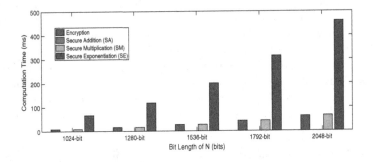

Fig. 2. Running time of our computing framework

We evaluate the performance of our privacy-preserving computing framework. Based on the ciphertexts under different parameters, we test the performance of our secure protocols - Secure Addition Protocol (**SA**), Secure Multiplication Protocol (**SM**), Secure Exponentiation Protocol (**SE**). For each secure protocol, we run 1000 times and record the average execution time. The results are shown in Fig. 2. Overall, the running time of each secure protocol increases with the bit-length of N. We observe that SE is the most time-consuming. On the contrary, the running time for SA is the minimum. SA takes less than 0.1 ms and thus the computation time of SA is negligible in Fig. 2.

7.2 Comparison with Existing Schemes

In our analysis, we focus on exponentiation operations and unfixed number of multiplications while ignoring fixed number of additions or multiplications (following the same setting as in [9]).

We compare our scheme with [9, 12] in terms of computational complexity. The results are shown in Table 1, where N/A means not applicable. Our scheme is obviously more efficient than [9] and there is no implementation of SE in [9]. Rong et al. [12] design two sets of building blocks. The first set of building blocks relies on additive homomorphism and the second set depends on multiplicative homomorphism. Therefore, we observe that SA is inversely correlated with SM in [12]. Moreover, SM and SE in [12] (first set) require Discrete Logarithm (Dlog) operation. A lookup table is maintained to facilitate decryption, which also restricts the size of message space. In terms of computational complexity, our scheme outperforms [9, 12].

We compare the communication overhead of our scheme with [9, 12] in Table 2, where μ is the bit-size of the plaintext. Overall, our scheme is better than [9]. As the underlying cryptosystem is different, our SE protocol causes more interactions than [12]. The advantage of our scheme is that it can be easily extended to support various operations, which we will present in our future work.

Table 1. The comparison of computational complexity

Operation	Our Scheme	[9]	[12] (First set)	[12] (Second set)
SA	2Mul	$33|N|$	2Mul	15Exp+20Mul
SM	$21|N|$	$72|N|$	9Exp+2Mul+1Dlog	2Mul
SE	$18|N|$	N/A	95Exp+27Mul+11Dlog	16Exp+13Mul

Table 2. The comparison of communication overhead (bits)

Operation	Our Scheme	[9]	[12] (First set)	[12] (Second set)
SA	0	$16|N|$	0	$18|G|$
SM	$12|N|$	$36|N|$	$6|G|$	0
SE	$O(\mu^2)$	N/A	$18|G|$	$10|G|$

$|G|$: the order of a multiplicative group G of the ElGamal cryptosystem

8 Conclusions

In this paper, we propose a privacy preserving computing framework (PCMFK), which supports addition, multiplication and exponentiation on encrypted data under different keys. Compared to the existing schemes, our computing framework performs better in terms of computational complexity and communication overhead. Based on our computing framework, various privacy preserving data mining systems can be constructed. Moreover, different data mining algorithms are allowed to run on the same encrypted dataset. One limitation of our scheme is that it runs only on the integer domain. Our future work is to extend our computing framework to include more data mining operations.

References

1. Bresson, E., Catalano, D., Pointcheval, D.: A simple public-key cryptosystem with a double trapdoor decryption mechanism and its applications. In: Laih, C.-S. (ed.) ASIACRYPT 2003. LNCS, vol. 2894, pp. 37–54. Springer, Heidelberg (2003). https://doi.org/10.1007/978-3-540-40061-5_3
2. Catalano, D., Fiore, D.: Using linearly-homomorphic encryption to evaluate degree-2 functions on encrypted data. In: ACM SIGSAC CCS, pp. 1518–1529 (2015)
3. Chen, H., Dai, W., Kim, M., Song, Y.: Efficient multi-key homomorphic encryption with packed ciphertexts with application to oblivious neural network inference. In: ACM SIGSAC CCS, pp. 395–412 (2019)
4. Chen, H., Dai, W., Kim, M., Song, Y.: Efficient multi-key homomorphic encryption with packed ciphertexts with application to oblivious neural network inference. In: ACM SIGSAC CCS, pp. 395–412. ACM (2019)
5. Dai, W., Qiu, M., Qiu, L., Chen, L., Wu, A.: Who moved my data? privacy protection in smartphones. IEEE Commun. Maga. **55**(1), 20–25 (2017)
6. Gai, K., Qiu, M., Zhao, H.: Security-aware efficient mass distributed storage approach for cloud systems in big data. In: 2016 IEEE 2nd International Conference on Big Data Security on Cloud (BigDataSecurity), pp. 140–145 (2016)
7. Li, C., Ma, W.: Comments on "an efficient privacy-preserving outsourced calculation toolkit with multiple keys". IEEE Trans. Inf. Forensics Secur. **13**(10), 2668–2669 (2018)
8. Li, L., Lu, R., Choo, K.K.R., Datta, A., Shao, J.: Privacy-preserving-outsourced association rule mining on vertically partitioned databases. IEEE Trans. Inf. Forensics Secur. **11**(8), 1847–1861 (2016)
9. Liu, X., Deng, R.H., Choo, K.K.R., Weng, J.: An efficient privacy-preserving outsourced calculation toolkit with multiple keys. IEEE Trans. Inf. Forensics Secur. **11**(11), 2401–2414 (2016)
10. López-Alt, A., Tromer, E., Vaikuntanathan, V.: On-the-fly multiparty computation on the cloud via multikey fully homomorphic encryption. In: ACM Symposium on Theory of Computing, pp. 1219–1234. ACM (2012)
11. Peter, A., Tews, E., Katzenbeisser, S.: Efficiently outsourcing multiparty computation under multiple keys. IEEE Trans. Inf. Forensics Secur. **8**(12), 2046–2058 (2013)
12. Rong, H., Wang, H.M., Liu, J., Xian, M.: Efficient privacy-preserving building blocks in cloud environments under multiple keys. J. Inf. Sci. Eng. **33**(3), 635–652 (2017)
13. Wang, B., Li, M., Chow, S.S., Li, H.: A tale of two clouds: computing on data encrypted under multiple keys. In: IEEE Conference on Communications and Network Security (CNS), pp. 337–345. IEEE (2014)

14. Zhang, J., He, M., Yiu, S.-M.: Privacy-preserving elastic net for data encrypted by different keys - with an application on biomarker discovery. In: Livraga, G., Zhu, S. (eds.) DBSec 2017. LNCS, vol. 10359, pp. 185–204. Springer, Cham (2017). https://doi.org/10.1007/978-3-319-61176-1_10
15. Zhang, J., He, M., Zeng, G., Yiu, S.M.: Privacy-preserving verifiable elastic net among multiple institutions in the cloud. J. Comput. Secur. **26**(6), 791–815 (2018)
16. Zhang, J., Wang, X., Yiu, S.M., Jiang, Z.L., Li, J.: Secure dot product of outsourced encrypted vectors and its application to SVM. In: Proceedings of the Fifth ACM International Workshop on Security in Cloud Computing, pp. 75–82 (2017)
17. Zou, Y., et al.: Highly Secure Privacy-Preserving Outsourced k-Means Clustering under Multiple Keys in Cloud Computing. Secur. Commun. Netwo. (2020)

Deep Reinforcement Learning Based on Spatial-Temporal Context for IoT Video Sensors Object Tracking

Panbo He[1], Chunxue Wu[1(✉)], Kaijun Liu[1], and Neal N. Xiong[2]

[1] University of Shanghai for Science and Technology, Shanghai 200093, China
wcx@usst.edu.cn
[2] Department of Mathematics and Computer Science, Northeastern State University, Tahlequah, OK 74464, USA

Abstract. The Internet of Things (IoT) is the upcoming one of the major networking technologies. Using the IoT, different items or devices can be allowed to continuously generate, obtain, and exchange information. The new video sensor network has gradually become a research hotspot in the field of wireless sensor network, and its rich perceptual information is more conducive to the realization of target positioning and tracking function. This paper presents a novel model for IoT video sensors object tracking via deep Reinforcement Learning (RL) algorithm and spatial-temporal context learning algorithm, which provides a tracking solution to directly predict the bounding box locations of the target at every successive frame in video surveillance. Crucially, this task is tackled in an end-to-end approach. Considering the tracking task can be processed as a sequential decision-making process and historical semantic coding that is highly relevant to future decision-making information. So a recurrent convolutional neural network is adopted acting as an agent in this model, with the important insight that it can interact with the video overtime. In order to maximize tracking performance and make a great use the continuous, inter-frame correlation in the long term, this paper harnesses the power of deep reinforcement learning (RL) algorithm. Specifically, Spatial-Temporal Context learning (STC) algorithm is added into our model to achieve its tracking performance more efficiently. The tracking model proposed above demonstrates good performance in an existing tracking benchmark.

Keywords: Recurrent network · Reinforcement learning · Spatial-temporal context · IOT · Video sensors · Visual object tracking

1 Introduction

The Internet of Things (IOT) is one of the integrated parts of the Internet in the future. The wide range of applications of video sensors target tracking in the behavior identification, video surveillance, human-computer interaction in, makes it a very popular field of computer vision research topics [1]. At present, target tracking algorithm based on deep learning adopts some measures to solve this problem. In the case where the training data

© Springer Nature Switzerland AG 2021
M. Qiu (Ed.): SmartCom 2020, LNCS 12608, pp. 226–235, 2021.
https://doi.org/10.1007/978-3-030-74717-6_24

of the object tracking is very limited, the auxiliary non-tracking training data is used for pre-training to obtain the general representation of the characteristics of the object [2]. That large-scale classification database such as ImageNet is directly used to be trained, and deep convolutional neural network (CNN) such as VGG-Net can be used to obtain the representation of the object. Then the observation model is used to classify to obtain tracking results [3, 4]. This approach not only avoids lacking of the sample when large-scale CNN is directly trained in the tracking process, but also take full advantage of the ability of a strong characterization of deep learning.

However, there remains great difference between image classification task and object tracking. Here one kind of model called MDNet [5] process that the tracking video data can be directly used to pre-train CNN, so that general representation of the characteristic of the object can be obtained. In recent years, Recurrent Neural Network (RNN), especially Long Short-Term Memory (LSTM), GRU, etc., shows prominent performance in the timing task. [6] introduces a novel measure to model and exploit the reliable part useful for overall tracking, and finally solved the problem of tracking drift caused by the accumulation of prediction errors and propagation. In addition, different from the RTT with RNN modeling two-dimensional plane association in the model, The framework region proposal + CNN [7–10] replaces the method with sliding window and manual design feature used for traditional target detection. In terms of slow object detection, other improvement models [11, 12] have been proposed for to solve the problem.

Avoiding manually extracting features, CNN has been becoming popular and dominated the field of object tracking in recent years [13–15]. To a large extend, the superior performance of the model depends on the capability of CNN to be able to learn good character representation of the objects. Although the good performance in capturing spatial feature, it's influenced by the limited manual temporal constraints added to the frameworks. To this end, in order to harness the power of deep-learning models to automatically learn both spatial and temporal constraints, especially with longer-term information aggregation and disambiguation, a novel visual tracking approach based on recurrent convolutional networks [16]. Although simple the idea is, it's powerful enough. Spatial-temporal context learning is outstanding in some extensive applications such as visual tracking [17], action detection [18, 40], action classification [19], action recognition [20].

This paper proposes an end-to-end approach to predict the bounding box of the target object and maximize it in the long run. Two methods which are respectively reinforcement learning (RL) and Spatial-temporal context learning (STC) are put into use when the model is trained, contributing to the efficient target object localization. RL and STC are designed in our framework and there is no need to get ground truth, so that more data can be reasonably used.

2   Related Works

2.1   Visual Object Tracking

Traditional Object Tracking. Traditional tracking algorithm can be divided into two categories: generation model and discriminative model. The generation method uses the generation model to describe the apparent characteristics of the target, and then minimizes the reconstruction error by searching the candidate target. The production method focuses on the characterization of the target itself, ignoring the background information, and is prone to drift when the target itself changes violently or is obscured. Generation model is generally divided into three categories: mode-based models [21–23], subspace-based models [24, 25] and the models based on sparse representation [26–31]. In contrast, the discriminative model distinguishes between the target and the background by training the classifier. This method is also often called tracking-by-detection. And STRUCK [37] achieve the adaptive scale during tracking using Haar with structured output SVM. TLD [38] is a classic long-term tracker, even if the effect is not particularly good. VTS [39] tracks a target robustly by searching for the appropriate trackers in each frame. In recent years, various machine learning algorithms have been applied to discriminant methods, among which there are multiple instances learning, boosting and structured SVM, etc.

Target Tracking Method Based on Deep Learning. One of the magic of deep learning comes from the effective learning of a large number of annotated training data, while the target tracking only provides the first frame of the bounding-box as training data. In this case, it is difficult to track a deep learning model at the start of tracking in terms of the current target. Here are some approaches to deal with the problem. [2] use the auxiliary image data pre-training deep learning model, and fine-tuning on-line tracking. [3, 4] extract features via the CNN classification network pre-trained by existing large-scale classification data set. [5] uses the tracking sequence to pre-train the model, and fine-tuning when online tracking. [6] propose a novel idea of harnessing the power of the recurrent neural network for target object tracking.

Recurrent-neural-network trackers. Currently, several researches have explored to apply recurrent neural network to visual tracking. [34] proposed a spatially supervised recurrent convolutional neural network. What's more, a YOLO is exploited to achieve object detection on each frame and a recurrent network is put into use to regress the detections from YOLO network [39].

Spatial-temporal-context trackers. [17] proposes a propose a multi-channel features spatial-temporal context (MFSTC) learning algorithm with an improved scale adaptive scheme. [35] introduces an improved spatial-temporal context tracking algorithm based on a dual-object model to solve the problem that the object model is wrongly updated and hard to be recovered after long term occlusion.

2.2 Deep Reinforcement Learning

Reinforcement learning [41, 42] acts as a sequential decision-making problem, requiring continuous selection of some actions, and get the greatest benefits from the completion of these actions as the best results. It does not need to have the correct input/output pair, nor does it need to precisely correct the suboptimal behavior. Agent can perform an action, receive the observation of current environment, and also receive the reward after it performs an action.

Reinforcement learning has been applied to learn task-specific policies such as image classification [32, 43], image caption generation [33], etc. [16] specially tailored for solving the visual tracking problem by combing CNN, RNN and RL algorithms. Inspired by the successful works, this paper explores a more effective strategy to develop a novel target object tracking approach. RNN and CNN are combined to pay attention to spatial and temporal features. In addition, the full framework is trained via RL and spatial-temporal context learning algorithm. This paper will describe the framework in detail in Sect. 3.

3 The Existing Popularity About Deep RL and STC Tracker (DRST)

3.1 Framework of DRST

Shown as Fig. 1, At each time step t, the feature extraction network takes the image x_t from input sequences. The visual features are generated from feature extraction network in the second stage. In order to obtain spatial-temporal features, firstly the visual features are incorporated by STC and the recurrent network. Then spatial-temporal feature c_t acting as the ground-truth and hidden state h_t are extracted from STC and the Recurrent network respectively. Specially, the Recurrent network also receives previous hidden state h_{t-1} as input. The last stage is to generate predicted location l_t of target object at each time-step, which is directly extracted from the last four elements of hidden state of the Recurrent network. The model is trained with reinforcement learning to make the predicted location as accuracy as possible. Specifically, the reward r_t is defined for each prediction during training process with RL to update network parameters. The aim of DRST model is to maximize the cumulative rewards, such that the tracking performance is maximized.

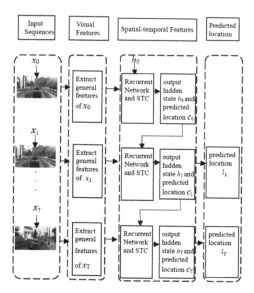

Fig. 1. Framework of DRST

3.2 DRST Networks

3.2.1 Feature Extraction Network

In this part, general features are extracted from feature extraction network f_c parameterized by w_c such as CNN. The learned weights contribute to a generalized understanding of visual objects. Specifically, the convolutional neural network takes every video frame and a location vector s_t as its input. Encoding the video frame into a feature vector i_t of size 4096, CNN processes the location vector and the feature vector as a combo denoted as at each time-step. The value of s_t is directly set to be the normalized location coordinate at the first frame in a given sequence. Otherwise, s_t is assigned a value of zero and only the feature vector i_t under these circumstances is fed into the recurrent network.

3.3 STC and RL Tracking

Based on the summary mentioned above, the spatial-temporal context (STC) and Reinforcement Learning (RL) is applied into the networks to run at frame-rates beyond real-time and to maximize the tracking performance.

3.3.1 STC Tracking

STC is applied in our framework to train the model for better performance, whose tracking result, that is, the location of target object, can be considered as the ground-truth at every video frame. And then it can correct the tracking results in the training process with RL algorithm.

In the field of tracking, the local context is composed of the target object with the background within a certain area near it. Thus, the local context has a strong spatial-temporal relationship in successive frames, which can help to predict the location of target object in the next frame. In general, time contexts help to guarantee target location; spatial contexts can provide more accurate information to help us distinguish between target object and background.

3.3.2 RL Tracking

At every timestep during training, the agent in Reinforcement learning is set to execute an action, and then receive a corresponding reward r_t from the environment. In our research, two kinds of rewards are defined. The first reward used in the early stage is:

$$r_t = -\rho max(|l_t - c_t|) - (1 - \rho)avg(|l_t - c_t|) \tag{1}$$

Where c_t means the predicted location of target object outputted by the STC model, l_t outputted by the recurrent network, ρ represents the parameter, the operators $avg(\cdot)$ and $max(\cdot)$ respectively represent the mean and maximum of the pixel-wise. The second reward used in the late stage is:

$$r_t = \frac{|l_t \cap c_t|}{|l_t \cup c_t|} \tag{2}$$

where the calculation of the reward in Eq. 13 can be described as the intersection region divided by the union one (IOU) between l_t and c_t.

c_t outputted by STC model is adopted as an auxiliary to help RL track the target object more fast and robustly. As shown in Eq. 12 and Eq. 13, the reward acts as a measure of the closeness between the predicted location l_t from recurrent network and the other one c_t from STC model.

3.4 Algorithm Implementation

In the framework, this paper leverages the power of CNN to extract feature representation, and it's followed by LSTM as recurrent network in our model for its great representation of temporal feature. Specifically, our model can be divided into two parts: feature extraction network and recurrent network. The feature extraction network takes every video frame x_t as input. The feature o_t encoded from feature vector x_t acts as the input to the recurrent network and STC at the same time. In addition, the recurrent network combines with the previous hidden state h_{t-1} and outputs the next hidden state h_t which is capable of extracting the predicted location of target object. Then the reward r_t in RL is calculated based on o_t and l_t. After that, the gradient is updated and finally the weight parameter w_c and w_r can be easily computed by standard backpropagation. RL and STC algorithm are applied during training process where RL trains the model with labeled data which means our model takes every video frame with target ground-truth location when trained via RL algorithm, while STC unlabeled data.

4 Experiments and Performance Evaluation

The proposed algorithm is implemented in Python, running at 9 frames per second on an Intel i5 Core machine with 2.4 GHz CPU and 10 GB RAM.

For comparison, this paper runs 3 state-of-art algorithms including STRUCK, TLD, VTS and our proposed approach DRST with the same video datasets. In order to compare the robustness of illumination of those four trackers, this paper conducts the experiment on three videos to evaluate the performance of DRST, where two taxies and one bus were driving under different light conditions and traffic conditions. What's more, an online target object tracking benchmark [1] including most highly representative tracking scenarios such as occlusions, different scale, illumination variations, high speed, etc. is also applied in our experiments. Experiments conducted on the online target object tracking benchmark is to evaluate the performance of DRST compared with STRUCK, TLD and VTS based on one-pass evaluation (OPE). The feature information in observed video frames is obtained by a YOLO network [11]. The network is fine-tuned on the PASCAL VOC dataset, capable of detecting target objects of 20 classes. This paper chooses a subset of 15 videos from the benchmark, which belong to different categories. The video sequences mentioned above are summarized in Table 1.

4.1 Comparison Experiments Based on DRST

This paper conducts the experiment on three videos to evaluate the performance of DRST, where two taxies and one bus were driving under different traffic conditions and

Table 1. Summary of Average Overlap Scores (AOS) results for all 12 trackers. The best and second best performance are respectively in redand bluecolors.

Sequence	DRST	STRUCK	TLD	VTS
Human 2	0.517	0.599	0.334	0.129
Gym	0.583	0.066	0.142	0.111
Skater	0.540	0.334	0.297	0.359
BlurBody	0.519	0.661	0.384	0.270
CarScale	0.595	0.347	0.434	0.429
Dancer2	0.683	0.771	0.655	0.717
BlurCar1	0.609	0.760	0.587	0.214
Dog	0.388	0.262	0.592	0.263
Jump	0.561	0.113	0.088	0.053
Woman	0.049	0.680	0.109	0.132
CarDark	0.636	0.849	0.473	0.702

three light conditions in varying degrees, which are respectively darker light, moderate light and brighter light.

The tracking results based on the challenging factors are shown on Fig. 2, Fig. 3, Fig. 4 respectively.

#35   #87   #144

Fig. 2. Tracking in the different light

In these three sequences, illumination and scale affect the performance of some trackers in the whole process. The STRUCK, TLD, VTS fail at frame around 9, 17, 55, 80, 131, 204 and 375 respectively. The proposed DRST can track the target correctly all the time.

It demonstrates that our model shows a more accurate prediction of the target location compared with other trackers, and hardly drift in the tracking process to the background. As a result of the combination about high-level feature and temporal sequence, our framework can handle challenging situations such as deformation, illumination variation, rotation, etc.

5 Conclusions

To conclude, this paper has proposed the improved model of video sensors object tracking based on deep reinforcement learning and spatial-Temporal context learning in videos. To the best of our knowledge, Reinforcement learning (RL) and spatial-temporal-context algorithm (STC) are applied in our framework to solve the target object tracking problem, achieving a good tracking performance in the long-term. Our model is trained end-to-end, allowing it run faster than real-time. In general, the main contributions of our work can be listed as follows:

1. This paper proposes an end-to-end approach to predict the bounding box of the target object and maximize it in the long run.
2. Two methods which are respectively reinforcement learning (RL) and Spatial-temporal context learning (STC) are put into use when the model is trained, contributing to the efficient target object localization.
3. RL and STC are designed in our framework and there is no need to get ground truth, so that more data can be reasonably used.

Acknowledgements. This research was supported by Shanghai Science and Technology Innovation Action Plan Project (16111107502, 17511107203) Shanghai key lab of modern optical systems.

References

1. Wu, Y., Lim, J., Yang, M.H.: Online object tracking: a benchmark. IEEE Conf. Comput. Vis. Pattern Recogn. **9**(4), 2411–2418 (2013)
2. Wang, N., Yeung, D.Y.: Learning a deep compact image representation for visual tracking. Int. Conf. Neural Inf. Process. Syst. **1**, 809–817 (2013)
3. Wang, L., Ouyang, W., Wang, X.: Visual tracking with fully convolutional networks. In: IEEE International Conference on Computer Vision, pp. 3119–3127. IEEE (2016)
4. Wu, P.F., Xiao, F., Sha, C., Huang, H.P., Wang, R.C., Xiong, N.: Node scheduling strategies for achieving full-view area coverage in camera sensor networks. Sensors **17**(6), 1303–1307 (2017)
5. Nam, H., Han, B.: Learning multi-domain convolutional neural networks for visual tracking. In: 2016 IEEE Conference on Computer Vision and Pattern Recognition (CVPR), pp. 4293–4302 (2016)
6. Cui, Z., Xiao, S., Feng, J., Yan, S.: Recurrently target-attending tracking. In: IEEE Conference on Computer Vision & Pattern Recognition, pp. 1449–1458. IEEE Computer Society (2016)
7. Girshick, R., Donahue, J., Darrell, T.: Region-based convolutional networks for accurate object detection and segmentation. IEEE Trans. Pattern Anal. Mach. Intell. **38**(1), 142–158 (2015)
8. Gui, J., Hui, L., Xiong, N.X.: A game-based localized multi-objective topology control scheme in heterogeneous wireless networks. IEEE Access **5**, 2396–2416 (2017)
9. Xia, Z., Wang, X., Sun, X., Liu, Q., Xiong, N.: Steganalysis of LSB matching using differences between nonadjacent pixels. Multimed. Tools Appl. **75**, 1947–1962 (2016). https://doi.org/10.1007/s11042-014-2381-8

10. Gao, L., Yu, F., Chen, Q., Xiong, N.: Consistency maintenance of do and undo/redo operations in real-time collaborative bitmap editing systems. Clust. Comput. **19**(1), 255–267 (2015). https://doi.org/10.1007/s10586-015-0499-8

11. Liu, W., et al.: SSD: single shot MultiBox detector. In: Leibe, B., Matas, J., Sebe, N., Welling, M. (eds.) Computer Vision – ECCV 2016. ECCV 2016. Lecture Notes in Computer Science, vol. 9905, pp. 21–37. Springer, Cham (2016). https://doi.org/10.1007/978-3-319-46448-0_2

12. Bertinetto, L., Valmadre, J., Henriques, J.F., Vedaldi, A., Torr, P.H.S.: Fully-convolutional siamese networks for object tracking. In: Hua, G., Jégou, H. (eds.) ECCV 2016. LNCS, vol. 9914, pp. 850–865. Springer, Cham (2016). https://doi.org/10.1007/978-3-319-48881-3_56

13. Fang, W., Li, Y., Zhang, H., Xiong, N., Lai, J., Vasilakos, A.V.: On the through put-energy trade off for data transmission between cloud and mobile devices. Inf. Sci. **283**, 79–93 (2014)

14. Lu, X., Chen, S., Xiong, N.: ViMediaNet: an emulation system for interactive multimedia based telepresence services. J. Super Comput. (SCI Indexed) **73**, 3562–3578 (2017)

15. Zhang, D., Maei, H., Wang, X., Wang, Y.F.: Deep Reinforcement Learning for Visual Object Tracking in Videos, p. 10. arXiv preprint (2017)

16. Zhou, X., Liu, X., Yang, C., Jiang, A., Yan, B.: Multi-channel features spatio-temporal context learning for visual tracking. IEEE Access **5**, 12856–12864 (2017)

17. Baek, S., Kim, K.I., Kim, T.: Real-time online action detection forests using spatio-temporal contexts. In: 2017 IEEE Winter Conference on Applications of Computer Vision (WACV), Santa Rosa, CA, pp. 158–167(2017)

18. Girdhar, R., Ramanan, D., Gupta, A., Sivic, J., Russell, B.: ActionVLAD: learning spatio-temporal aggregation for action classification. In: 2017 IEEE Conference on Computer Vision and Pattern Recognition (CVPR), pp. 3165–3174 (2017)

19. Lee, H., Jung, M., Tani, J.: Recognition of visually perceived compositional human actions by multiple spatio-temporal scales recurrent neural networks. IEEE Trans. Cogn. Dev. Syst. **10**(4), 1058–1069 (2018)

20. Wang, Y., et al.: Dynamic propagation characteristics estimation and tracking based on an EM-EKF algorithm in time-variant MIMO channel. Inf. Sci. **408**, 70–83 (2017)

21. Lu, Z., Lin, Y.-R., Huang, X., Xiong, N., Fang, Z.: Visual topic discovering, tracking and summarization from social media streams. Multimed. Tools Appl. **76**(8), 10855–10879 (2016). https://doi.org/10.1007/s11042-016-3877-1

22. He, S., Yang, Q., Wang, J., Yang, M.H.: Visual tracking via locality sensitive histograms. In: Computer Vision and Pattern Recognition. IEEE 2013, pp. 2427–2434 (2013)

23. Shu, L., Fang, Y., Fang, Z., Yang, Y., Fei, F., Xiong, N.: A novel objective quality assessment for super-resolution images. Int. J. Signal Process. Image Process. Pattern Recogn. **9**(5), 297–308 (2016)

24. Xu, T., Feng, Z.H., Wu, X.J., Kittler, J.: Learning adaptive discriminative correlation filters via temporal consistency preserving spatial feature selection for robust visual object tracking. IEEE Trans Image Process. **28**(11), 5596–5609 (2019)

25. Zhang, T.Z., Liu, S., Yan, S.C., Ghanem, B., Ahuja, N., Yang, M.H.: Structural sparse tracking. In: Proceedings of the 2015 IEEE Conference on Computer Vision and Pattern Recognition, pp. 150–158. IEEE (2015)

26. Xiong, N., Liu, R.W., Liang, M., Liu, Z., Wu, H.: Effective alternating direction optimization methods for sparsity-constrained blind image deblurring. Sensors **7**, 174–182 (2017)

27. Zhang, H., Liu, R.W., Wu, D., Liu, Y., Xiong, N.N: Non-convex total generalized variation with spatially adaptive regularization parameters for edge-preserving image restoration. J. Internet Technol. **17**(7), 1391–1403 (2016)

28. Xia, Z., Xiong, N.N., Vasilakosc, A.V., Sun, X.: EPCBIR: an efficient and privacy-preserving content-based image retrieval scheme in cloud computing. Inf. Sci. **387**, 195–204 (2017)

29. Fang, Y., Fang, Z., Yuan, F., Yang, Y., Yang, S., Xiong, N.N.: Optimized Multi-operator Image Retargeting Based on Perceptual Similarity Measure. IEEE Transactions on Systems, Man, and Cybernetics: Systems **47**, 1–11 (2016)

30. Zhang, C., Wu, D., Xiong, N., et al.: Non-local regularized variational model for image deblurring under mixed gaussian-impulse noise. J. Internet Technol. **16**(7), 1301–1320 (2015)

31. Ba, J., Mnih, V., Kavukcuoglu, K.: Multiple object recognition with visual attention. Comput. Sci. 1–10 (2014)

32. Xu, K., Ba, J., Kiros, R.: Show, attend and tell: neural image caption generation with visual attention. In: Computer Science, pp. 2048–2057 (2015)

33. Ning, G., et al.: Spatially supervised recurrent convolutional neural networks for visual object tracking. In: IEEE International Symposium on Circuits and Systems. IEEE, pp. 1–4 (2017)

34. Zhang, H.Y., Zheng, X.: Spatio-temporal context tracking algorithm based on dual-object model. Optics Preci. Eng. **24**(5), 1215–1223 (2016)

35. Williams, R.J.: Simple statistical gradient-following algorithms for connectionist reinforcement learning. In: International Conference on Learning Representation. ICLR, pp. 1095–32 (2015)

36. Hare, S., Saffari, A., Torr, P.H.S.: Struck: structured output tracking with kernels. In: IEEE International Conference on Computer Vision, ICCV 2011, pp. 6–11 (2011)

37. Kalal, Z., Mikolajczyk, K., Matas, J.: Tracking-learning-detection. IEEE Trans. Pattern Anal. Mach. Intell. **34**(7), 1409–1422 (2012)

38. Kwon, J., Lee, K.M.: Tracking by sampling trackers. In: IEEE International Conference on Computer Vision. IEEE, pp. 1195–1202 (2011)

39. Shahzad, A., et al.: Real time MODBUS transmissions and cryptography security designs and enhancements of protocol sensitive information. Symmetry **7**(3), 1176–1210 (2015)

40. Huang, K., Zhang, Q., Zhou, C., Xiong, N., Qin, Y.: An efficient intrusion detection approach for visual sensor networks based on traffic pattern learning. IEEE Trans. Syst. Man Cybern. Syst. **47**(10), 2704–2713 (2017)

41. Wu, W., Xiong, N., Wu, C.: Improved clustering algorithm based on energy consumption in wireless sensor networks. IET Netw. **6**(3), 47–53 (2017)

42. Chunxue, W., et al.: UAV autonomous target search based on deep reinforcement learning in complex disaster scene. IEEE Access **7**, 117227–117245 (2019)

43. Ling-Fang Li, X., Wang, W.-J., Xiong, N.N., Yong-Xing, D., Li, B.-S.: Deep learning in skin disease image recognition. a review. IEEE Access **8**, 208264–208280 (2020)

Imputation for Missing Items in a Stream Data Based on Gamma Distribution

Zhipeng Sun[1,2], Guosun Zeng[1,2(✉)], and Chunling Ding[3]

[1] Department of Computer Science and Technology, Tongji University, Shanghai 200092, China
{1710351,gszeng}@tongji.edu.cn
[2] Tongji Branch, National Engineering and Technology Center of High Performance Computer, Shanghai 200092, China
[3] College of Chemical Science and Engineering, Tongji University, Shanghai 200092, China
dingcl@tongji.edu.cn

Abstract. During the collection of real-time data, data missing is a common phenomenon in some stream data, which leads to the difficulty of such stream analysis and knowledge mining. In this paper, we study an approach of data imputation based on gamma distribution, which determines the number of missing data items in each time interval and the values of missing data items in a stream. We also present some metrics, such as fitting degree, credibility, matching degree, to evaluate the effectiveness of data imputation. Experimental results show that our approach improves the credibility by 15.0% to 20.0% compared to EMI, a widely adopted approach of data imputation, and outperforms traditional techniques significantly.

Keywords: Stream computing · Real-time data · Data missing · Gamma distribution · Probability Density Function (PDF) · Data imputation

1 Introduction

With the rapid development of the Internet of things and smart terminals, applications from various domains, such as social network, e-commerce, health care and many others, produce lots of stream data [1]. However, due to irresistible reasons including sensor failure, network interruption and virus attack, data missing may occur frequently [2, 3]. For example, traffic monitoring devices of the intelligent transport system are running all day in San Antonio, USA. The missing data detected range from 5 to 25% in most days [4]. Unfortunately, if some important data are missing, serious problems will arise. For example, in real-time traffic accident detection, many cameras are installed in traffic accident-prone areas. If some parts of a video are missing due to malfunction of equipment, traffic policemen may be hard to identify the person who is responsible for this accident. To analyze and find out the causes of a traffic accident, policemen need to work hard to complete missing parts for such video by collecting information related to the accident. Therefore, what we need is very complete sampling data even in the era of big data [5]. Once incomplete and missing data sets occur, we must do everything possible to complete them.

© Springer Nature Switzerland AG 2021
M. Qiu (Ed.): SmartCom 2020, LNCS 12608, pp. 236–247, 2021.
https://doi.org/10.1007/978-3-030-74717-6_25

In this paper we present a novel approach based on gamma distribution to imputing missing data in a stream. Our main contributions can be summarized as follows:

- Based on gamma distribution, we calculate the number of missing data items in a stream, and obtain the value of each missing data item using Lagrange's mean value theorem.
- We present some metrics, like the fitting degree, credibility, quantile relative error, and matching degree. These metrics evaluate not only the effectiveness of our proposed approach but also the accuracy of missing data items imputed.
- We conduct experiments both on a synthetic dataset and a real-world dataset. Experimental results show that our approach improves the credibility by 15.0% to 20.0% on the synthetic dataset and the matching degree by 38.2% on the real-world dataset compared to EMI.

The rest of this paper is structured as follows. Section 2 investigates the related work. Section 3 presents some concepts and metrics. Section 4 proposes a novel approach to imputing missing data items in a stream data and makes a theoretical analysis. Some experiments are conducted to show the advantage of our approach over traditional techniques in Sect. 5. Section 6 concludes this paper.

2 Related Work

Data imputation usually fills missing data with observed information. Scholars have made a lot of efforts on data imputation approaches. The traditional imputation methods can generally be divided into two categories: Imputation based on non-model and model-based imputation. The former method such as random imputation (RI) and mean imputation (MI) [6] do not use any model to estimate missing data. RI method is the simplest method which is similar to max/min imputation methods [7], but their results are very inaccurate. In order to improve accuracy, missing data can also be imputed by the mean of all observed values. Unfortunately, the above methods do not take into account the relationship between data, leading to poor results. The latter method, model-based imputation method [8], is based on the construction of a special data model. Such models can be either predictive or descriptive [7]. Imputation methods based on a predictive model impute data by making a prediction over the observed data where a predictive model can be a linear regression, a decision tree etc. The other model-based imputation methods are based on a descriptive model which identifies patterns or relationship among data. For example, the missing data can be imputed by the most frequent data of the k-most similar data [9]. Obviously, the effect of data completion is affected by parameter k, and k is hard to choose and determine. In order to obtain higher accuracy of imputation, scholars adopt the maximum likelihood method. Expectation maximization imputation (EMI) [10] is just such a method which computes the best possible imputed values based on the available dataset. Since imputing one data item may underestimate standard error, multiple imputation method [11] imputes missing data with a set of n possible values. This method uses these possible values n times to generate n complete data sets. However, multiple imputation method creates lots of data sets, which leads to high computing overheads. As traditional imputation approaches have high computational complexity, it is not suitable to impute missing data for application scenarios where real-time response is particularly required. To this end, many researchers study data imputation combined statistics-based models with artificial intelligence techniques.

In fact, data items in a stream follow a certain distribution law. So it is reasonable to impute missing data based on its distribution. Dick et al. [12] proposed that miss data can be imputed based on Gaussian distribution. Vellido [13] designed a method of data imputation by a constrained mixture of distribution models, and pointed out that student t-distribution was a robust alternative to Gaussian distribution for mixture models. Deng et al. [14] investigated the relationship between the data distribution and the topology of wireless sensor networks, studied the generation of missing data by using Poisson stochastic process, and discussed maximum missing data rates in term of different topology networks. Demirtas [15] proposed a multiple imputation method based on the generalized lambda distribution.

Gamma distribution is a very important distribution in probability theory and statistics. Exponential distribution, Erlang distribution, and chi-squared distribution are special cases of the gamma distribution [16]. In recent years, a few literatures begin to study data imputation methods based on gamma distribution. Luo et al. [17] estimated the parameters of gamma in the case of incomplete data. Li et al. [18] proposed the empirical Bayes bilateral test of shape parameter for gamma distribution on a dataset with random missing data. Ma et al. [19] established the mixture-Gamma distribution model to describe the lifetime distribution of aircraft components in the case of missing data, and forecasted failure time of such components.

The above works focus on missing data in static context, but there is little study on data imputation for a stream that is a dynamic context. Stream data are large in amount and should be quickly processed in time. Imputation methods mentioned in static context cannot meet these requirements. In addition, data imputation method based on gamma distribution is still in its infancy. Therefore, this work focuses on missing data in a stream, and presents an imputation method based on gamma distribution for missing data in a stream.

3 Stream Data and Data Distribution

3.1 Concepts

Definition 1 (Stream Data). A sequence of data tuples continuously arriving over time is referred to a stream data. Let $S(D, T)$ be the dataset of a stream data. $T = \{t_1, t_2, \ldots, t_i, \ldots\}$ is an infinite set of time intervals. $D = \{D(t_1), D(t_2), \ldots, D(t_i), \ldots\}$ is all data in this stream. More concretely, $D(t_i) = \{d_1, d_2, \ldots, d_m\}$ is these data coming over time interval t_i. Data tuple $d_j = (d_{j1}, d_{j2}, \ldots, d_{jr})$, $j \in [1,m]$, contains r attributes, and an attribute is called a data item whose data type can be integer, character, and others.

For convenience of discussion, without losing the generality, this paper assumes that $r = 1$ in Definition 1 and each data is specified as a numerical data. In this way, multi-attribute data processing is reduced to be single-attribute data processing. Further, as mentioned in Sect. 2, we do suppose that not only the number of data items in each time interval t_i but also the values of data items in each time interval follow gamma distribution whose probability density function (PDF) [20] is $f(x; \alpha, \beta) = \frac{\beta^\alpha}{\Gamma(\alpha)} e^{-\beta x} x^{\alpha-1}$, $x > 0$, as shown in Fig. 1, where $\alpha > 0$ is a shape parameter, $\beta > 0$ is a rate parameter, and $\Gamma(x) = \int_0^\infty t^{x-1} e^{-t} dt$ is the gamma function. For $\alpha \leq 1$, the PDF of gamma distribution is strictly decreasing, $f(x; \alpha, \beta) \to \infty$ as $x \to 0$, and $f(x; \alpha, \beta) \to 0$ as $x \to \infty$. For $\alpha > 1$, the PDF of gamma distribution is unimodal with the mode at $(\alpha - 1)/\beta$.

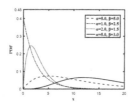

Fig. 1. Probability density functions of gamma distribution

3.2 Data Missing in a Stream

Definition 2 (Data Missing). In a stream data, if part of data or even the whole may be lost at some moment. This phenomenon is called data missing, and these lost data called missing data items in such a stream.

The motivation of this study is to make up for these missing data items and try to restore the original appearance of a stream data.

Definition 3 (Missing Ratio, R). The missing ratio is a proportion of the number of missing data items to the total amount, which indicates the degree of data missing in a stream. Let N' be the number of missing data items and N the total number of data items. The miss ratio is $R = N'/N$, $R \in [0,1]$.

In a real time computing environment, missing ratio R in each time interval may be different. If $R = 1$, all the data items are missing, and it is difficult to estimate missing data items because of without any information. Usually, part of data items in a stream is missing, namely, $R < 1$. Generally speaking, there is a close relationship between data imputation method and missing ratio. The higher the missing ratio is, the harder the selection of data imputation methods will be. The ideal situation is $R = 0$. It means that no data item is missing in a time interval, and the whole data items are real and complete, so there is no need to do any imputation work.

3.3 Metrics of Data Imputation

To evaluate the effect of data imputation in a stream, we present some metrics defined as follows.

Definition 4 (Fitting Degree, τ). Let $X = \{x_1, x_2, \ldots, x_n, \}$ be the original dataset with its PDF being $g(x)$, and there exist missing data items in this dataset. If these missing data items are imputed by a fitted PDF $f(x; \varphi)$ where φ is the parameter of data distribution, the average residual (a special error) is obtained by $u_1 = \sum_{i=1}^{n} |f(x_i; \varphi) - g(x_i)|/n$. Let the set of each data item imputed into X be denoted as $X' = \{x_1', x_2', \ldots, x_{n'}'\}$ where $n' = |X'|$ and $X^* = X \cup X'$ be the complete set after data imputation with probability density function $g^*(x)$, the average residuals associated with X and X' are obtained by $u_2 = \sum_{i=1}^{n} |f(x_i; \varphi) - g^*(x_i)|/n$ and $v = \sum_{i=1}^{n'} |f(x_i; \varphi) - g^*(x_i')|/n'$, respectively. Thus, the average residual of X^* equals to $(u_2 n + v n')/(n + n')$. The fitting degree τ is defined as the difference of average residuals before and after data imputation, that is

$$\tau = u_1 - (u_2 n + v n')/(n + n') \tag{1}$$

If $\tau > 0$, it means that after data imputation the data quality of a dataset has a good improvement. Otherwise, if $\tau \le 0$, data imputation makes something worse.

Definition 5 (Credibility, η). As defined in Definition 4, let the residual of each $x_i' \in X'$ be $e_i' = |f(x_i'; \varphi) - g^*(x_i')|$, and x_i' satisfies $e_i' \le u_2$, $i = 1, 2, \ldots, n'$. The set of all such x_i' is denoted as C, $k = |C|$. The credibility is referred to

$$\eta = \frac{k}{n'} \tag{2}$$

Since $k \in [0, n']$, we know $\eta \in [0, 1]$. Especially, if $k = 0$, the residual of each data item after imputation is larger than average residuals u_2 and data imputation has the lowest credibility. If $k = n'$ and $\eta = 1$, it means data imputation is the most credible.

Definition 6 (Quantile Relative Error, λ). Considering a dataset X, it is rearranged in ascending order and its probability distribution range is divided into Q equivalents. Let q_i be the ith quantile of X. The ith quantile of the complete set X^* after data imputation is denoted as q_i^*. Let c_i be the ith quantile of data that fits a PDF $f(x; \varphi)$ where $i = 1, 2, \ldots, Q-1$. The sum of quantile error before and after data imputation are given by $\sum_{i=1}^{Q-1} |q_i - c_i|$ and $\sum_{i=1}^{Q-1} |q_i^* - c_i|$, respectively. The quantile relative error λ is defined by

$$\lambda = \frac{\sum_{i=1}^{Q-1} |q_i^* - c_i| - \sum_{i=1}^{Q-1} |q_i - c_i|}{\sum_{i=1}^{Q-1} |q_i - c_i|} \tag{3}$$

If $\sum_{i=1}^{Q-1} |q_i - c_i| = 0$, λ is meaningless and it indicates that the fitted PDF is exactly the same with the real distribution. However, such a situation is too idealistic to achieve. We focus on the condition of $q_i \ne c_i$ in this paper. If $\lambda > 0$, the quantile error after data imputation is larger than that one before data imputation. But if $\lambda < 0$, the quantile relative error becomes smaller after imputing missing data items, which implies that the fitted distribution function reflects the real situation more accurately.

Definition 7 (Matching Degree, γ). For a dataset X, we use the PDF $f(x;\varphi)$ to impute missing data. Let $X^* = \{x_1^*, x_2^*, \ldots, x_{n^*}^*\}$ be the complete set after data imputation and the actual PDF of X^* be $g^*(x)$. The mean value of X^* is denoted as $\bar{x}^* = \frac{\sum_{i=1}^{n^*} x_i^*}{n^*}$. Let $x_i^{\#}$ be the value via $f\left(x_i^{\#}; \varphi\right) = g^*\left(x_i^*\right)$, $i = 1, 2, \ldots, n^*$. The mean value of all $x_i^{\#}$ is denoted as $\bar{x}^{\#} = \frac{\sum_{i=1}^{n^*} x_i^{\#}}{n^*}$. Thus, the matching degree γ is presented as follows.

$$\gamma = 1 - \frac{\sum_{i=1}^{n^*} (x_i^* - x_i^{\#})^2}{\sum_{i=1}^{n^*} \left(\left|x_i^* - \bar{x}^*\right| + \left|x_i^{\#} - \bar{x}^{\#}\right|\right)^2} \tag{4}$$

The matching degree γ reflects whether the imputed data values are equal to the corresponding fitting function values. The larger γ is, the higher accuracy of the imputed data values are.

4 Data Imputation Based on Gamma Distribution

4.1 Gamma Distribution Parameters

Let $X = \{x_1, x_2, \ldots, x_n\}$ be a given dataset that is the dataset to be studied in this paper. We suppose X follow a gamma distribution with its PDF being $f(x; \alpha, \beta) = \frac{\beta^\alpha}{\Gamma(\alpha)} e^{-\beta x} x^{\alpha-1}$ where α and β are two distribution parameters. But α and β are unknown. Only when they are determined, making use of gamma distribution law becomes possible. Therefore, we have to firstly estimate α and β based on X. According to Reference [21], α and β can be calculated by maximum likelihood method as follows:

Let the likelihood function be $L(x_1, x_2, \ldots, x_n; \alpha, \beta) = \prod_{i=1}^n f(x_i; \alpha, \beta)$. Take natural logarithms on both sides of this equation and derive this equation with respect to α and β. Let $\psi(\alpha) = \frac{\partial[\Gamma(\alpha)]}{\partial \alpha}$. We estimate α and β by the following two equations [22]:

$$\begin{cases} -\frac{n\psi(\alpha)}{\Gamma(\alpha)} + n\ln\beta - \sum_{i=1}^n \ln x_i = 0 \\ \frac{\alpha}{\beta} - \sum_{i=1}^n x_i = 0 \end{cases} \tag{5}$$

4.2 Calculating the Number of Missing Data Items

Stream data is a continuous arrival time series data. Due to the huge volume, we cannot wait for all the data to arrive before processing. In addition, too much data puts a lot of pressure on memory. Thus, sliding window or snapshot model is usually employed to manage stream data.

Definition 8 (Sliding Window, SW). Sliding window is a window with a specified length of time. All data items in each sliding window are processed in a batch way. The whole stream is processed repeatedly one sliding window after the other.

This paper assumes that the size of a sliding window can be flexibly adjusted according to the actual demand. Our research work focus on a sliding window, namely, data imputation is done in a single sliding window.

Following the symbols in Definition 1, a sliding window SW may consist n time intervals that is $t_1, t_2, \ldots, t_n$. Let $N(t_i)$ be the number of data items in each time interval t_i. $W = \sum_{i=1}^{n} N(t_i)$ is the total number of data items in this SW.

As stated in Sect. 3.1, in a sliding window the number of data items arrived in each time interval follows a gamma distribution, we can calculate such the gamma distribution parameters α_w and β_w by Formula 5. Once α_w and β_w are determined, $N^\circ(t_i)$, the number of data items in time interval t_i, can be computed as follows.

$$N^\circ(t_i) = \left\lceil W \frac{\beta_w^{\alpha_w}}{\Gamma(\alpha_w)} e^{-\beta_w t_i} t_i^{\alpha_w - 1} \right\rceil, i = 1, 2, \ldots \ldots \tag{6}$$

Since $N^\circ(t_i)$ is an estimated value by a fitted gamma function, it may not be equal to $N(t_i)$, which is the number of the actually arriving data items in the same time interval t_i. Let $N'(t_i) = N^\circ(t_i) - N(t_i)$. If $N'(t_i) > 0$, it indicates that there are missing data items in time interval t_i, and the number of missing data items is $N'(t_i)$. In this case we must impute these missing data items, and this is one of our purposes of this paper.

4.3 Estimating the Values of Missing Data Items

After determining the number of missing data items in each time interval related to a SW, we need to estimate the values of missing data items in this time interval. Suppose that there are missing data items in time interval t_i, which means $N'(t_i) > 0, i = 1,2,\ldots,n$. Under the previous assumptions, our method about estimating for the values of missing data items is explained below.

As discussed above, $N'(t_i) > 0$ means that there are $N'(t_i)$ data items need to be imputed in time interval t_i. The missing ratio R related to time interval t_i is $R(t_i) = N'(t_i)/(N'(t_i) + N(t_i))$. Let $D(t_i) = \{d_1, d_2, \ldots, d_{N(t_i)}\}$ be all data items that should arrive within time interval t_i, and d'_j be a missing data item in time interval t_i, $j = 1, 2, \ldots, N'(t_i)$. To estimate d'_j and impute it into $D(t_i)$, we need to know the numerical range of d'_j. According to the confidence interval estimation [23], we allow the missing ratio R to be equal to the significance level θ, which usually is given beforehand in practical applications. Contrary to significance level θ, confidence level is equal to $1 - \theta$. Let the confidence interval be denoted as $[\underline{I}, \overline{I}]$, and $\underline{I} < \overline{I}$. The probability of the upper and lower bound are $P\left(\underline{I}\right) = f(\underline{I}; \alpha, \beta)$ and $P(\overline{I}) = f(\overline{I}; \alpha, \beta)$ respectively, where f(...) is the corresponding PDF. To this end, the confidence interval for missing data items is calculated by:

$$P\left(\underline{I} \leq d'_j \leq \overline{I}\right) = 1 - \theta \tag{7}$$

Let's use the assumption in Sect. 3.1 again. The numerical values of data items in each time interval t_i follow a gamma distribution. Like steps in Sect. 4.2, we can determined two gamma distribution parameters α_i and β_i. The values of I_{-i} and $\bar{I}_i$ can be easily calculated by Formula 7, $i = 1, 2, \ldots, n$. $[I_{-i}, \bar{I}_i]$ is qualified as the value range of all missing data items in time interval t_i.

Since a gamma PDF is continuous in the closed interval $[I_{-i}, \bar{I}_i]$ and differentiable on the open interval $(I_{-i}, \bar{I}_i)$, it can be known from Lagrange's mean value theorem [24] that in the open interval $(I_{-i}, \bar{I}_i)$, there exists a point $\varepsilon \in (I_{-i}, \bar{I}_i)$ such that $f(\bar{I}_i; \alpha_i, \beta_i) - f(I_{-i}; \alpha_i, \beta_i)$ $= \frac{\partial f}{\partial e}(\varepsilon; \alpha_i, \beta_i)(\bar{I}_i - I_{-i})$, where $\frac{\partial f}{\partial e}(\ldots)$ is the first derivative. More specifically, $\frac{\partial f}{\partial e}(\varepsilon; \alpha_i, \beta_i)$ $= \beta_i^{\alpha_i} e^{-\beta_i \varepsilon} \varepsilon^{\alpha_i - 2}(\alpha_i - 1 - \beta_i \varepsilon)/\Gamma(\alpha_i)$. So far, we can obtain ε according to Lagrange's mean value theorem.

To reasonably impute missing data items as much as possible, we divide the value range of data items into $N'(t_i)$ subintervals, and impute a value which is corresponding to a missing data item in each subinterval. Let $d = (\bar{I}_i - I_{-i})/N'(t_i)$ be the width of each subinterval. We can calculate each missing data item d'_j by the following Formula 8.

$$\frac{\partial f}{\partial d'_j}(d'_j; \alpha_i, \beta_i) = \frac{f\left(I_{-i} + jd; \alpha_i, \beta_i\right) - f\left(I_{-i} + (j-1)d; \alpha_i, \beta_i\right)}{d}, j = 1, 2, \ldots, N'(t_i)$$

(8)

4.4 Theoretical Analysis

Proposition 1. Considering a dataset $D(t_i) = \{d_1, d_2, \ldots, d_m\}$ in time interval t_i belonging to a sliding window, its actual PDF is g(d). By a proper gamma PDF $f(d; \alpha, \beta)$, missing data items, $D'(t_i) = \left\{d'_1, d'_2, \ldots, d'_n\right\}$, are imputed into D(ti). After data imputation, the complete dataset is $D^*(ti) = D(ti) \cup D'(ti)$ whose PDF is g*(d). Let the average residuals related to dataset D(ti) before and after data imputation be $\bar{e}_1$ and $\bar{e}_2$, and the residual of data item d'_i in $D'(t_i)$ be $e_{d'_i}$, $i = 1, 2, \ldots, n$. If $e_{d'_i} = \bar{e}_2$, then the credibility after imputation reaches its maximum.

Proof: By Definition 4, we have $\bar{e}_1 = \sum_{i=1}^{m}|g(d_i) - f(d; \alpha, \beta)|/m$ and $\bar{e}_2 = \sum_{i=1}^{m}|g^*(d_i) - f(d; \alpha, \beta)|/m$. From Formula 1, $\tau = \bar{e}_1 - (\bar{e}_2 m + e_{d'_i} n)/(m + n) = \bar{e}_1 - (\bar{e}_2 m + \bar{e}_2 n)/(m + n) = \bar{e}_1 - \bar{e}_2$. Since $e_{d'_i} = \bar{e}_2$, $i = 1, 2, \ldots, n$, the credibility $\eta = n/n = 1$ by Definition 5. Thus, the credibility reaches its maximum.

Proposition 2. Abide by these statements in Proposition 1, and let the average residual related to dataset $D^*(t_i)$ be $\bar{e}$. If the fitting degree $\tau > 0$ and $\bar{e} = 0$, then all data items in $D^*(t_i)$ obey gamma distribution completely, and $g^*(d) = f(d; \alpha, \beta)$.

Proof: By Definition 4, we have $\bar{e}_1 = \sum_{i=1}^{m}|f(d_i; \alpha, \beta) - g(d_i)|/m$ and $\bar{e} = \sum_{i=1}^{m+n}|f(d_i^*; \alpha, \beta) - g^*(d_i^*)|/(m + n)$. The fit degree $\tau = \bar{e}_1 - \bar{e}$ is calculated. If

$\tau > 0$, then $\bar{e}_1 > \bar{e}$. Since $\bar{e} = 0$, then $\bar{e}_1 > 0$. It indicates that even though the actual data distribution and the fitted gamma distribution for $D(t_i)$ are not identical, but for the complete dataset $D^*(t_i)$, the fitted gamma distribution exactly match with the actual distribution after data imputation, that is, $g^*(d) = f(d; \alpha, \beta)$.

Proposition 3. Abide by these statements in Proposition 1, and let (p, q) be the range of all imputed data items which is divided into n subintervals with the same width Δd. If there is a constant $\varsigma \in (0, 1)$ that satisfies $d_i' = p + (i - \varsigma)d$, i = 1,2,...,n, and Formula 8 holds for ever, then any two consecutive imputed data items have the same distance.

Proof: From Definition 3, the missing ratio R = n/(m + n), and the lower bound p and the upper bound q of imputed data items can be calculated by $P\left(p \leq d_i' \leq q\right) = 1 - R$. Data item d_i' is calculated by Formula 8. Since $d_i' = p + (i-\varsigma)d$, $d_{i+1}' - d_i' = p + (i + 1 - \varsigma)d - (p + (i-\varsigma)d) = d$. In this case, data imputation belongs to equidistant insertion.

Proposition 4. Suppose that dataset $D(t_i)$ are imputed via $f(d; \alpha, \beta)$ and the complete dataset is $D^*(t_i)$ after data imputation, $|D^*(t_i)| = n^*$. If the quantile Q^* of $D^*(t_i)$ is approaching n^* (i.e. $Q^* \to n^*$), and the quantile relative error $\lambda = -1$, then the matching degree reaches its maximum.

Proof: From Definition 6, since $Q^* \to n^*$, the Q^* quantiles divide all the data items of $D^*(t_i)$ into n^* equal parts, which means that the jth quantile is equal to data item d_j^*, namely, $Q_j^* = d_j^*$. Since the quantile relative error $\lambda = -1$, Q^* quantiles of dataset $D^*(t_i)$ match with the quantiles calculated by $f(d; \alpha, \beta)$. Let the fitted data item be $\widehat{d_i^*}$. Under the condition of $Q_j^* = d_j^*$, we have $d_j^* = \widehat{d_i^*}$. According to Definition 7, as $d_j^* = \widehat{d_i^*}$, the matching degree γ can be calculated by Formula 4, and $\gamma = 1$ is the maximum.

5 Experimental Evaluations

5.1 Experimental Environment and Setup

We conduct experiments with DELL Vostro 3668-R3428 computer whose configuration is Intel(R) Core(TM) i5-7500 CPU @3.40 GHz, 8 GB RAM, running MATLAB R2016a on Microsoft Windows10. Two kinds of datasets are employed to test. One is the synthetic dataset, named ANMs_Rand, which contains more than 2000 random data items generated via the ANMs method proposed by Peters [25]. The other is the real-world dataset, named Windspeed, which is downloaded from the web site https://dataju.cn/Dataju/web/datasetInstanceDetail/381. This is the open data of the international ground exchange station. Our experiments only extract the wind speed data observed in the last 1–12 months.

To evaluate the effectiveness of data imputation, we analyze experimental results by using random imputation method (RI), mean imputation method (MI), EM-based imputation method (EMI) [26], and our proposed method (GI). K sliding windows SW with different sizes are set in each group of experiments. It is advisable to set the size ratio of SW as $SW_1 : SW_2 : \cdots : SW_K = 1 : 2 : \cdots : K$. Specially, K = 10 in Experiment 1 and 2, while K = 12 in Experiment 3 and 4. We also assume the missing ratio R is the same in each SW, and R = 10% data items are randomly selected as missing data.

5.2   Experimental Results

Experiment 1. Comparison of fitting degree τ on dataset ANMs_Rand. Experimental result is shown in Fig. 2. It can be seen that the fitting degree τ of GI are generally larger than the other three imputation methods. When the size of the sliding window SW exceeds 1600, RI is better than GI in terms of τ. This is because RI imputes missing data items at random, which may get a larger value of τ. However, in the majority of cases RI is worse than GI. On the other hand, the τ values obtained by GI are larger and more stable than that by EMI, which indicates that GI has wide applicability to data imputation for stream computing.

Experiment 2. Comparison of credibility η on dataset ANMs_Rand. Experimental result is shown in Fig. 3. Obviously, all values of the credibility η obtained by GI are larger than that by the other three methods, GI improves the credibility by 15.0% to 20.0% compared to EMI, and GI exhibits very stable performance although at different sliding window sizes. In addition, it is worth mentioning that all η values obtained by MI are 0. Therefore, the MI method for data imputation is not credible.

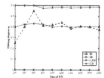

Fig. 2. The fitting degree on ANMs_Rand

Fig. 3. The credibility on ANMs_Rand

Fig. 4. Mean square error on Windspeed

Fig. 5. Matching degree on Windspeed

Experiment 3. Comparison of mean square error (MSE) on Windspeed. Experimental result is shown in Fig. 4. When the size of SW is less than 6000, these MSEs calculated by GI are slightly larger than that by MI, but smaller than that by RI or EMI. As the size of SW increases, these MSEs by GI become smaller and smaller. Further, they are very close to zero when the size of SW is 8000 or above. Therefore, data imputation by GI is very suitable for cleaning and repairing big data from a stream.

Experiment 4. Comparison of matching degree γ on dataset Windspeed. Experimental result is shown in Fig. 5. Obviously, all values of matching degree γ obtained by GI are larger than that by the other three methods. In addition, it is worth mentioning that η values oscillate at different sizes of SW. This is because MI does not consider the distribution law of data items in imputation operations. Instead, MI simply calculates the average value as a missing data item.

6 Conclusions

This paper focuses on missing data items in a stream, and proposes a data imputation method. Based on gamma distribution and Lagrange's mean value theorem, we are able to determine the value of each missing data item and add them to the original dataset. In addition, we present a series of metrics to evaluate the effect of data imputation. On this basis, theoretical analysis can be carried out in many aspects. Finally, experiments on a synthetic dataset and a real-world dataset with different sizes show that our proposed approach outperform the traditional methods for data imputation. Although the premise of our work is to assume that the stream data obeys gamma distribution, our methods presented in this paper also have the value of reference and imitation even if the stream data follow other distributions. Of course, for data imputation, it is better not to assume any premises, but to make use of historical data logs directly and analyze the possibility of data loss through machine learning, so as to realize data completion. This is our future research work.

Acknowledgments. This work was supported by the National Social Science Foundation of China under grant No. 17BTQ086; the National Key R&D Program of China under grant No. 2019YFB1704100; the National Natural Science Foundation of China under grant No. 62072337; and Excellent experimental project of Tongji University under grant No. 1380104112.

References

1. Xiao, Q., Chen, S., Zhou, Y., et al.: Estimating cardinality for arbitrarily large data stream with improved memory efficiency. IEEE/ACM Trans. Netw. **28**(2), 433–446 (2020)
2. Qiu, H., Noura, H., et al.: A user-centric data protection method for cloud storage based on invertible DWT. IEEE Trans. Cloud Comput. 1 (2019)
3. Security protection and checking for embedded system integration against buffer overflow attacks via hardware/software. IEEE Trans. Comput. **55**(4), 443–453 (2006)
4. Turner, S., Albert, L.: Archived intelligent transportation system data quality: preliminary analyses of San Antonio TransGuide data. Transp. Res. Rec. **1719**(1), 77–84 (2000)
5. Gai, K., Qiu, M., Zhao, H.: Privacy-preserving data encryption strategy for big data in mobile cloud computing. IEEE Trans. Big Data, 1 (2017)
6. Strike, K., Emam, K.E.: Software cost estimation with incomplete data. IEEE Trans. Softw. Eng. **27**(10), 890–908 (2001)
7. Valarmathie, P., Dinakaran, K.: An efficient technique for missing value imputation in microarray gene expression data. In: Proceedings of IEEE International Conference on Computer Communication and Systems, Chennai, pp. 073–080 (2014)
8. Jea, K.F., Hsu, C.W.: A missing data imputation method with distance function. In: 2018 International Conference on Machine Learning and Cybernetics, Chengdu, pp. 450–455. IEEE (2018)
9. Chang, G., Ge, T.: Comparison of missing data imputation methods for traffic flow. In: IEEE International Conference on Transportation, Mechanical, and Electrical Engineering, Changchun, pp. 639–642. IEEE (2011)
10. Dempster, A.P., Laird, N.M.: Maximum likelihood from incomplete data via the EM algorithm. J. Roy. Stat. Soc. B **39**(1), 1–22 (1977)

11. Murray, J.S.: Multiple imputation: a review of practical and theoretical findings. Stat. Sci. **33**(2), 142–159 (2018)
12. Dick, U., Haider, P.: Learning from incomplete data with infinite imputations. In: the 25th International Conference on Machine Learning, pp. 232–239. Association for Computing Machinery, Helsinki (2008)
13. Vellido, A.: Missing data imputation through GTM as a mixture of t-distributions. Neural Netw. **19**(10), 1624–1635 (2006)
14. Deng, C.J., Chen, D.: A recombination information process method of missing data in WSN. Int. J. Electron. **104**(6), 1063–1076 (2017)
15. Demirtas, H.: Multiple imputation under the generalized lambda distribution. J. Biopharm. Stat. **19**(1), 77–89 (2009)
16. Li, H.C., Hong, W.: On the empirical-statistical modeling of SAR images with generalized gamma distribution. IEEE J. Sel. Top. Sign. Proces. **5**(3), 386–397 (2011)
17. Luo, Q., Zhou, J.L.: Parameter estimation and hypothesis testing of two gamma populations with missing data. Math. Pract. Theory **47**(13), 196–201 (2017)
18. Li, N.Y.: The empirical Bayes two-sided test for the parameter of gamma distribution family under random censored. J. Syst. Sci. Math. Sci. **31**(4), 458–465 (2011)
19. Ma, X., Liu, X.D.: Lifetime distribution fitting method for a type of component with missing field data. Mech. Sci. Technol. Aerosp. Eng. **31**(7), 1136–1139 (2012)
20. Saulo, H., Bourguignon, M.: Some simple estimators for the two-parameter gamma distribution. Commun. Stat.-Simul. Comput. **48**, 13 (2018)
21. Thorn, H.C.S.: A note on the gamma distribution. Mon. Weather Rev. **86**(4), 117–122 (1958)
22. Shenton, L.R., Bowman, K.O.: Further remarks on maximum likelihood estimators for the gamma distribution. Technometrics **14**(3), 725–733 (1972)
23. Kyriakides, E., Heydt, G.T.: Calculating confidence intervals in parameter estimation: a case study. IEEE Trans. Power Delivery **21**(1), 508–509 (2005)
24. Rehmana, A., Ashraf, S.: New results on the measures of transitivity. J. Intell. Fuzzy Syst. **36**(4), 3825–3832 (2019)
25. Peters, J., Janzing, D.: Identifying cause and effect on discrete data using additive noise models. In: The Thirteenth International Conference on Artificial Intelligence and Statistics, pp. 597–604. JMLR.org, Sardinia (2010)
26. Dalca, A.V., Bouman, K.L.: Medical image imputation from image collections. IEEE Trans. Med. Imaging **38**(2), 504–514 (2019)

Electronic Stethoscope for Heartbeat Abnormality Detection

Batyrkhan Omarov[1,2(✉)], Aidar Batyrbekov[3], Kuralay Dalbekova[4],
Gluyssya Abdulkarimova[5], Saule Berkimbaeva[4], Saya Kenzhegulova[4],
Faryda Gusmanova[1], Aigerim Yergeshova[2], and Bauyrzhan Omarov[1]

[1] Al-Farabi, Kazkah National University, Almaty, Kazakhstan
[2] International Kazakh-Turkish University, Turkistan, Kazakhstan
[3] International Information Technology University, Almaty, Kazakhstan
[4] University of International Business, Almaty, Kazakhstan
[5] Abai Kazakh National Pedagogical University, Almaty, Kazakhstan

Abstract. Cardiovascular diseases (CVD) are one of the main causes of death and disability in most countries of the world. However, most countries do not currently have a comprehensive mass identification of risk factors and an overall assessment of the risk of developing CVD. Most heart diseases are related and are reflected by the sounds that the heart produces. Auscultation of the heart, defined as listening to the sound of the heart, was a very important method for early diagnosis of cardiac dysfunction. In this case, phonocardiogram (PCG) records heart sounds and noises that contain significant information about heart health. Analysis of the PCG signal has the potential to detect an abnormal heart condition. Traditional auscultation requires significant clinical experience and good listening skills. The advent of the electronic stethoscope paved the way for a new field of computer auscultation. This article discusses in detail the technology of an electronic stethoscope and the method of diagnosing heart rhythm disorders based on computer auscultation.

Keywords: Smart stethoscope · Machine learning · Classification · PCG · Abnormal heart sounds · Heartbeat

1 Introduction

One of the most common death in the world inclined toward heart diseases. It estimates that 17.9 million people dies each year from disorder of heart, blood vessels and coronary heart disease. Heart disease is one of the typical. Coronary illness is one of the average sorts of cardiovascular infections that cause huge discomfort and unfavorable impacts on the usefulness and long haul life of the patient [1]. Diagnosis plays a crucial function in reducing cardiovascular disease mortality. Electrocardiogram is a powerful and popular method of cardiac disease screening. It's relatively cheap, non-surgical and easy to use [2]. There are however, several drawbacks, one of which being the difficulties of identifying structural irregularities in the heart valves and cardiac noise defects [3]. At present, a magnetic and ultrasound scanner can take precise and even shifting images

© Springer Nature Switzerland AG 2021
M. Qiu (Ed.): SmartCom 2020, LNCS 12608, pp. 248–258, 2021.
https://doi.org/10.1007/978-3-030-74717-6_26

of the heart. An echocardiogram uses the Jumping Wave concept to construct a moving picture of the heart, providing detail on the scale, shape, configuration and function of the heart. Magnetic resonance imaging of the heart uses radio waves, magnets, and a computer to make images of the heart as it beats. The Magnetic resonance imaging test also includes major blood vessels and moving images of the heart [4]. Under other conditions, computed tomography (CT) of the heart is a procedure that uses x-rays to produce precise images of the heart and its blood vessels. Many different variety of advanced geometric processing techniques have recently been developed to recreate 3D and 4D cardiac models based on MRI and CT to better explain and imagine what we can't get from 2D static images [5–8]. But then again the key drawbacks of echo, MRI and CT are their high costs and the need for skilled workers to deal with sophisticated devices. Such technology is generally only accessible in big hospitals in big cities. According to [9], 85% of cardiovascular disease deaths range from low middle-income areas. The supply of medical imaging equipment in these countries is comparatively limited.

It is quite important to provide an efficient and reliable approach for early diagnosis of heart disease. Ultrasonography of the heart, described as reacting to and decoding heart sounds, was a very effective tool for early detection of heart disease by identifying abnormal cardiac sounds [10]. Phonocardiogram (FKG) is a high-precision chart that measures heart tones using a computer called a Phonocardiogram. Despite the benefits of low price and ease of operation, the ultrasonography of the heart has historically been limited to three criteria. First of all, since heart sounds include a combination of high-frequency and low-frequency sounds with low amplitude, it is essential that the stethoscope has a high sensitivity. Second, the sound of the heart captured by a stethoscope is often obscured by interference, that can disrupt with the precision and identifying diagnosis of heart disease. Third, the assessment of heart rhythms is very dependent mostly on the knowledge, expertise and listening ability of the doctor. The need to establish auscultation of the heart is therefore very clear.

Owing to the new technical advances, from the design of acoustic instruments, sophisticated optical signal processing to computer-based machine learning algorithms, cardiac malfunction identification systems can be automated using an electronic stethoscope.

2 Modules of the Electronic Stethoscope and Its System

There are three primary components, including a data collection module, a preprocessing module and a processing sound module, in a computer device for the diagnosis of cardiac disease using an electronic stethoscope system. Electronic stethoscope captures heart sounds and transform it to digital signals and transfer it to the preprocessing module. There, the signal will be filtered, normalized and segmented for further processing. The three main modules and their submodules are described in detail below.

2.1 Data Collection Module

The heart sound collection stage generates automated heart sound data for further analysis.

Electronic Sensor Stethoscope. Sounds are obtained directly from patients using an electronic stethoscope. Such widely used stethoscope detectors are microphones, piezoelectric sensors, and so on. (Figure 1) The heart sounds are translated into analog electrical signals.

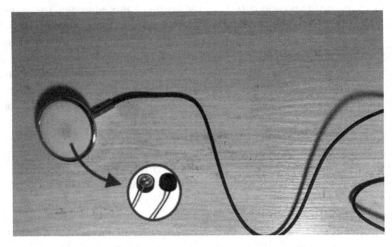

Fig. 1. Electronic sensor stethoscope

Amplifier and Filter. Boost and filtering are the two primary features of any signal acquisition system. Usually a low-gain preamplifier is used to block 50 to 60 Hz interference. The anti-aliasing filter is then used to eliminate anti-aliasing effects. In certain device architectures, the filter section consists of a bandpass filter circuit which has the midrange frequencies of most heart signals. Bandpass filter with the right spectrum selection not only avoids overlapping of the frequency, but also removes any noise beyond the bandwidth. Afterwards the filtered signal is amplified to the level needed by the analog-to-digital converter with corresponding amplification.

Analog-to-Digital Converter. An augmented and processed analog signal is digitized by means of an analog to a digital converter.

2.2 Heart Tone Preprocessing Module

At this point, noise reduction, normalization, and segmentation will be applied to the digital audio signal of the heart.

Noise Reduction Device. A digital filter is often used to generate a signal from noise data. In order to equip the device with even better noise-removal properties, certain specialized anomaly removal methods are typically used so that the output signal-to-noise ratio can be more increased.

Normalization and Segmentation. When gathering data, various sampling locations typically affect the signal. Thus the heart signals are standardized to a certain level, so that the predicted signal amplitude does not depend on the location of the sample group and the various samples. After receiving normalized signals, the sound is separated into intervals that are prepared to retrieve the features.

2.3 Heart Sound Signal Processing Module

At this stage, features are extracted and classified.

Extraction of Special Characteristics. Signal processing takes place to transform raw data to some kind of parametric representation. This parametric representation, referred to as a feature, that is used for further study and processing.

Classification. A classification model trained using derived features is being used to identify the data and to assist a medical professional in making clinical diagnostic decisions. Followed by a detailed studies, research into the automatic detection of various pathological conditions and heart disorders by audio signal focuses primarily on three stages: the cardiac tone acquisition and sensor design, the elimination of noise and cardiac segmentation, and the extraction of specific functions and the automatic analysis of cardiac tone.

2.4 User Interface

Getting all the processed information and sending the results to the device showing the final result.

3 Scheme for Building Machine Learning for Classification

3.1 Data Acquisition

The database is a mandatory requirement in the study of the patterns, thresholds, time and details of the intensity of the heart tones. The new database was created using samples collected at the hospital of the city's cardiology center in Almaty. An electronic stethoscope was used to obtain the necessary samples of heart tones (Fig. 2). For one subject, five samples were obtained from the heart apex regions (Fig. 3).

The accuracy of the database was vital because it was used to train the detection system. Therefore, only the heart tones of patients who have already been diagnosed with diseases by echocardiogram were recorded.

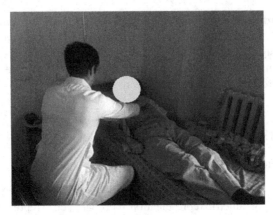

Fig. 2. Data acquisition process

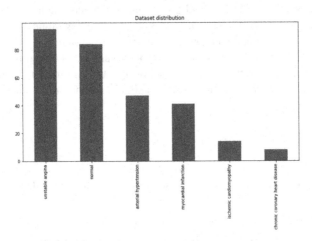

Fig. 3. Working principle of the smart stethoscope

3.2 Data Preprocessing

Feature detection is used to highlight the characteristic features of the heart tone, which will allow you to identify noises and added heart tones. The feature extraction algorithm was developed in Python, which was used to implement signal processing operations. The feature detection algorithm uses 8 steps to obtain the highest possible accuracy and has the ability to collect an acceptable number of properties from a single heart tone, since they can represent each patient individually during the optimal processing time. The feature detection algorithm uses the number of statically obtained resolutions and thresholds to perform heart tone analysis.

Most literature sources present methods for identifying the main components of heart tones (first heart tone (S1) and second heart tone (S2)) and calculating the boundaries of S1, S2, systole and diastole (Fig. 4).

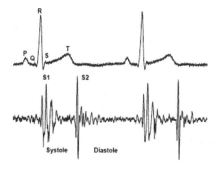

Fig. 4. Working principle of the smart stethoscope [13]

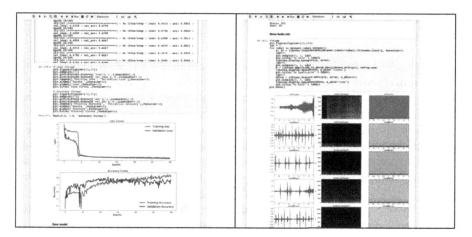

Fig. 5. Working principle of the smart stethoscope

Phase 1: Get an uniform root-mean-square curve with a length of 10 s. Since cardiac tones are non-stationary, 10-s-long signals have been derived from the original captured sound to prevent losing special features that are irregularly scattered over abnormal cardiac tones. Standardized RMS energy curves of the above obtained waves were then achieved to improve the visibility of the S1 and S2 peaks by reducing noise interference.

Phase 2: Locate the high peak. Finding high peaks between the tones of S1 and S2. Parameter resolution was used to evaluate the peaks, and if the intervals between the peaks found were not within an appropriate range, the procedure of determining high peaks would be replicated over and over again, adjusting the resolution to achieve an acceptable number of continuous peaks. The principle behind this repetitive algorithm [11] is that the frequency of any heart tone can vary from person to person. The required resolution for each heart tone is then connected to the corresponding heart rhythm of the patient. As a result, each heart signal must loop through various resolutions before an exact match is found.

Phase 3: Segmentation of the highest peak. Next, all zero wave crossings were evaluated for the calculated threshold and mapped to the estimated peaks for the peak boundary. Then after the verification process, if the number of correctly segmented peaks was not at an appropriate amount, the whole process was returned to classify a high peak with a different resolution.

Phase 4: Define the shortest peak. Waves among correctly segmented, continuous limits of high peaks were obtained, and the highest point in the interval was taken as the short peak of interest. The test then measured the distances between the defined short peaks and the related high peaks and tested for appropriate standard deviations. Otherwise the short peak, the path of which is the highest deviation, was replaced by the peak, which was at the same interval, but induced the lowest deviation effect. Following the same method, the highest continuous short peaks is reached and if the amount was not appropriate, the entire process returns to the detection of high peaks [12].

Phase 5: Short peak segmentation. Zero intersections of the extracted waves between the border of the high peaks were measured for the presumed threshold. After, they were compared to the measured short peaks. There, if the limits of the short peaks were wrongly determined or if there was any overlap, the segmentation of the short peaks would be replicated by adjusting the threshold.

Phase 6: Final measurement of the time frame and final verification. Ultimately, all results obtained above were used to measure the ranges of peak values from high to low and peak values from short to high in different arrays. When tested, any peak or interval that induced overlap or unacceptable deviation was omitted to ensure high precision. Even if the measured number of peaks or periods is not adequate, the procedure has been restored to classify high peaks, but in any case, the process was disrupted by passing all the specified resolutions in the high-pitch identification to ensure 100% accuracy of the results. This termination happens only for cardiac tones that are heavily influenced by noise.

Phase 7: S1, S2, category of systoles and diastoles. In consistent with the observation that the systolic duration is shorter than the diastolic period for heart tones,' all the parts and periods collected have been listed as S1, S2, systole and diastole.

Phase 8: Extract functions. Extracted features were followed only by samples which passed final verification.

3.3   Characteristics of Sound Identification

Based on the characteristic features obtained using the algorithm, and using some derived features, such as the square of the fourth wavelet detail coefficient, a statistical analysis was performed to identify patterns that differ depending on diseases. The square of the fourth wavelet detail coefficient was calculated for each systole and diastole separately. A range of this value with a confidence level of 87% was obtained for each type of disease.

4   A Review of the App with a Stethoscope

With the rapid development of smartphone technology, smartphones can support every-day practice for health services. Applications include using mobile devices to collect health data, deliver medical information to practitioners, researchers, and patients, monitor patients' vital signs in real time, and provide direct medical care.

The device is designed as simple as possible using the smallest components: a stethoscope, a mobile application and a small device. An electret microphone is inserted into the stethoscope tube to produce sound. The hose is blocked at all other ends except the reception area to eliminate the noise factor. Figure 6 illustrates components of the smart stethoscope.

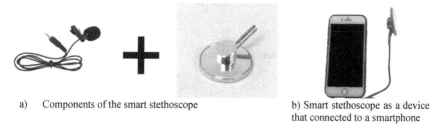

a)   Components of the smart stethoscope   b) Smart stethoscope as a device
                                                                      that connected to a smartphone

Fig. 6. Components of the smart stethoscope

Figure 7 shows heart abnormality detection process using smartphone after getting the heartbeat sounds through the stethoscope. The first is a smart stethoscope that studies the sound received from the stethoscope. The second is a trained algorithm that analyzes extraneous noise. The third is the classification process. Moreover, the output shows your potential diagnosis.

Fig. 7. Heartbeat abnormality detection

After processing and eliminating the noises in signals, we start to identify heat sounds. After detecting normal and abnormal heart sounds, the result were given in Table 2 and Table 3. As we notified before, we divided the dataset into two parts, as 200 abnormal to 200 normal heart sounds. In addition training and testing data divided in 80% to 20% proportion.

The results of detecting heartbeat abnormalities using machine learning techniques are concidered true positive (TP), false positive (FP), false negative (FN), and true negative (TN). A true positive means the total number of abnormalities in heart sounds that are actually identified, while a false positive represents the total number of heartbeat abnormalities that are incorrectly detected by the model. A false negative value is the total number of heartbeat abnormalities that are not identified at all. These results are used to calculate the results of evaluating the detection of heart rhythm disorders using various machine learning classifiers, which include sensitivity (SN), positive Predictivity (PP), and overall accuracy (OA). The equations below show the calculation for SN, PP, and OA. Table 1 and Table 2 demonstrate segmentation result of normal and abnormal heart sounds.

Table 1. Segmentation of normal cardiac sounds

Normal sounds	True positive	True negative	False negative	False positive	Sensitivity	Accuracy
S1	187	2	6	5	96.89	94.5
S2	185	3	5	7	97.37	92.5
Total	372	5	11	12	97.12	93

Table 2. Segmentation of abnormal cardiac sounds

Normal sounds	True positive	True negative	False negative	False positive	Sensitivity	Accuracy
S1	188	3	4	5	97.91	94
S2	185	4	5	6	97.36	92.5
Total	373	7	9	11	97.64	93.25

$$SN = \frac{TP}{TP + FN} \times 100 \tag{1}$$

$$PP = \frac{TP}{TP + FP} \times 100 \tag{2}$$

$$OA = \frac{TP + TN}{TP + TN + FP + FN} \times 100 \tag{3}$$

5   Discussion and Conclusion

The article presents a model for detecting pathological patterns on phonocardiograms to detect heart obstruction. Audio signal analysis is mostly related with problems such as signal weakness, narrow band bandwidth, contamination by various types of noise, randomness, the need to separate the combined signals into a single signal, etc. Heart sounds as complex signals are also subject to that rule. Heart valve sounds, the filling and emptying of ventricles and arteries, and the interference between these sounds and a narrow frequency range make it difficult to detect heart-related problems based on sound analysis.

On the other hand, the presence of reliable signals, such as ECGs, has led experts to rely on them for the diagnosis of heart disease. While the PCG signal can be used as an independent method to detect many improper disorders and cardiac arrest functions in an inexpensive and fast way.

To be able to use the phonocardiogram signal as a diagnostic tool in the field of heart disease, we must first provide a method for segmenting and detecting heart sounds as clinically significant segments and identifying their correspondence to heart cycles. The value of this algorithm and method increases when the signal is independently used as a diagnostic method, rather than still affecting the ECG signal. Of course, sometimes it is necessary to use either of these methods, since the complementary method is used together.

The main advantages of the proposed stethoscope are speed – 15 s is enough for analysis. Locality-Smart stethoscope+ does not send data to servers, so everything is stored in your phone. Audio recording can be provided to the doctor without problems. Accuracy – the accuracy reached about 90%.

References

1. Cardiovascular Diseases. https://www.who.int/health-topics/cardiovascular-diseases/#tab=tab_1. Retrieved from www.who.int
2. Leng, S., Tan, R.S., Chai, K.T.C., et al.: The electronic stethoscope. BioMed. Eng. OnLine **14**, 66 (2015). https://doi.org/10.1186/s12938-015-0056-y
3. Molcer, P.S., Kecskes, I., Delic, V., Domijan, E., Domijan, M.: Examination of formant frequencies for further classification of heart murmurs. In: Proceedings of the International Symposium on Intelligent Systems and Informatics, pp. 575–578 (2010)
4. How is heart disease diagnosed? http://www.nhlbi.nih.gov/health/health-topics/topics/hdw/diagnosis.html. Accessed 30 Mar 2015
5. Zhong, L., Zhang, J.M., Zhao, X.D., Tan, R.S., Wan, M.: Automatic localization of the left ventricle from cardiac cine magnetic resonance imaging: a new spectrum-based computer-aided tool. PLoS ONE **9**(4), (2014)
6. Tan, M.L., Su, Y., Lim, C.W., Selvaraj, S.K., Zhong, L., Tan, R.S.: A geometrical approach for automatic shape restoration of the left ventricle. PLoS ONE **8**(7), (2013)
7. Lim, C.W., et al.: Automatic 4D reconstruction of patient-specific cardiac mesh with 1-to-1 vertex correspondence from segmented contours lines. PLoS ONE **9**(4), (2013)
8. Cui, H.F., et al.: Coronary artery segmentation via hessian filter and curve skeleton extraction. In: Proceeding of 36th International Conference IEEE Engineering in Medicine and Biology Society (EMBS) (2014)
9. Cardiovascular diseases (CVDs). http://www.who.int/mediacentre/factsheets/fs317/en/. Retrieved from www.who.int
10. Omarov, B., et al.: Applying face recognition in video surveillance security systems. In: Mazzara, M., Bruel, J.-M., Meyer, B., Petrenko, A. (eds.) TOOLS 2019. LNCS, vol. 11771, pp. 271–280. Springer, Cham (2019). https://doi.org/10.1007/978-3-030-29852-4_22
11. Maitra, S., Dutta, D.: A systemic review on the technological development in the field of phonocardiogram. In: 2019 Devices for Integrated Circuit (DevIC), pp. 161–166. IEEE, March 2019

12. Real-Time Smart-Digital Stethoscope System for Heart Diseases Monitoring (2019). https://www.mdpi.com/1424–8220/19/12/2781/htm

13. Debbal, S.M.: Computerized Heart Sounds Analysis, Discrete Wavelet Transforms - Biomedical Applications, Hannu Olkkonen, IntechOpen, 12 September 2011. https://doi.org/10.5772/23700, https://www.intechopen.com/books/discrete-wavelet-transforms-biomedical-applications/computerized-heart-sounds-analysis

14. Dia, N., Fontecave-Jallon, J., Gumery, P.Y., Rivet, B.: Heart rate estimation from phonocardiogram signals using non-negative matrix factorization. In: ICASSP 2019–2019 IEEE International Conference on Acoustics, Speech and Signal Processing (ICASSP), pp. 1293–1297. IEEE, May 2019

15. Banerjee, R., et al.: Time-frequency analysis of phonocardiogram for classifying heart disease. In: 2016 Computing in Cardiology Conference (CinC), pp. 573–576. IEEE, September 2016

16. Durand, L.G., Pibarot, P.: Most recent advancements in digital signal processing of the phonocardiogram. Critical Reviews™ Biomed. Eng. **45**, 1–6 (2017)

17. Clifford, G.D., et al.: Recent advances in heart sound analysis. Physiol. Meas. **38**(8), E10–E25 (2017)

Geometric Surface Image Prediction for Image Recognition Enhancement

Tanasai Sucontphunt$^{(\boxtimes)}$

National Institute of Development Administration, Bangkok, Thailand
tanasai@as.nida.ac.th

Abstract. This work presents a method to predict a geometric surface image from a photograph to assist in image recognition. To recognize objects, several images from different conditions are required for training a model or fine-tuning a pre-trained model. In this work, a geometric surface image is introduced as a better representation than its color image counterpart to overcome lighting conditions. The surface image is predicted from a color image. To do so, the geometric surface image together with its color photographs are firstly trained with Generative Adversarial Networks (GAN) model. The trained generator model is then used to predict the geometric surface image from the input color image. The evaluation on a case study of an amulet recognition shows that the predicted geometric surface images contain less ambiguity than their color images counterpart under different lighting conditions and can be used effectively for assisting in image recognition task.

Keywords: Surface reconstruction · Geometric normal · Photometric stereo · Few-short image recognition · GAN

1 Introduction

From a photograph, a human can intuitively recognize an object from different lighting. Even objects are very similar to each other, an expert can also tell them apart from a photograph. This is partly because the human can guess the surface of the objects from the photograph. Since there are many self-shadows from ridge and valley in the photograph, a photograph can be ambiguous for a machine to comprehend. Several images with different lighting directions and views are typically required to add variance to the training set. Thus, an image recognition algorithm requires a large photograph dataset to train a model in order to register and recognize the target object. There is still an on-going research on how many and different images are enough for the job [14]. A Few-Shot learning [21] attempts to mitigate this issue by re-using the pre-trained model, augmenting image samples, or performing automate learning. For objects with visually different in shapes and colors such as different types of amulets, using pre-trained model with perceptual loss e.g. VGG-16 [16] can give very satisfied results. However, if the target objects are very different from the pre-trained dataset, a new

© Springer Nature Switzerland AG 2021
M. Qiu (Ed.): SmartCom 2020, LNCS 12608, pp. 259–268, 2021.
https://doi.org/10.1007/978-3-030-74717-6_27

dataset of the target objects, which is typically very large, are required to add to the training model. Furthermore, if the target objects are very similar to each other such as human faces or same-shape amulets which can be only differentiated by minor details, the pre-trained model is still needed to be re-trained with a large dataset. For example, to utilize the pre-trained model for very similar faces, the model must be re-trained with several image pairs of the same and different faces using triplet-loss function on Siamese networks [8]. Several pair examples are required to induce the feature space so that images of the same object are close together. Nonetheless, for many image recognition tasks, obtaining a large dataset of image pairs at the beginning is very difficult. In this work, a simple Few-Shot image recognition framework is developed where large examples are not require. The recognition is performed by firstly predicting a geometric surface from the input photograph, then extracting features by an existing pre-trained network, and finally recognizing by linear classifiers. To prove the concept, in this work, a pre-trained network model of VGG-16 is used as a perceptual feature extraction. SVM is then used as a linear classifier to recognize the object.

An object's surface details i.e. a geometric surface image is used instead of its pure color image in order to reduce the lighting condition dependency. The geometric surface image is a 2D image that contains surface's normal vectors. Thus, the geometric surface image should be a better representation since it does not contain ambiguity from self-shadow i.e. ridge and valley as appearing in the color image. Since each object contain only one geometric image independently from lighting conditions, the geometric surface image can be used as an intrinsic representation of the object. However, capturing object's surface image requires a special equipment e.g. a photometric scanner. For reasonable quality, a DIY photometric stereo can capture finer surface details than using a stereo vision or a depth camera. The photometric stereo uses images from different lighting directions to generate a geometric normal image (a.k.a. geometric surface image) as shown in Fig. 1. The more sophisticated the system, the higher precision the surface details can be captured. Even though the captured normals may not contain the precise normals but they can be used for image recognition effectively. The geometric normals represent only normal directions from camera view but it does not contain the 3D surface geometry. The normal image can be also used to estimate height map in order to obtain its 3D surface geometry.

Capturing a geometric normal image by the photometric stereo scanner is not practical for image recognition tasks. In this work, the geometric normal image is synthesized by a generative model from its photograph instead. Firstly, a DIY photometric scanner with eight equally spacing light sources and one high resolution camera is used to create a training dataset (Fig. 1). The eight images taking from the same camera position but in different light source directions are then used to reconstruct a geometric normal map/image. This reconstruction process takes each image pixel at the same location from eight images with known light source directions to solve for its 3D normal vector at the pixel location. Secondly, to be able to construct a geometric normal image from a

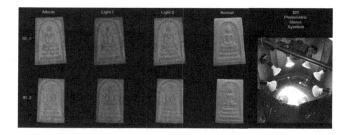

Fig. 1. From left to right: Phra Somdej Thai Amulet' albedo image, its color image with lighting direction 1, its color image with lighting direction 2, its normal image. Each row represents different amulet. The red dot boxes show ambiguity surface area from the color image which does not appear in its normal image. Rightmost: a Photometric Stereo Surface Scanning System. (Color figure online)

single photograph, Generative Adversarial Networks (GAN) is used to map a relation between 2 images from different domains i.e. a geometric normal image and a color photograph image. GAN is a popular method to randomly generate a novel image from example images such as human faces, cartoons, and scenery views.

Lastly, ambiguity evaluation on different lightings of color photographs and predicted normal images are illustrated. Also, Few-Shot image recognition evaluations compare photograph, geometric normal image, and a combination of both for their accuracy and robustness. These evaluations shows that the predicted geometric normal images highly benefit the Few-Shot image recognition in different lighting conditions.

2 Related Work

Advanced 3D surface reconstruction systems aim to reassembling information of a target object from cameras or sensors. For example, Smith and Fang [17] use photometric stereo to reconstruct normal map as well as the height of the surface from photometric ratio. These systems require intensive efforts in sensor setting as well as capturing process. To synthesize a surface image without developing a capturing system, several works combines Bidirectional Reflectance Distribution Function (BRDF) and Photometric Stereo with Convolutional Neural Network (CNN) to generate reflectance maps [12], outdoor surface normal maps [4], or Physically Based Rendering (PBR) maps with U-Net [1]. Multiple images from known illumination directions can also be used to reconstruct their normal maps [18] which is similar to this work except known illumination directions is not required in this work. Many works also focus on using CNN to generate a depth map from a single image such as using global and local layers [3] or with Conditional Random Field (CRF) [10]. Furthermore, coarse normal map [19] with albedo [13] or with label [2] can also be generated with CNN based techniques from a single image.

Currently, Generative Adversarial Networks (GAN) can be used to translate an image from one domain to another domain with higher details and less artifacts than CNN in general. Also, using GAN in 3D reconstruct gains popularity recently due to its robustness. Special devices can also be used to capture images to train with GAN to reconstruct a surface normal such as using NIR images [23] or RGBD images [20]. Sketches can also be used to reconstruct their surface normals using GAN [6]. Face normal map and its albedo map can also be generated with GAN using reconstructed surface as a training set [15]. In this work, conditional GAN (cGAN) [7] is used as the main learning model to generate a surface normal from an image. From current GAN family, DualGAN [22] can also produce similar results to cGAN with less artifacts. However, cGAN is preferred in this work because pixel-to-pixel condition is required to translate from RGB of a color space to its directional XYZ of a normal space precisely.

3 Approach

The main contributions of this work compose of Photometric Data Preparation and GAN-Based Model Training for Normal Map Prediction.

3.1 Photometric Data Preparation

Photometric stereo system [5] is a popular surface reconstruction using images from different illuminations. In this work, photometric stereo system with eight light directions and $45°$ equally apart is used to capture illumination images as shown in Fig. 1. This technique reconstructs surface normals from each pixel of all eight illumination images. A surface normal map represents tangent space normals of each captured pixel which is view-dependent to the camera view. In this work, a normal map and its albedo map (pure material colors) are reconstructed without 3D point clouds because the normal image is the main character to be used for image recognition.

Under Lambertian reflectance assumption (ideal diffuse reflection), surface reflecting light as a color pixel (I) on a camera could be derived by $I = k(L \cdot n)$ where k is albedo reflectivity, L is a light direction, and n is surface normal at the point. With known I and L (from light calibration e.g. with mirror ball), at least three light directions (eight in this work) can be used to solve for unknown kn with least-square based method because I is the size of three (RGB). Since n is a unit vector, the magnitude of kn is the k and its unit direction is the n. Each n is a 3D vector with ranging value X of $[-1, 1]$, Y of $[-1, 1]$, Z of $[-1, 1]$. These values will then be used to train in GAN-Based model directly. Normally, a camera is pointing perpendicular to the surface, thus, Z values is limited to $[0, 1]$ which pointing toward the camera. For practical usage, these normal values are stored in UV coordinate of texture map as a normal map. The X,Y,Z values are then stored in the texture map as R of $[0, 255]$, G of $[0, 255]$, B of $[128, 255]$ making the flat surface purple in color e.g. XYZ direction of $(0, 0, 1)$ is mapping to RGB color of $(128, 128, 255)$.

3.2 GAN-Based Model Training for Normal Map Prediction

The goal of using GAN-Based model is to imitate the way photometric stereo calculate for the normal image e.g. with $I = k(L \cdot n)$. Instead of fitting the parameters in least square sense, the GAN-Based model attempts to non-linearly generate each normal image from each color image of different light source. Without specific reconstruction equation, the generated normal image will find relationship between given shading patterns and normal directions implicitly. However, typical GAN-based model generates an image to fit only to its training distribution without exact pixel locations. To give a condition (controllable parameters) for GAN to generate a proper image, conditional Generative Adversarial Networks (cGAN) appends a condition image in the training pipeline to constraint the output image. In this case, the color photographs from different lighting directions are used as condition images to generate their geometric normal image. Figure 2 (Training) shows an overview of the cGAN training process in this work. The input are photometric-stereo color images and their normal image. The photometric-stereo color image is the condition to feed to both Generator and Discriminator. Since one reconstructed normal image is the result of eight photometric-stereo color images from different light directions, each of eight photometric-stereo color images will be trained with the same normal image. This condition will train the network to capture the reflective nature of the surface. Even though the dataset is small, cGAN can generate a reasonable result without data augmentation since the training set contains highly structure data.

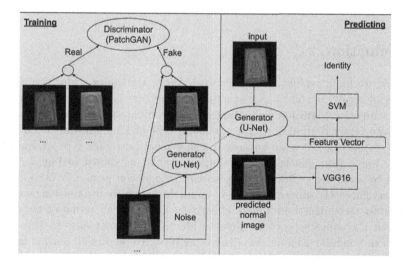

Fig. 2. The training and predicting processes in this work.

An objective function (G^*) for cGAN using in this work is shown in Eq. 1 where x is a color image, y is a normal image, and z is a noise vector. The training loop starts with a Discriminator (D), then a Generator (G) is trained with the Discriminator's loss together with its loss in iterative. In this work, the PatchGAN, with 32×32 patches, is used as the Discriminator in order to discriminate each patch for fake (0) or real (1). The Discriminator of C64-C128-C256-C512, where C represents Con2D-BatchNorm (except the 1st layer)-LeakyReLU block, is trained directly by detecting if the normal image is fake (predicted normal image from Generator) or real (real normal image) typically with binary-cross-entropy loss. However, in this work, mean-squared-error loss is used instead as it helps the Discriminator learn better in our experiment.

$$G^* = \arg \min_{G} \max_{D} (\mathbb{E}_{x,y}[\log D(x, y)] + \mathbb{E}_{x,z}[\log(1 - D(x, G(x, z)))]) + \lambda Cos(G)$$

(1)

To generate fine details of normal maps, U-Net, which is an encoder-decoder with skip connections, is used as the Generator to keep resolutions of the output similar to the input. The U-Net of C64-C128-C256-C512-C512-C512-C512-C512-reverse-back is trained with composite loss of an adversarial loss from the Discriminator and a regularization loss from the generated image. All input images are firstly re-scaled to 512×512 pixels. While training the Generator, the Discriminator will not be trained but it will evaluate for the adversarial loss. To keep generated normal image similar to the real normal image, typically L1 loss is used as a regularization. However, in this work, a cosine similarity $(Cos(G))$ loss (with weight of $\lambda = 100$) is used instead since the normal image contains only normal vectors. The trainings are optimized using Adam solver with learning rate of 0.0002 and momentum parameters of $\beta_1 = 0.5$ and $\beta_2 = 0.999$ with batch size of 1 for 200 epochs.

4   Evaluation

There are two main evaluations in this work. First, ambiguity evaluation aims to compare ambiguity of color images versus their predicted normal images. Second, Few-Shot image recognition evaluation aims to compare recognition accuracy of color image versus normal image. A particular type of Thai amulets called "Phra Somdej" is used as the main subject in the evaluations. This type of amulet is typically made from plaster in the shape of tablet as shown in Fig. 1. There are numerous research works on Thai amulet recognition using CNNs to classify or authenticate [11] different types of amulets in which feature extraction is sufficient to recognize the amulets effectively [9]. However, none of the works focuses on the same type of amulets to identify the exact one. In this work, there are very similar 23 amulets of the same type. To predict all normal images, first 11 amulets are predicted using cGAN trained from another 12 amulets and vice versa. Figure 3 shows the input color images comparing to their predicted normal image from cGAN, ground-truth normal image, and the color differences between them. From the figure, the color images are relatively different from lighting directions but their predicted normal images remain resemble to each other. The heat-maps of differences between the predicted normal images and the ground-truth normal image reveal that the predicted normals can mostly represent the real surface structure of the amulet except some crisp silhouettes.

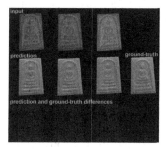

Fig. 3. The input color images (from different light directions) and their predicted normal images in comparison. Bottom: the color differences between a ground-truth normal image and its predicted normal images (Color figure online)

4.1   Ambiguity Evaluation

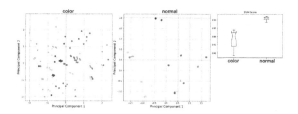

Fig. 4. Ambiguity measurement comparing color images and their predicted normal images. Left: color PCA-space. Middle: normal PCA-space. Right: SSIM score box plot. (Color figure online)

In this work, the ambiguity measurement evaluates if the predicted normal image contained less ambiguity than its color image. There are two experiments which are an arrangement visualization and a similarity measurement. To visualize how images are arranged, each type of images are projected to its own domain space using Principal Component Analysis (PCA). The PCA spaces of color and normal images are built from all images of each type. Figure 4 shows 2D points of principle-component-1 and principle-component-2 of each image where different label colors means different amulets. There are 8 images from different light directions for each amulet. Figure 4 (left and middle) shows color PCA space and normal PCA space. From this figure, normal images obviously group together closer than color images.

Second, to evaluate if the predicted normal images are similar to each other than their color image inputs, the predicted normal images are measured against their average. As well, the color image inputs of different light directions are measured against their average. The measurement is quantified by Structure Similarity (SSIM) score. Figure 4 (right) shows a box plot of SSIM score comparing the normal images and the color images. The box plot reveals that the predicted normal images are significantly less sensitive to different light directions.

4.2 Few-Shot Image Recognition Evaluation

From Fig. 2 (Predicting), each amulet has eight images from different light directions. For recognition evaluation, half of them (4 images) is used for registration and another half (4 images) is used for recognition. For registration, the images are transformed to their feature vectors of 4096 dimensions by using VGG-16's second-last fully connected layer. Then, they are trained with SVM to classify for their identities. To add variations, all test images are first transformed by minor rotations and translations before predicted for their normal images. Color images, normal images, and combination of both (by concatenating their feature vectors) are evaluated. To compare which input types are more robust for the recognition, image variations by downgrading of brightness, contrast, color, and blurriness are gradually applied to the test color images before predicting for their normal images.

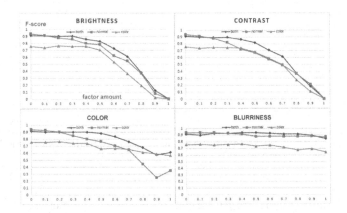

Fig. 5. Recognition accuracy by F-score of color, predicted normal, and combination of both images on applying brightness, contrast, color, and blurriness image downgrading. (Color figure online)

Figure 5 shows recognition accuracy by F-score of each image variation with downgrading amounts. The downgrading amount is a real number from 0 to 1.0 where 0 returns original image and 1.0 returns fullest image downgrading, i.e. worsen quality from the original image, which means total black image for brightness, gray image for contrast, black/white image for color, and blurred image for blurriness. From the figure, using color together with its predicted normal for image recognition mostly outperforms color or normal image alone since color and normal features can compensate each other's drawback. However, when the downgrading amounts are less (close to original images), the predicted normal images give higher accuracy than other types. This is because the predicted normal images are accurately predicted from highest quality input color images. For the original images, F-score of color, predicted normal, and combination of both images are 0.75837, 0.93913, and 0.91498 respectively.

5 Conclusion and Future Work

In this work, a normal image of target surface is predicted from a color image by using a GAN-Based model which is trained with photometric stereo images captured by a scanner system. The evaluation illustrates that the predicted normal image from this work contain less ambiguity from different illuminations comparing to their color image counterpart and can be used for Few-Shot image recognition effectively. Also, from the evaluation, using both color and predicted normal images together for image recognition give highest accuracy in general. The main reason is that the color information is still important for image recognition. This work does not intend to reconstruct exact surface details. Instead, it tries to create a proper representation for image recognition by predicting its possible surface.

In the future, more types of target objects such as ancient-heritage objects and more variations of materials will be collected. GAN architecture as well as its loss function can also be adjusted in order to achieve better resolution. Also, this work focuses only on lighting conditions of image ambiguity. However, another important factor which is camera parameters (intrinsic and extrinsic) will need to be addressed. In this case, a 3D model together with its normal map can be used to predict for a better representation. Finally, if there are enough collection of scanned objects, fully automated object identification which robust for real world environments (e.g. taking a photograph from mobile phone under different lighting directions) can be developed.

Acknowledgment. This project is funded by Research Awards for Complete Research from National Institute of Development Administration.

References

1. Deschaintre, V., Aittala, M., Durand, F., Drettakis, G., Bousseau, A.: Single-image SVBRDF capture with a rendering-aware deep network. ACM Trans. Graph. **37**(4), 1–5 (2018)
2. Eigen, D., Fergus, R.: Predicting depth, surface normals and semantic labels with a common multi-scale convolutional architecture. In: ICCV, pp. 2650–2658, December 2015
3. Eigen, D., Puhrsch, C., Fergus, R.: Depth map prediction from a single image using a multi-scale deep network. In: NIPS (2014)

4. Hold-Geoffroy, Y., Sunkavalli, K., Hadap, S., Gambaretto, E., Lalonde, J.F.: Deep outdoor illumination estimation. In: CVPR (2017)
5. Horn, B.K.P.: Obtaining Shape from Shading Information, pp. 123–171. MIT Press, Cambridge (1989)
6. Hudon, M., Grogan, M., Pagés, R., Smolic, A.: Deep normal estimation for automatic shading of hand-drawn characters. In: ECCV Workshops (2018)
7. Isola, P., Zhu, J.Y., Zhou, T., Efros, A.: Image-to-image translation with conditional adversarial networks. In: CVPR, pp. 5967–5976, July 2017
8. Koch, G.R.: Siamese neural networks for one-shot image recognition (2015)
9. Kompreyarat, W., Bunnam, T.: Robust texture classification using local correlation features for Thai Buddha amulet recognition. In: Advanced Engineering Research. Applied Mechanics and Materials, vol. 781, pp. 531–534 (9 2015)
10. Liu, F., Shen, C., Lin, G.: Deep convolutional neural fields for depth estimation from a single image. In: CVPR, pp. 5162–5170 (2014)
11. Mookdarsanit, L.: The intelligent genuine validation beyond online Buddhist amulet market. Int. J. Appl. Comput. Technol. Inf. Syst. 2(9), 7–11 (2020)
12. Rematas, K., Ritschel, T., Fritz, M., Gavves, E., Tuytelaars, T.: Deep reflectance maps. In: CVPR, pp. 4508–4516 (2015)
13. Sengupta, S., Kanazawa, A., Castillo, C.D., Jacobs, D.W.: SfSNet: learning shape, reflectance and illuminance of faces 'in the wild'. In: CVPR, pp. 6296–6305 (2018)
14. Shorten, C., Khoshgoftaar, T.: A survey on image data augmentation for deep learning. J. Big Data 6, 1–48 (2019)
15. Shu, Z., Yumer, E., Hadap, S., Sunkavalli, K., Shechtman, E., Samaras, D.: Neural face editing with intrinsic image disentangling. In: CVPR (2017)
16. Simonyan, K., Zisserman, A.: Very deep convolutional networks for large-scale image recognition (2014)
17. Smith, W., Fang, F.: Height from photometric ratio with model-based light source selection. Comput. Vis. Image Underst. 145(C), 128–138 (2016)
18. Taniai, T., Maehara, T.: Neural inverse rendering for general reflectance photometric stereo. In: ICML, pp. 4864–4873 (2018)
19. Trigeorgis, G., Snape, P., Kokkinos, I., Zafeiriou, S.: Face normals "in-the-wild" using fully convolutional networks. In: CVPR, pp. 340–349 (2017)
20. Wang, X., Gupta, A.: Generative image modeling using style and structure adversarial networks. In: Leibe, B., Matas, J., Sebe, N., Welling, M. (eds.) ECCV 2016. LNCS, vol. 9908, pp. 318–335. Springer, Cham (2016). https://doi.org/10.1007/978-3-319-46493-0_20
21. Wang, Y., Yao, Q., Kwok, J.T., Ni, L.M.: Generalizing from a few examples: a survey on few-shot learning. ACM Comput. Surv. 53(3), 1–34 (2020)
22. Yi, Z., Zhang, H., Tan, P., Gong, M.: Dualgan: Unsupervised dual learning for image-to-image translation. In: 2017 IEEE International Conference on Computer Vision (ICCV), pp. 2868–2876 (2017)
23. Yoon, Y., Choe, G., Kim, N., Lee, J.-Y., Kweon, I.S.: Fine-scale surface normal estimation using a single NIR image. In: Leibe, B., Matas, J., Sebe, N., Welling, M. (eds.) ECCV 2016. LNCS, vol. 9907, pp. 486–500. Springer, Cham (2016). https://doi.org/10.1007/978-3-319-46487-9_30

A Novel Approach for Robust and Effective Pose Estimation via Visual-Inertial Fusion

Kun Tong[✉], Yanqi Li[✉], Ningbo Gu[✉], Qingfeng Li[✉], and Tao Ren[✉]

Hangzhou Innovation Institute, Beihang University, Hangzhou, Zhejiang, China
{tongkun,liyanqi_07,guningbo,liqingfeng,taotao_1982}@buaa.edu.cn

Abstract. This paper proposes a visual-inertial sensor fusion method to perform fast, robust and accurate pose estimation. The fusion framework includes two major modules: orientation fusion and position fusion. The orientation fusion is performed via the combination of vision and IMU (Inertial Measurement Unit) measurement. The position fusion is implemented via the combination of visual position measurement and accelerometer measurement using a proposed adaptive complementary filter. The proposed framework is robust to visual sensor failures from a poor illumination, occlusion or over-fast motion and is efficient in computation due to the adoption of complementary filters. Another important advantage is the error-reducing feature: the direction of the optical axis can be automatically compensated by Madgwick filter with inclination taken as magnetic distorsion. The performance is evaluated with a dual-arm manipulator. The results show a better pose estimation than visual sensor alone in terms of accuracy and robustness to vision failures.

Keywords: Sensor fusion method · Pose estimation · Virtual Magnetic Field · IMU measurements

1 Introduction

IMU fusion, with accelerometer and gyroscope or optionally plus magnetometer, is a technology for estimating orientation of a rigid body and used in many applications like human motion tracking [1], range of motion monitoring [2], human motion tracking in rehabilitation [3], human navigation in free-living environments [4] and fall detection system for elderly people [5].

Although IMU has been applied in many daily scenarios beyond the above mentioned, it is yet to be improved in terms of efficiency, stability and accuracy. The well-known Kalman Filter is widely used to estimate the orientation of IMU fusing angular rate and acceleration (optionally plus magnetic sensor) [1]. The application of Kalman Filter is very successful for the advantage of effective rejection of noise. However, the algorithm is not computing efficient enough, especially for some real-time demanding case [17]. Besides, for full pose estimation,

Supported by the Key R&D Program of Zhejiang Province (2020C01026).

M. Qiu (Ed.): SmartCom 2020, LNCS 12608, pp. 269–280, 2021.
https://doi.org/10.1007/978-3-030-74717-6_28

an Extended Kalman Filter is typically used which doesnot guerranty the convergence. To improve the computing efficiency, fuzzy processing [6] or frequency domain filters [7] were used previously but under limited operating conditions. An alternative method is using complementary filter which can achieve effective performance with less computation burden [8,9]. Madgwick et al. applied the complementary filter and used an analytically [19] derived and optimised gradient descent algorithm to compute the direction of the gyroscope measurement error as a quaternion derivative [10]. This dramatically reduced the computing load with the estimation accuracy still comparable to a Kalman Filter. Even in low sampling frequency (at 10 Hz), the filter still works well.

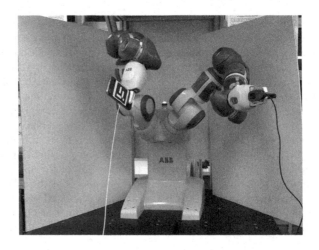

Fig. 1. Hardware setup for pose estimation

This paper proposes a solution to improving the performance of pose estimation using monocular vision and IMU. The main idea is like this: (1) visual orientation measurement and IMU data are fused by Madgwick complementary filter for orientation estimation (Madgwick algorithm was not implemented with vision before) which was proved to be orders of magnitude faster than Kalman Filter without hurting the accuracy; (2) visual position measurement and IMU's accelerometer measurement are fused by an adaptive complementary filter which can handle vision failures and inconsistency of update frequencies between vision and IMU; (3) a small 2D marker is used to help tracking in absolute coordinate of camera frame which doesnot result in accumulated drift; (4) optical axis is used as virtual magnetic field which doesnot introduce any IMU fusion errors from rotations around this axis. This solution can be used meters off the camera for purpose of object manipulation or human-robot interaction. Furthermore, the hardware used is simply a phone (with Gyro and Acc. sensors) attached with a marker on the back and a usb web camera. These components are accessible just off the shelf. Compared to Kinect, this solution is more low-cost and accessible, able to identify subtle movements which Kinect still cannot perceive very well [11]. Furthermore, it can handle short-time occlusion.

The remaining of the paper is organized as follows: Sect. 2 breifly reviews the closely related works and the existing problems. Section 3 introduces the proposed pose estimation method. Section 4 presents the experimental results in senario of moving with a dual-arm manipulator (as shown in Fig. 1). The paper is concluded in Sect. 5.

2 Related Works

This section briefly reviews the closely related works and merits/existing problems with the methods used therein.

2.1 Orientation Estimation with IMU

Kalman Filter (KF) is one of the most often used approaches for fusing sensor informations [18]. As the orientation estimation problem is nonlinear, so Extended Kalman Filter (EKF) is often chosen, such as in [12,13]. However, EKF implementation imposes a huge computation burden due to many recursive formulas. Furthermore, the optimality could be broken for the linear approximations and unknown statistical nature of noise.

Another widely used method for estimating IMU orientation is using Complementary Filter (CF). Mahony et al. formulate the filtering problem as a deterministic observer imposed on a special orthogonal group termed 'explicit complementary filter' [9]. Madgwick et al. propose a computationally efficient algorithm applying a gradient descent method to CF and obtain a comparible accuracy to EKF with much higher computing efficiency [10]. A lot of comparison studies are performed. cavallo et al. systematically compares the EKF and complementary filters (both Madgwick version and Mahony version) and show that EKF performs best of all but is 5 to 25 times computation-expensive than complementary filters [14], although the EKF is better in terms of generality which avoids redesign of filter when new sensor is added.

Although the IMU can be used to estimate orientation very efficiently, there exists drift for long term applications especially for the heading estimation. Magnetic field measurement can assist in correcting heading direction but the relative orientation estimation with respect to other object other than earth frame is not feasible. In a lot of cases, we need a pose estimation in a static (or moving) local reference frame which might have uncertain pose with respect to earth frame. Vision-based method is a good solution to achieve this goal. It is presented in next sub-section.

2.2 Pose Estimation by Fusing Visual and IMU Measurements

IMU can only be used to estimate orientation. To estimate the full pose of IMU-attached object, we need combine visual sensors. On one hand, visual information can not only provide orientation but also location estimation. On the other hand, IMU is normally with a higher sampling rate which can predict fast movements and free of occlusion problem of vision method. Dual to the complementary nature between vision and IMU, it is reasonable to fuse both together to provide fast and accurate full pose estimation.

Armesto et al. propose an ego-motion estimation system which fuses vision and inertial measurements using EKF and UKF (Unscented Kalman Filter) [12]. A constant linear acceleration model and constant angular velocity model is used to formulate a non-linear state model with linear measurement equations. The results show an significant performance improvement over standalone vision or IMU measurements and the capability of handling system failures. However, the use of EKF for IMU is not computation efficient considering the fast update of IMU measurements. [15] proposes a method for global pose estimation fusing IMU, vision and UWB (Ultra Wide Band) sensors. This method doesnot need any fiducial markers while natural landmarks are estimated using SLAM (simultaneous localization and mapping). But the implementation is more complex and EKF as its fusing method could limit the IMU/UWB processing speed.

3   Fast and Accurate Pose Estimation Fusing Vision and IMU Measurements

This section proposes a novel method for vision and IMU fusion - Virtual Magnetic Field (VMF) fusion. The main idea is representing the optical axis of static/moving camera in the IMU-attached local frame which can be taken as a virtual magnetic field measurement - so called VMF. The VMF can be feeded to a complementary filter to be fused with IMU sensor. VMF can correct the heading direction estimation of IMU with center of camera as virtual magnetic north pole. The camera orientation with respect to odom frame can be automatically estimated and compensated by considering fusion and standalone vision-estimated orientation. The location estimation can also be made by fusing acceleration and visual translation measurements which are fused by an adaptive complementary filter.

3.1   Madgwick Algorithm

Madgwick et al. propose an IMU and MARG orientation estimation algorithm which uses quaternion to describe the coupled nature of orientations in three-dimensions without incurring any singularities [10]. The algorithm incorporates magnetic distorsion correction and fuses the IMU and magnetic sensors to obtain a best pose estimation by analytically solving a nonlinear optimization problem leveraging gradient descent method.

The coordinate frames are defined as: local frame S attached to the IMU (optionally plus magnetic sensor) and global frame E attached to earth. $^S\hat{\omega}$ is the quaternion expression of angular rate vector (by padding a zero at the beginning). $^S_E\hat{q}$ is a quaternion corresponding to a rotation from frame S to frame

E (with the hat $\hat{}$ indicating a quaternion in this paper). The other quaternions are defined the same way with pre-superscripts denoting parent frames and pre-subscripts denoting child frames. The post-subscript ω denotes the prediction is calculated from angular rate ω. The post-subscript 'est' denotes the estimation result from the fusion. At time point t, the prediction of quaterion derivative and quaternion is performed by (1) and (2).

$$S_E\dot{\hat{q}}_{\omega,t} = \frac{1}{2}{}^S_E\hat{q}_{est,t-1} \otimes {}^S\hat{\omega}_t \tag{1}$$

$$S_E\hat{q}_{\omega,t} = {}^S_E\hat{q}_{est,t-1} + {}^S_E\dot{\hat{q}}_{\omega,t}\triangle t \tag{2}$$

The reference direction of a (gravitational or magnetic) field is represented as a versor ${}^E\hat{d}$ defined in global earth frame by augmenting the vectorwith an initial zero. The pose estimation problem is formulated as (3) and (4). Equation (4) changes the view point from earth frame to sensor frame by applying a quaternion to reference versor ${}^E\hat{d}$ and calculate the error from sensor measurement ${}^S\hat{s}$. The Eq. (3) can be solved to find the optimal estimation with minimal error. The gradient descent method as (5) can be used to find the minimum, where ∇f defined by (6) denotes the gradient of f depending on the Jacobian J.

$$\min_{{}^S_E\hat{q}\in\Re^4} f\left({}^S_E\hat{q}, {}^E\hat{d}, {}^S\hat{s}\right) \tag{3}$$

where

$$f\left({}^S_E\hat{q}, {}^E\hat{d}, {}^S\hat{s}\right) = {}^S_E\hat{q}^* \otimes {}^E\hat{d} \otimes {}^S_E\hat{q} - {}^S\hat{s} \tag{4}$$

$$S_E q_{k+1} = {}^S_E q_k - \mu\frac{\nabla f\left({}^S_E\hat{q}_k, {}^E\hat{d}, {}^S\hat{s}\right)}{\|\nabla f\left({}^S_E\hat{q}_k, {}^E\hat{d}, {}^S\hat{s}\right)\|} \tag{5}$$

$$\nabla f\left({}^S_E\hat{q}_k, {}^E\hat{d}, {}^S\hat{s}\right) = J^T\left({}^S_E\hat{q}_k, {}^E\hat{d}\right) f\left({}^S_E\hat{q}_k, {}^E\hat{d}, {}^S\hat{s}\right) \tag{6}$$

where $k = 0, 1, 2\cdots n$. μ is a variable step proportional to quaternion change rate ${}^S_E\dot{\hat{q}}_\omega$.

3.2   New Concept – Virtual Magnetic Field

Considering the hardware setup as Fig. 1. An ArUco marker [16] of $5\,cm \times 5\,cm$ is attached to the back of an iPhone. An RGB webcamera is used to perceive the visual posture of the marker. The embedded IMU sensors in the phone are used for inertia measurement. The coordinate frames are defined in Fig. 2. The world frame Σ_W is defined with z pointing vertically at opposite gravity direction and the horizontal heading is determined by optical axis of camera which passes through x–z plane and crosses with x or/and z axes. The camera frame Σ_C is defined with z axis pointing out camera along the optical axis allowing any rotation around z axis. The odometry frame Σ_O is determined by the fusion output of IMU and vision with assumption that the z axis always points at measured gravity direction. Note that the heading of odometry frame cannot be measured accurately with only IMU. Marker frame Σ_M and IMU sensor frame Σ_S are binded together by the phone with π rad difference around y axis (indicated by dash line between them in Fig. 2). Note that the origins of Σ_C and Σ_O coincide with Σ_W although they are taken apart in Fig. 2 for clarity.

Assume the Virtual Magnetic Field vector $^C\hat{\boldsymbol{v}}$ is along the optical axis, i.e., z axis of camera frame Σ_C. Therefore $^C\hat{\boldsymbol{v}}$ is expressed as (7).

$$^C\hat{\boldsymbol{v}} = (0,0,0,1) \tag{7}$$

The rotation of IMU sensor frame Σ_S w.r.t. camera frame Σ_C can be calculated given the orientation measurement of marker frame Σ_M w.r.t. camera frame Σ_C, as depicted in Fig. 2, formulated as (8).

$$^C_S\hat{\boldsymbol{q}} = {}^M_S\hat{\boldsymbol{q}} \otimes {}^C_M\hat{\boldsymbol{q}} \tag{8}$$

With (8), the Virtual Magnetic Field measurement $^C\hat{\boldsymbol{v}}$ can be transformed into IMU sensor frame Σ_S, formulated as (9)

$$^S\hat{\boldsymbol{v}} = {}^C_S\hat{\boldsymbol{q}} \otimes {}^C\hat{\boldsymbol{v}} \otimes {}^C_S\hat{\boldsymbol{q}}^* \tag{9}$$

where * denotes the quaternion conjugate operator. The $^S\hat{\boldsymbol{v}}$ can be used to replace the real magnetometer measurement.

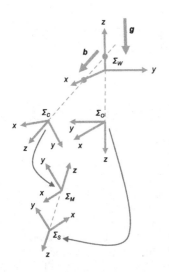

Fig. 2. Coordinate frame definition for Vision and IMU sensor fusion

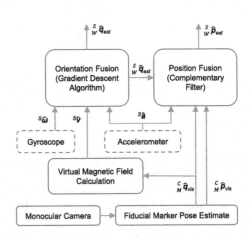

Fig. 3. Vision and IMU sensor fusion based on Virtual Magnetic Field

3.3 Proposed Fusion Method – Virtual Magnetic Field Based Fusion

The visual-inertial fusion method is depicted as Fig. 3. The whole framework includes two main modules as Orientation Fusion and Position Fusion. The gyroscope and accelerometer measures the angular rate and acceleration which are used to predict the rotation and estimate the gravity field. These two sensors alone can enforce a good estimation of inclination angle w.r.t. z axis of world frame Σ_W. To get an accurate estimation of horizontal heading, monocular camera sensor is incorporated which measures the raw image which is fed to a Fidual Marker Pose estimation module taken as a black box here. The generated rotation estimation $_{M}^{C}\hat{\boldsymbol{q}}_{vis}$ is fed to Virtual Magnetic Field Calculation module to compute the VMF vector $^{S}\hat{\boldsymbol{v}}$ expressed in IMU sensor frame Σ_S and position estimation $_{M}^{C}\hat{\boldsymbol{p}}_{vis}$ is fed to Position Fusion module to be fused with acceleration measurement from accelerometer. The Orientation Fusion is implemented with Madgwick Algorithm and Position Fusion implemented with an adaptive complementary filter.

The VMF measurement $^{C}\hat{\boldsymbol{v}}$ in camera frame Σ_C and $^{S}\hat{\boldsymbol{v}}$ in IMU sensor frame Σ_S is substituted in (4) and (6) for $^{E}\hat{\boldsymbol{d}}$ and $^{S}\hat{\boldsymbol{s}}$.

The global rotation of camera can be calculated by comparing orientation estimation from fusion and vision standalone, as formulated by (10). The inclination of optical axis (taken as magnetic field) in world frame Σ_W is taken care by the magnetic distorsion compensation of Madgwick Algorithm [10]. Note that the rotation around optical axis doesnot affect IMU orientation estimation.

$$
\begin{aligned}
_{C}^{W}\hat{\boldsymbol{q}} &= {}_{C}^{M}\hat{\boldsymbol{q}} \otimes {}_{M}^{S}\hat{\boldsymbol{q}} \otimes {}_{S}^{W}\hat{\boldsymbol{q}} \\
&= {}_{S}^{C}\hat{\boldsymbol{q}}^{*} \otimes {}_{S}^{W}\hat{\boldsymbol{q}}_{est}
\end{aligned}
\tag{10}
$$

with rotation estimation of IMU sensor frame Σ_S w.r.t. world frame Σ_W formulated as (11)

$$
_{S}^{W}\hat{\boldsymbol{q}}_{est} = {}_{S}^{O}\hat{\boldsymbol{q}}_{est} \otimes {}_{O}^{W}\hat{\boldsymbol{q}}
\tag{11}
$$

where $_{S}^{C}\hat{\boldsymbol{q}}$, i.e. the relative orientation of IMU sensor frame Σ_S w.r.t. camera frame Σ_C, is calculated by (8). The $_{O}^{W}\hat{\boldsymbol{q}}$ is the orientation estimation of odometry frame Σ_O w.r.t. world frame Σ_W, which is constant as depicted in Fig. 2 when the orientation fusion gets stable (it can be more accurately estimated by averaging over a period if the camera is fixed like in this work). The $_{S}^{O}\hat{\boldsymbol{q}}_{est}$ is the orientation estimation of IMU sensor in odometry frame Σ_O output from Madgwick Algorithm (i.e. Orientation Fusion module in Fig. 3).

Note that the world frame is not identical to earth frame as the horizontal heading is defined by camera optical axis. Also note that only the optical axis is used in orientation fusion. It means the visual posture error of rotation around the axis doesnot affect the orientation fusion accuracy. In other words, in terms of visual measurement, only the direction of the camera optical axis contributes to the orientation estimating accuracy.

The adaptive complementary filter for position estimation of IMU sensor frame Σ_S relative to world frame Σ_W is designed as (12).

$$
{}_{S}^{W}\hat{\boldsymbol{p}}_{est,t+\triangle t} = \frac{1}{\lambda+1}\left(\lambda \; {}_{C}^{W}\hat{\boldsymbol{q}}_{t}^{*} \otimes {}_{M}^{C}\hat{\boldsymbol{p}}_{t} \otimes {}_{C}^{W}\hat{\boldsymbol{q}}_{t} \right.
$$

$$
+ {}_{S}^{W}\hat{\boldsymbol{p}}_{est,t} \tag{12}
$$

$$
\left. + \int_{t}^{t+\triangle t}\int_{t}^{t+\triangle t} {}_{S}^{W}\hat{\boldsymbol{q}}_{est,\tau}^{*} \otimes \left({}^{S}\hat{\boldsymbol{a}}_{\tau} + {}^{S}\hat{\boldsymbol{c}}_{a}\right) \otimes {}_{S}^{W}\hat{\boldsymbol{q}}_{est,\tau}d\tau dt\right)
$$

where subscripts t and $t + \triangle t$ are attached to discriminate current time point and future time point of IMU updating. $\triangle t$ denotes the time step. Subscript 'est' means the fusion result. ${}^{S}\hat{\boldsymbol{a}}$ denotes the acceleration measurement in IMU sensor frame Σ_S. ${}^{S}\hat{\boldsymbol{c}}_{a}$ denotes the bias of accelerometer measurement (assuming gaussian noise and constant bias, hand-tuned in this paper and to be automized in future). λ is a gain parameter to tune the weights of IMU and vision, defined by (13).

$$
\lambda = \begin{cases} 0, & if \; T_0 < t < T \\ c, & else \; (i.e. \; t = T_0 \; or \; T) \end{cases} \tag{13}
$$

with $0 < c \in \Re$. T_0 and T are last and next time points with vision updating. In this way, λ can be switched between 0 and c by checking whether visual update is there or not at current time point.

As above described, in position estimation, there is an adaptive λ to handle the irregular update frequency of visual pose estimation and the synchronization mismatch from the IMU sensor. Also note that, in orientation estimation, the complementary gain is constant (for the less drift by only first order derivative). Within a vision updating cycle, the previous sampled vision measurement is used and the fast updating of IMU sensor can predict a short time (several hundred ms to several seconds) pose estimation without vision update. Even if the vision only occasionally works during several seconds, a reasonable estimation can still be obtained.

4   Experimental Results

An experiment is designed to verify the feasibility of the proposed visual-inertial sensor fusion framework for pose estimation.

4.1   Experimental Setup

The whole system is shown as Fig. 1. An ArUco marker [16] of size 5 cm × 5 cm is attached to the back of an iPhone 4, which is connected to PC via an USB cable and the raw data from embedded inertial sensors are streamed to PC by an UDP server app on the phone named Sensor Streamer 100 Hz (wifi-connected by an wireless router is also ok to achieve almost the same sampling frequency in average but the cycle is not regular in practice, to be further tested in future). The camera Logitech HD WEBCAM C270 with resolution of 1280 × 720 pixels is used as the visual sensor. The Yumi dual-arm robot, which has an position repeatibility accuracy of 0.02 mm, is used to provide reference measurements. As shown in Fig. 1, the iPhone and the camera are attached to the grippers with the marker facing to the camera initially. The robot base is calibrated to be level.

The programs are all running on Ubuntu 14.04 operating system. The visual pose estimation is executed by Ar_Sys ROS package [20] which in practice is found with tracking range of 30–120 cm (for 5 cm × 5 cm markers), noise of ±0.1 pixel at 30 cm distance and sampling frequency 30 Hz. This package can resume tracking at the maximum distance when lost tracking. The orientation fusion is mainly performed by Madgwick complementary filter algorithm. And the position fusion is executed by a ROS node running an adaptive complementary filter in real time. The orientation update frequency 100 Hz consistent with IMU sampling frequency. Position update frequency is 100 Hz compared 30 Hz of vision processing speed.

4.2 Results

The pose estimation results are benchmarked with Yumi robot pose measurements. As shown in Fig. 4, the extrinsic Enler angles estimated by the proposed fusion method and by only vision are presented. The black dot lines are the corresponding reference measured from Yumi robot. There are periods (e.g. 10 to 20 s) when the camera are blocked for rotating the device. It is shown that the proposed fusion method can always get a better estimate even when vision failes as from 10 to 20 s and the result sticks to the reference.

The camera pose can also be estimated by the fusion method. As shown in Fig. 6, the three Enler angles are almost constant although the pose is calculated from IMU sensor pose estimated by fusion and the visual marker pose w.r.t. camera frame. It can be seen, the camera rotates around x, y, z axes of world frame with around 1.52 rad, 0 rad, and 1.51 rad in sequence. The heading of the world frame is decided by camera frame and IMU odometry frame. This means the world frame is depending on measurements like odometry frame. When the fusion gets stabilized, the estimated relation between world frame and IMU odometry and camera frames are approaching constant. The relationships are demonstrated in Fig. 2.

The position estimation results are shown in Fig. 5. The position estimation based on the fusion method is denoted by the solid lines. The dotted line is the reference from Yumi robot. The worst accuracy happens when vision fails about 10 s to 17 s in Fig. 5 where the error roars upto 7 cm, even though the motion is almost cyclic. The big drift mainly results from the second integration of acceleration. When the vision recovers, the estimation is fairly good, as shown from 17 s to 25 s and from 34 s to 41 s.

In the Fig. 4, 5 and 6, the periods with straight lines indicate vision failures. It is designed to occlude visual marker during the motion to test the robustness. As can be seen in Fig. 5, in the vision active periods, the IMU motion is very slow, this is partly because the camera only captures the marker in slow motion (for our configuration). In the future, it is necessary to use IMU to boost visual tracking.

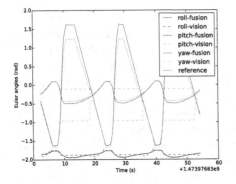

Fig. 4. Orientation estimation results

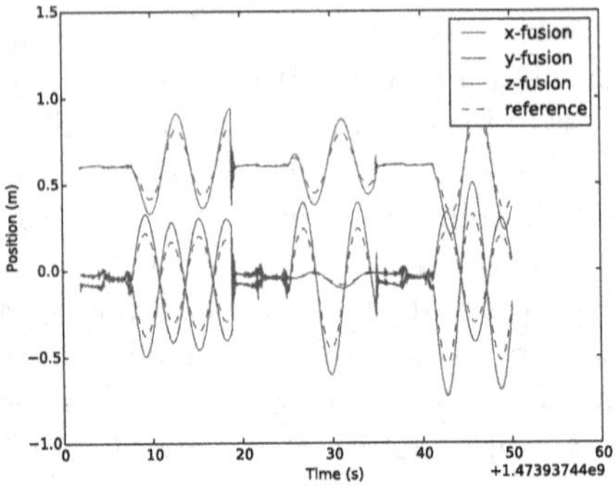

Fig. 5. Position estimation results

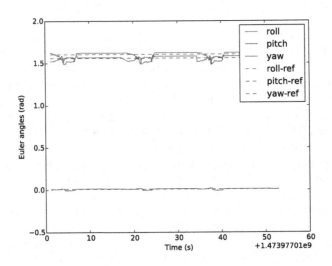

Fig. 6. Camera orientation estimation results

5   Conclusions

This paper proposes a novel visual-inertial sensor fusion method for pose estimation. The optical axis is taken as VMF (Virtual Magnetic Field) and is transformed into IMU sensor frame as simulated magnetometer measurement. The
orientation estimation is performed by fusing IMU and VMF with Madgwick
complementary filter. The estimated orientation is also used for calculating the
(static) camera orientation which is important for the consistency of reference
frame between visual position estimation and accelerometer measurement, which
are fused by an adaptive gain complementary filter. The advantages of the proposed fusion method are as follow:

- Compared with using visual sensor alone, the pose estimation is more fast and robust to a short term failure of vision, which frequently happens when the illumination is bad, marker is too far or moving too fast as well as there is occlusion for turnning marker to the side blocked by the device.
- The VMF method uses only the optical axis to estimate orientation. The rotation around optical axis doesnot contribute to IMU orientation errors.
- Not only the IMU orientation but also the camera rotation in world frame can be estimated. This feature is used to align the visual position and IMU accelerometer in the same (world) frame.
- The fusion is very efficient for using complementary filters. It is applicable to embedded system with restricted computing power.
- The implementation is very easy and straightforward as the whole framework is modulized. The orientation fusion module (Madgwick Algorithm) can almost be kept intact. The visual pose estimation and position estimation can be replaced with any other methods to gain a better performance.

The disadvantage is the occurrence of singularity when camera faces vertically.

Future work will improve the position estimation performance and stability of vision tracking.

References

1. Ligorio, G., Sabatini, A.M.: A novel Kalman filter for human motion tracking with an inertial-based dynamic inclinometer. IEEE Trans. Biomed. Eng. **62**(8), 2033–2043 (2015)
2. Lee, W.W., Yen, S.-C., Tay, E.B.A., Zhao, Z., Xu, T.M., Ling, K.K.M., Ng, Y.-S., Chew, E., Cheong, A.L.K., Huat, G.K.C.: A smartphone-centric system for the range of motion assessment in stroke patients. IEEE J. Biomed. Health Inform. **18**(6), 1839–1847 (2014)
3. Daponte, P., De Vito, L., Riccio, M,. Sementa, C.: Experimental comparison of orientation estimation algorithms in motion tracking for rehabilitation. In: Medical Measurements and Applications (MeMeA), 2014 IEEE International Symposium on IEEE pp. 1–6 (2014)
4. Tian, Y., Hamel, W.R., Tan, J.: Accurate human navigation using wearable monocular visual and inertial sensors. IEEE Trans. Instrum. Measur. **63**(1), 203–213 (2014)
5. Pierleoni, P., Belli, A., Palma, L., Pellegrini, M., Pernini, L., Valenti, S.: A high reliability wearable device for elderly fall detection. IEEE Sens. J. **15**(8), 4544–4553 (2015)
6. Hong, S.K.: Fuzzy logic based closed-loop strapdown attitude system for unmanned aerial vehicle (UAV). Sens. Actuators A Phys. **107**(2), 109–118 (2003)
7. Hyde, R.A., Ketteringham, L.P., Neild, S.A., Jones, R.J.: Estimation of upper-limb orientation based on accelerometer and gyroscope measurements. IEEE Trans. Biomed. Eng. **55**(2), 746–754 (2008)
8. Bachmann, E.R., McGhee, R.B., Yun, X., Zyda, M.J.: Inertial and magnetic posture tracking for inserting humans into networked virtual environments. In: Proceedings of the ACM Symposium on Virtual Reality Software and Technology, pp. 9–16. ACM (2001)

9. Mahony, R., Hamel, T., Pflimlin, J.-M.: Nonlinear complementary filters on the special orthogonal group. IEEE Trans. Autom. Control **53**(5), 1203–1218 (2008)
10. Madgwick, S.O., Harrison, A.J., Vaidyanathan, R.: Estimation of IMU and MARG orientation using a gradient descent algorithm. In: 2011 IEEE International Conference on Rehabilitation Robotics, pp. 1–7. IEEE (2011)
11. Galna, B., Barry, G., Jackson, D., Mhiripiri, D., Olivier, P., Rochester, L.: Accuracy of the microsoft kinect sensor for measuring movement in people with Parkinson's disease. Gait Posture **39**(4), 1062–1068 (2014)
12. Armesto, L., Tornero, J., Vincze, M.: Fast ego-motion estimation with multi-rate fusion of inertial and vision. Int. J. Robot. Res. **26**(6), 577–589 (2007)
13. Simanek, J., Reinstein, M., Kubelka, V.: Evaluation of the EKF-based estimation architectures for data fusion in mobile robots. IEEE/ASME Trans. Mechatron. **20**(2), 985–990 (2015)
14. Cavallo, A., et al.: Experimental comparison of sensor fusion algorithms for attitude estimation. IFAC Proc. Vol. **47**(3), 7585–7591 (2014)
15. Nyqvist, H.E., Skoglund, M.A., Hendeby, G., Gustafsson, F.: Pose estimation using monocular vision and inertial sensors aided with ultra wide band. In: Indoor Positioning and Indoor Navigation (IPIN), 2015 International Conference on IEEE, pp. 1–10 (2015)
16. Garrido-Jurado, S., Muñoz-Salinas, R., Madrid-Cuevas, F.J., Marín-Jiménez, M.J.: Automatic generation and detection of highly reliable fiducial markers under occlusion. Pattern Recogn. **47**(6), 2280–2292 (2014)
17. Qiu, M., Jia, Z., Xue, C., Shao, Z., Sha, E.H.-M.: Voltage assignment with guaranteed probability satisfying timing constraint for real-time multiproceesor DSP. J. VLSI Signal Process. Syst. Signal Image Video Technol. **46**(1), 55–73 (2007)
18. Zhang, Q., Huang, T., Zhu, Y., Qiu, M.: A case study of sensor data collection and analysis in smart city: provenance in smart food supply chain. Int. J. Distrib. Sens. Netw. **9**(11), 382132 (2013)
19. Chen, M., Zhang, Y., Qiu, M., Guizani, N., Hao, Y.: SPHA: smart personal health advisor based on deep analytics. IEEE Commun. Mag. **56**(3), 164–169 (2018)
20. Sahloul, H.: Ros Wiki: 3D pose estimation ROS package using ArUco marker boards (2016). http://wiki.ros.org/ar_sys

A Novel Image Recognition Method Based on CNN Algorithm

Yuwen Guo[1](✉), Xin Zhang[1](✉), Qi Yang[1], and Hong Guo[1,2]

[1] College of Computer Science and Technology, Wuhan University of Science and Technology, Wuhan, China
[2] Hubei Province Key Laboratory of Intelligent Information Processing and Real-Time Industrial System, Beijing, China

Abstract. Handwritten digital image recognition is a special problem in digital recognition, which is much more difficult due to uncertainties such as writer and writing environment. But the accuracy of handwritten digital recognition is particularly important in some critical areas such as public security and finance. In this paper, we used machine learning and deep learning technology, to identify MNIST handwritten digital data sets. And the support vector machine and convolutional neural network were used to image recognition, among them, the support vector machine compares the accuracy of the identification of the two kernel functions. Convolutional neural networks are improved based on the classic LeNet-5 model. The Experiments result shown that the identification accuracy of Gaussian kernel function model and linear kernel function model is 94.03% and 90.47% for support vector machines, respectively. In contrast, Gaussian kernel function performs better, but the overall accuracy is low. The best performance is based on the classic convolutional neural network LeNet-5 improved model, the test set recognition rate can reach 99.05% .

Keywords: Digital picture recognition · Support vector machine · Linear kernel function · Gaussian kernel function · Convolutional neural network · LeNet-5

1 Introduction

With the advent of the information age, people are having higher and higher requirements for information processing. It not only requires higher and higher precision, but also needs faster and faster processing speed. The technology of digital image recognition will play a more important role in the development of information technology in the future. In the field of digital image recognition, handwritten digital image recognition is a very important branch, which is widely used in real life. China has a large population, the amount of data that people generate when they fill out reports is also rising synchronously, resulting in a large number of digital information. In the past, manual entry was done by manual means, This will not only lead to high labor costs, low efficiency and error prone output, but also no longer applicable in today's era of rapid information development.

The recognition of character and handwriting is one of the most difficult problems in pattern recognition and artificial intelligence. Unlike computer-generated characters,

© Springer Nature Switzerland AG 2021
M. Qiu (Ed.): SmartCom 2020, LNCS 12608, pp. 281–291, 2021.
https://doi.org/10.1007/978-3-030-74717-6_29

handwriting numbers vary in size, width, and orientation due to human-written factors. The uniqueness and diversity of different people's writing habits will affect the formation and appearance of numbers, increasing the difficulty of correct character recognition. For completely unrestricted handwritten numbers, it's almost certain that simple solutions do not result in high recognition and accuracy [1]. In normal life, errors are often generated due to similarities between numbers, such as the numbers 0 and 6, 1 and 7.

In this context, the MNIST data set is used as an experimental training test set for research on digital picture recognition. The accuracy of handwriting digital recognition is improved by the research of supporting vector machine and convolutional neural network.

The rest of this paper is organized as follows. The second part introduces the current research of digital image recognition. The third part proposes the specific recognition process of the two algorithms. The fourth part explains the environment and the framework used in the development of this paper first, and then analyzes and compares the recognition accuracy of the two recognition algorithms, the last part is the summary of the paper.

2   Related Work

Handwritten digital image recognition can be used to automatically read bank check information, zip codes on envelopes, data in some documents, and so on. At present, there are some handwritten digital recognition products in the market, such as postcode automatic identification system, bank check automatic processing system and other practical systems [2], these have been very good applications.

Currently, the main research methods for handwriting digital recognition are statistics, clustering and neural networks. The minimum distance classification algorithm is one of the most traditional and simplest methods of pattern recognition, but it cannot be applied to handwritten fonts. K-Nearest Neighbor (KNN) algorithm is a statistically based classification algorithm first proposed by Cover and Hart in 1968. As the simplest method of machine learning, theoretical research is relatively mature [3]. However, these methods above may have some difficulties in feature extraction, and the recognition results are not satisfactory. The idea of deep learning makes this problem better solved.

In deep learning, the convolutional neural network is the first model that is more in line with its definition, enabling the establishment of a multi-layered structure for learning [4]. The close connection and spatial information between the layers in the convolutional neural network make it particularly suitable for the recognition of handwritten digital images, and can automatically extract rich correlation characteristics from the image.

In this paper, the support vector machine recognition algorithm and the convolutional neural network recognition algorithm are introduced in view of the problem of handwritten digital recognition of MINIST data set. The activation function, pooling method and padding method of convolutional neural network are analyzed. Finally, the identification accuracy of the above two methods is compared and analyzed.

However, because of the factors of handwritten numbers themselves, handwriting digital recognition is the most challenging problem in the field of digital character recognition [5], even if for decades, researchers have been proposing new recognition methods, but so far, it can be said that no handwritten digital recognition algorithm has achieved perfect recognition results. Therefore, to further improve the recognition rate and recognition accuracy of handwriting digital recognition algorithm is the goal of every digital recognition worker.

3 Architecture

3.1 Support Vector Machine Recognition Algorithm

Construction of the SVM Multiple Classifier
Handwritten numeric recognition has a sample of 0 to 9, and it's a multi-category issue. The SVM classifier is a two-classifier [6], and to identify 10 types of numbers, multiple SVM classifiers must be combined to form a multi-classifier.

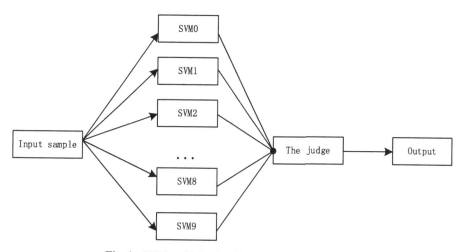

Fig. 1. SVM multiple classifier recognizer framework

The principle of constructing a multi-class classifier is as follows: For an N-class classification problem, we need to construct N two-class classifiers based on support vector machines, each two-class classifier distinguishes each class in the N class from the rest classes, runs the N classifier once we identify it, then compares the identification results of the N classifiers, and selects the highest likely output for each input variable [7].

In this experiment, there are 10 number categories, so we need to design 10 two-class classifiers, such as Classifier 1, which separates 1 from the other 9 numbers. During identification, enter the input samples into 10 SVM classifiers, the resulting judge compares the classification results of these classifiers to determine which category the input vector belongs to, and finally outputs the results [8], the composition framework diagram is shown in Fig. 1.

The SVM Identification Process

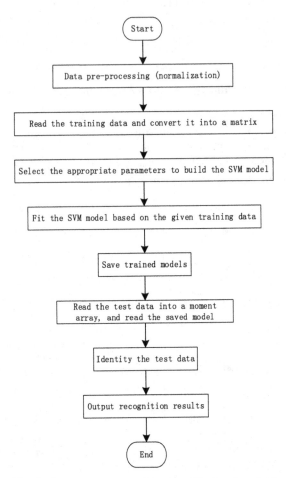

Fig. 2. Supports vector machine identification flowcharts

The implementation of 10 classifiers is involved, the workload is relatively large, so this experiment is based on the sklearn. Coding with SVM module, and the SVM model using different kernel functions is compared, and finally output a detailed report, giving the overall accuracy of the SVM model under different parameters, prediction accuracy and so on. The identification process is shown in Fig. 2.

3.2 Convolutional Neural Network Recognition Algorithm

LeNet-5 Classic Model

The LeNet-5 [9] model is a convolutional neural network designed by Yann LeCun et al. in 1998 to identify handwritten numbers and documents. LeNet-5 contains a total of 8 layers of input and output layers. The schema is shown in Fig. 3, and the structure of each layer is shown in Table 1.

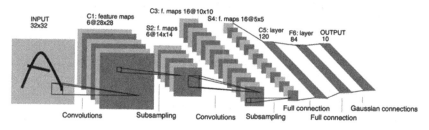

Fig. 3. LeNet-5 architecture 14

Table 1. LeNet-5 per-layer structure

Layer	Layer type	Feature maps	Size	Kernel size	Stride	Activation
Input	Image	1	32×32	–	–	–
1	Convolution	6	28×82	5×5	1	tanh
2	Average pooling	6	14×14	2×2	2	tanh
3	Convolution	16	10×10	5×5	1	tanh
4	Average pooling	16	5×5	2×2	2	tanh
5	Convolution	120	1×1	5×5	1	tanh
6	Fully connected	–	84	–	–	tanh
Output	Fully connected	–	10	–	–	softmax

The original LeNet-5 model used 32×32 picture data size, this paper is using the MNIST dataset, the size of each picture is 28×28, without changing the overall structure of LeNet-5 to implement LeNet.

LeNet-5 Model Improvements

The classic leNet-5 model has three major improvements:

(1) Activation function

Replace the original sigmoid activation function with the relu activation function [10, 11]. The sigmoid activation function, as shown in the formula (1), has a value of z from negative to positive infinity, and a has a value between 0 and 1. The closer the value of z is to 0, the larger the derivative, and when z is greater or less than 0, the derivative value is close to 0. Because the value of a range is between 0 and 1, the sigmoid activation function is also often used in two-category problems.

$$a = \sigma(z) = \frac{1}{1 + e^{-z}} \tag{1}$$

The relu activation function, also known as the function that corrects the linear unit, the mathematical formula is shown in the Eq. (2), as long as z is positive, a = z, the derivative is constantly equal to 1, when z is negative, a = 0, the derivative is constantly equal to 0.

$$a = \max(0, z) \tag{2}$$

(2) Pooling method

Replace average pooling with maximum pooling. According to Boureau et al.'s research, the characteristic extraction of the maximum pooling in the sparse matrix performs better than the average pooling performance [12]. The input sample used in the article is a grayscale picture of the handwritten number, most of which is 0 and a small number is 1, which is eligible. The average pooling averages features within a quadrant and may lose some specific features. Therefore, the recognition effect will be better if we replace average pooling with maximum pooling.

(3) Padding method

Change the Valid padding method in the convolution operation to Same padding [13]. Valid padding way is not to fill before doing convolution operations, so there will be two disadvantages, the first disadvantage is that each time you finish the convolution operation, the image will be reduced, after a few times, the image may become very small, not conducive to the later recognition operation. The second disadvantage, is that the pixels at the edge of the corner are touched or used by only one output when doing convolution operations. But in the middle of the pixel, there are many areas that overlap with it. Therefore, pixels in corners or edge areas are used less in the output, which means that information about the edge position of many images is lost.

4 Experiments

4.1 The Dataset

The experiment was conducted with MNIST data set. The dataset is based on many scanned document datasets obtained by the National Institute of Standards and Technology (NIST).

The MNIST dataset is an Arabic handwritten digital dataset written by 250 different people. It contains 70,000 images, of which 60,000 are training samples and 10,000 are test samples. Each sample is a 28×28-pixel grayness image, i.e. 784 features per sample.

The MNIST dataset is available in IDX format. This IDX file format is a simple format that is convenient for working with vectors and high-dimensional matrices with different numeric types.

Pixels in image and label file sets are arranged in lines by RGB color code, ranging from 0 to 255. The background is white (the value from RGB is 0) and the foreground is black (the value from RGB is 255). Digital labels, ranging from 0 to 9. To identify better reads by the algorithm, convert the IDX format of the file to PNG format.

4.2 Environment Construction

All of the code in this article is written in the Python language, and some of the code is implemented using the Keras [14] framework.

The Keras framework is one of the advanced neural network APIs. It was written in Python and supports multiple back-end neural network computing engines, such as Tensorflow or Theano. Keras, which supports both CPUs and GPUs, has packaged the embedded Keras API as tf.keras in the API version of TensorFlow 2.0.

Keras has two main model types: the sequential model and the functional API model. This paper uses the Sequential model, which is linearly stacked by multiple network layers.

4.3 Environment Configuration

This article uses Anaconda, an open source Python release that contains more than 180 science packages and their dependencies, including conda, Python, and more. The Anaconda installed version is Anaconda 3–5.3.1 and Python version 3.7. To avoid dependency package version conflicts, TensorFlow runs in a separate virtual environment created by Anaconda. TensorFlow installs a version of TensorFlow 2.1.0 because it uses the integrated keras framework in TensorFlow2, a feature that is only available in TensorFlow2. Open the command-line window under Anaconda, Anaconda Prompt, create a virtual environment called TensorFlow and specify Python version 3.6. TensorFlow supports Python versions 3.5–3.7. Once the environment is created, activate the environment, be sure to check the version of the pip before installing TensorFlow, pip must meet 19.0 or later to successfully install TensorFlow, and upgrade the pip package if it is below 19.0.

4.4 Comparison of SVM Recognition Results

Linear kernel function recognition results, as shown in Table 2, have an average accuracy of 90.47%.

The Gaussian kernel function recognition results, as shown in Table 3, have an average accuracy of 94.03%.

By comparison, it can be found that the recognition rate of the Gaussian kernel function SVM model is much higher than that of the linear kernel function SVM model. For both models, the numbers 0 and 1 are better recognized, and the numbers 5, 8, and 9 are poorly recognized.

Table 2. Linear kernel function recognition results

Number	Number of test	Correct quantity	Error quantity	Error rate
0	980	938	42	0.042857
1	1135	1112	23	0.020264
2	1032	922	110	0.106589
3	1010	910	100	0.099010
4	982	907	75	0.076375
5	892	759	133	0.149103
6	958	895	63	0.065762
7	1028	925	103	0.100195
8	974	827	147	0.150924
9	1009	866	143	0.141724

Table 3. The Gaussian kernel function recognition results

Number	Number of test	Correct quantity	Error quantity	Error rate
0	980	968	12	0.012245
1	1135	1120	15	0.013216
2	1032	961	71	0.068798
3	1010	949	61	0.060396
4	982	931	51	0.051935
5	892	807	85	0.095291
6	958	927	31	0.032359
7	1028	957	71	0.069066
8	974	879	95	0.097536
9	1009	912	97	0.096135

4.5 Comparison of Convolutional Neural Network Results

The identification of the MNIST dataset on the original LeNet-5 model, using cross-validation entropy and rmsprop optimization, it can be found after 5 iterations, the loss of the training set was reduced to 0.1346 and the accuracy of the training set was 95.79%. The accuracy of test set recognition was 96.22%.

Following the improved LeNet-5 model as described in the optimization method above, it can be found that after 5 iterations, the loss of the training set is reduced to 0.0124 and the accuracy of the training set is 99.59%. And the accuracy of test set recognition was 99.05%.

As shown in Fig. 4, there are also five iterations, and the improved loss is much smaller than the loss before the improvement. The improved network starts with the first iteration and the loss value is small. The accuracy of the final test set was also increased from 96.22% to 99.05%, and the error rate was reduced by 94.9%.

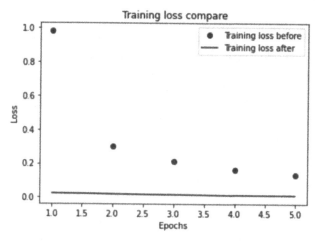

Fig. 4. Improve the before-and-after loss comparison chart

4.6 Two Ways to Compare

The accuracy of the two identification algorithm test sets is shown in Table 4, and the convolutional neural network performs best when it is used to identify MNIST datasets, with an accuracy rate of 99.05%.

Table 4. Two ways to compare results

The recognition algorithm	The accuracy of the identification
Support vector machine	94.03%
Convolutional neural network	99.05%

5 Conclusion

Based on the Python language, this paper identifies handwritten digital pictures in digital pictures, uses MNIST data sets as experimental data, uses two learning algorithms (support vector machines, convolutional neural networks) to analyze and identify, and uses Python language, Keras framework programming implementation. The main contributions and innovations in this article are as follows:

(1) This paper compares the accuracy of the identification of different kernel function models that support vector machines, and it is found that the Gaussian kernel function performs better than the linear kernel function in the problem of MNIST data set. The numbers 0 and 1 of both kernel functions are more accurate than the numbers 8 and 9. The average accuracy of identification is 90.47% and 94.03%, respectively.

(2) In this paper, the classic convolutional neural network LeNet-5 model is analyzed and improved, the relu activation function is used to improve convergence speed, the maximum pooling replaces the average pooling, and the corner letter is retained by same padding interest. Recognition accuracy increased from 96.22% to 99.59%. Through the experimental results, it can be found that this optimization method is feasible, the recognition rate is improved to a certain extent, and achieve the desired results.

References

1. Bian, Z., Zhang, X.: Pattern Recognition, pp. 17–20. Tsinghua University Press, Beijing (2000)
2. Cao, D., Yang, C., Zhang, W.: Advance on handwritten digital recognition. Comput. Knowl. Technol. **5**(03), 688–695 (2009)
3. Wang, Y., Xia, C., Dai, S.: Handwritten digital recognition optimization method based on LeNet-5 model. Comput. Dig. Eng. **47**(12), 3177–3181 (2019)
4. Song, X., Wu, X., Gao, S., et al.: Simulation study on handwritten numeral recognition based on deep neural network. Sci. Technol. Eng. **19**(05), 193–196 (2019)
5. Lu, H., Yang, Y., Su, J., Zhou, X.: Handwritten numerals recognition based on image processing. Instrum. Anal. Monit. **03**, 13–15 (2005)
6. Gai, K., Qiu, M., Zhao, H: Security-aware efficient mass distributed storage approach for cloud systems in big data. In: 2016 IEEE 2nd International Conference on Big Data Security on Cloud (2016)
7. Meng, G., Fang, J.: System design of recognition of handwritten digits by support vector machine. Comput. Eng. Des. **06**, 1592–1594+1598 (2005)
8. Wang, J., Cao, Y.: The application of support vector machine in classifying large namber of catalogs. J. BeiJing Inst. Technol. **02**, 225–228 (2001)
9. LeCun, Y., Bottou, L., Bengio, Y., Haffner, P.: Gradient-based learning applied to document recognition. In: Proceedings of IEEE, pp. 2278–2324. IEEE, USA (1998)
10. Boureau, Y.L., Ponce, J., Lecun, Y.: Atheoretic alanalys is of feature pooling invisual recognition. In: International Conference on Machine Learning, vol. 32, no. 4, pp. 111–118 (2010)
11. Qiu, M., Jia, Z., Xue, C., Shao, Z., Sha, E.H.M.: Voltage assignment with guaranteed probability satisfying timing constraint for real-time multiproceesor DSP. J. VLSI Signal Process. Syst. Signal Image Video Technol. **46**, 55–73 (2007)
12. Zhang, Q., Huang, T., Zhu, Y., Qiu, M.: A case study of sensor data collection and analysis in smart city: provenance in smart food supply chain. Int. J. Distrib. Sensor Netw. **9**(11), 382132 (2013)

13. Chen, M., Zhang, Y., Qiu, M., Guizani, N., Hao, Y.: SPHA: smart personal health advisor based on deep analytics. IEEE Commun. Mag. **56**(3), 164–169 (2018)
14. El-Sawy, A., EL-Bakry, H., Loey, M.: CNN for handwritten arabic digits recognition based on LeNet-5. In: Hassanien, A.E., Shaalan, K., Gaber, T., Azar, A.T., Tolba, M.F. (eds.) AISI 2016. AISC, vol. 533, pp. 566–575. Springer, Cham (2017). https://doi.org/10.1007/978-3-319-48308-5_54

Author Index

Printed in the United States
by Baker & Taylor Publisher Services